CW01506922

Astrophysical Techniques
Fourth Edition

Astrophysical Techniques

Fourth Edition

C R Kitchin
University of Hertfordshire Observatory

Institute of Physics Publishing
Bristol and Philadelphia

British Library Cataloguing-in-Publication Data

A catalogue record for this book is available from the British Library.

ISBN 0 7503 0946 6

Library of Congress Cataloging-in-Publication Data are available

First edition 1984, reprinted 1988
Second edition 1991, reprinted 1998
Third edition 1998, reprinted 2002

Commissioning Editor: John Navas
Production Editor: Simon Laurenson
Production Control: Sarah Plenty
Cover Design: Frédérique Swist
Marketing: Nicola Newey and Verity Cooke

Published by Institute of Physics Publishing, wholly owned by The Institute of Physics, London

Institute of Physics Publishing, Dirac House, Temple Back, Bristol BS1 6BE, UK

US Office: Institute of Physics Publishing, The Public Ledger Building, Suite 929, 150 South Independence Mall West, Philadelphia, PA 19106, USA

Typeset by Academic + Technical, Bristol
Printed in the UK by MPG Books Ltd, Bodmin, Cornwall

For Christine

Contents

Standard symbols

Most of the symbols used in this book are defined when they are first encountered and that definition then applies throughout the rest of the section concerned. Some symbols, however, have acquired such a commonality of use that they have become standard symbols among astronomers. Some of these are listed below, and the symbol will not then be separately defined when it is encountered in the text.

amu	atomic mass unit $= 1.6605 \times 10^{-27}$ kg
c	velocity of light in a vacuum $= 2.9979 \times 10^8$ m s^{-1}
e	charge on the electron $= 1.6022 \times 10^{-19}$ C
e^-	symbol for an electron
e^+	symbol for a positron
eV	electron volt $= 1.6022 \times 10^{-19}$ J
G	gravitational constant $= 6.670 \times 10^{-11}$ N m^2 kg^{-2}
h	Planck's constant $= 6.6262 \times 10^{-34}$ J s
k	Boltzmann's constant $= 1.3806 \times 10^{-23}$ J K^{-1}
m_e	mass of the electron $= 9.1096 \times 10^{-31}$ kg
n	symbol for a neutron
p$^+$	symbol for a proton
U, B, V	magnitudes through the standard UBV photometric system
$^{12}_{6}$C	symbol for nuclei (normal carbon isotope given as the example); the superscript is the atomic mass (in amu) to the nearest whole number and the subscript is the atomic number
γ	symbol for a gamma ray
ε_0	permittivity of free space $= 8.85 \times 10^{-12}$ F m^{-1}
λ	symbol for wavelength
μ	symbol for refractive index (also, n is widely used)
μ_0	permeability of a vacuum $= 4\pi \times 10^{-7}$ H m^{-1}
ν	symbol for frequency (also, f is widely used).

Preface

The aim of this book is to provide a coherent state-of-the-art account of the instruments and techniques used in astronomy and astrophysics today. Whilst every effort has been made to make it as complete and up-to-date as possible, the author is only too well aware of the many omissions and skimpily treated subjects throughout the work. For some types of instrumentation it is possible to give full details of the instrument in its finally developed form. However, for the 'new astronomies' and even some aspects of established fields, development is occurring at a rapid pace and the details will change between the writing and publishing of this edition. For those areas of astronomy therefore, a fairly general guide to the principles behind the techniques is given and this should enable the reader to follow the detailed designs in the scientific literature.

With this fourth edition, many new instruments and techniques are included for the first time, and some topics have been eliminated on the grounds that they have not been used by either professional or amateur astronomers for many years. Other topics, while no longer employed by professional observers for current work, are included because archive material that is still in use was obtained using them and/or because amateur astronomers use the techniques. A few references to web sites have been included but not many because the sites change so frequently and search engines are now so good. However, this resource is now the point of first call for most scientists when they have a query, and much material, such as large sky surveys, is only available over the web. Furthermore it is used for the operation of some remote telescopes and it forms the basis of 'virtual' telescopes as discussed in the last chapter.

Another aim has been to try to reduce the trend towards fragmentation of astronomical studies. The new techniques that are required for observing in exotic regions of the spectrum bring their own concepts and terminology with them. This can lead to the impression that the underlying processes are quite different, when in fact they are identical but are merely being discussed from differing points of view. Thus, for example, the Airy disc and its rings and the polar diagram of a radio dish do not at first sight look related, but they are simply two different ways of presenting the same data. As far as possible, therefore, broad regions of the spectrum are dealt with as a single area, rather than as many smaller disciplines. The underlying unity of all of astronomical observation is also emphasized by the layout of the book; the pattern of detection–imaging–ancillary instruments has been adopted so that one stage of an observation is encountered together with the similar stages required for all other information carriers. This is not an absolutely hard and fast rule, however, and in some places it seemed appropriate to deal with topics out of sequence, either in order to prevent a multiplicity of very small sections, or to keep the continuity of the argument going.

The treatment of the topics is at a level appropriate to a science-based undergraduate degree. As far as possible the mathematics and/or physics background that may be needed for a topic is developed or given within that section. In some places it was felt that some astronomy background would be needed as well, so that the significance of the technique under discussion could be properly realized. Although aimed at an undergraduate level, most of the mathematics should be understandable by anyone who has attended a competently taught mathematics course in their final years at school and some sections are non-mathematical. Thus many amateur astronomers will find aspects of the book of use and interest. The fragmentation of astronomy, which has already been mentioned, means that there is a third group of people who may find the book useful, and that is professional astronomers themselves. The treatment of the topics in general is at a sufficiently high level, yet in a convenient and accessible form, for those professionals seeking information on techniques in areas of astronomy with which they might not be totally familiar.

Lastly I must pay a tribute to the many astronomers and other scientists whose work is summarized here. It is not possible to list them by name, and to stud the text with detailed references would have ruined the intention of making the book readable. I would, however, like to take the opportunity afforded by this preface to express my deepest gratitude to them all.

C R Kitchin

Chapter 1

Detectors

1.1   Optical and infrared detection

Introduction

In this and the immediately succeeding sections, the emphasis is upon the detection of the radiation or other information carrier, and upon the instruments and techniques used to facilitate and optimize that detection. There is inevitably some overlap with other sections and chapters, and some material might arguably be more logically discussed in a different order from the one chosen. In particular in this section, telescopes are included as a necessary adjunct to the detectors themselves. The theory of the formation of an image of a point source by a telescope, which is all that is required for simple detection, takes us most of the way through the theory of imaging extended sources. Both theories are therefore discussed together even though the latter should perhaps be in chapter 2. There are many other examples such as x-ray spectroscopy and polarimetry that appear in section 1.3 instead of sections 4.2 and 5.2. In general the author has tried to follow the pattern detection–imaging–ancillary techniques throughout the book, but has dealt with items out of this order when it seemed more natural to do so.

The optical region is taken to include the near infrared and ultraviolet, and thus roughly covers the region from 10 nm to 100 μm. The techniques and physical processes employed for investigations over this region bear at least a generic resemblance to each other and so may be conveniently discussed together.

Detector types

In the optical region, detectors fall into two main groups, thermal and quantum (or photon) detectors. Both these types are incoherent: that is to say, only the amplitude of the electromagnetic wave is detected; the phase information is lost. Coherent detectors are common at long wavelengths

1

(section 1.2), where the signal is mixed with that from a local oscillator (heterodyne principle). Only recently have optical heterodyne techniques been developed in the laboratory, and these have yet to be applied to astronomy. We may therefore safely regard all optical detectors as incoherent in practice. With optical aperture synthesis (section 2.5), some phase information may be obtained provided that three or more telescopes are used.

In quantum detectors, the individual photons of the optical signal interact directly with the electrons of the detector. Sometimes individual detections are monitored (photon counting); at other times the detections are integrated to give an analogue output like that of the thermal detectors. Examples of quantum detectors include the eye, photographic emulsion, photomultiplier, photodiode and charge-coupled devices.

Thermal detectors, by contrast, detect radiation through the increase in temperature that its absorption causes in the sensing element. They are generally less sensitive and slower in response than quantum detectors, but have a very much broader spectral response. Examples include thermocouples, pyroelectric detectors and bolometers.

The eye

This is undoubtedly the most fundamental of the detectors to a human astronomer although it is a very long way from being the simplest. It is now rarely used for primary detection, though there are still a few applications in which it performs comparably with or possibly even better than other detection systems. Examples of this could be very close double-star work and planetary observation, in which useful work can be undertaken even though the eye may have been superseded for a few specific observations by, say, interferometry, speckle interferometry and planetary space probes. Much more commonly it is necessary to find and/or guide on objects visually whilst they are being studied with other detectors. Plus there are, of course, the millions of people who gaze into the skies for pleasure—and that includes most professional astronomers. Thus there is some importance in understanding how the eye and especially its defects can influence these processes.

It is vital to realize that a simple consideration of the eye on basic optical principles will be misleading. The eye and brain act together in the visual process, and this may give better or worse results than, say, an optically equivalent camera system depending upon the circumstances. Thus the image on the retina is inverted and suffers from chromatic aberration (see below) but the brain compensates for these effects. Conversely, high contrast objects or structures near the limit of resolution, such as planetary transits of the sun and the Martian 'canals', may be interpreted falsely.

The optical principles of the eye are shown in figure 1.1.1 and are too well known to be worth discussing in detail here. The receptors in the retina (figure 1.1.2) are of two types: cones for colour detection, and rods

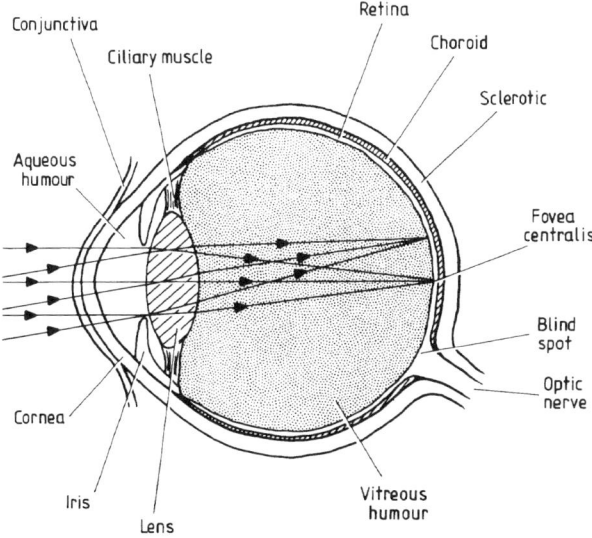

Figure 1.1.1. Optical paths in a horizontal cross-section of the eye.

for black and white reception at higher sensitivity. The light passes through the outer layers of nerve connections before arriving at these receptors. In the rods a pigment known as rhodopsin or, from its colour, visual purple, absorbs the radiation. It is a complex protein with a molecular weight of

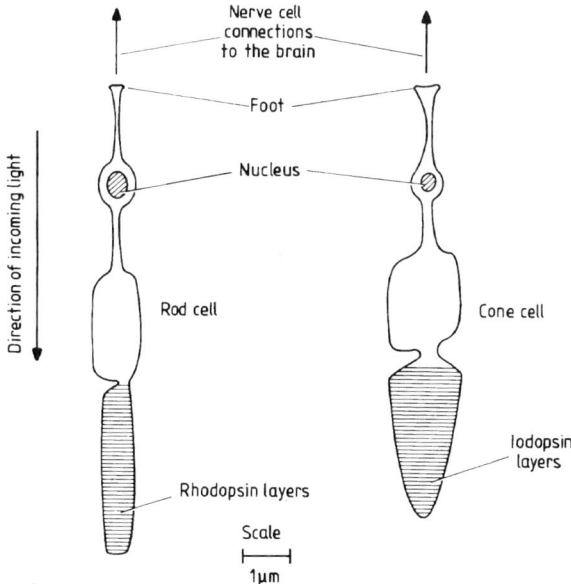

Figure 1.1.2. Retinal receptor cells.

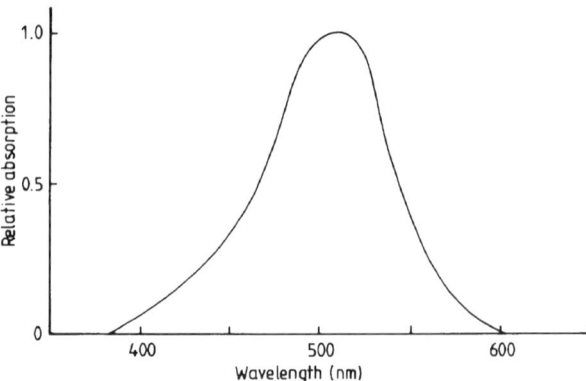

Figure 1.1.3. Rhodopsin absorption curve.

about 40 000 amu, and its absorption curve is shown in figure 1.1.3. It is arranged within the rods in layers about 20 nm thick and 500 nm wide, and may comprise up to 35% of the dry weight of the cell. Under the influence of light a small fragment of it will split off. This fragment, or chromophore, is a vitamin A derivative called retinal or retinaldehyde and has a molecular weight of 286 amu. One of its double bonds changes from a *cis* to a *trans* configuration (figure 1.1.4) within a picosecond of the absorption of the photon. The portion left behind is a colourless protein called opsin. The instant of visual excitation occurs at some stage during the splitting of the rhodopsin molecule, but its precise mechanism is not yet understood. The reaction causes a change in the permeability of the receptor cell's membrane to sodium ions and thus a consequent change in the electrical potential of the cell. The change in potential then propagates through the nerve cells to the brain. The rhodopsin molecule is slowly regenerated. The response of the cones is probably due to a similar mechanism. A pigment known as iodopsin is found in cones and this also contains the retinaldehyde group. Cone cells, however, are of three varieties with differing spectral sensitivities, and this seems to arise from the presence of much smaller quantities of other pigments. The absorption curves of these other pigments are shown

Figure 1.1.4. Transformation of retinaldehyde from *cis* to *trans* configuration by absorption of a photon near 500 nm.

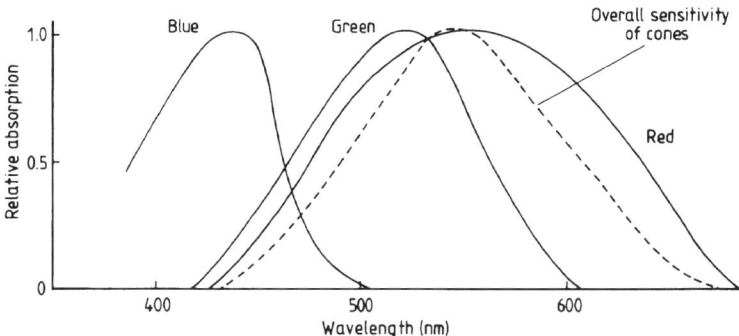

Figure 1.1.5. Absorption curves of pigments involved in cone vision.

in figure 1.1.5, together with the overall sensitivity curve of the cones for comparison.

In bright light much of the rhodopsin in the rods is split into opsin and retinaldehyde, and their sensitivity is therefore much reduced. Vision is then provided primarily by the cones, although their sensitivity is only about 1% of the maximum for the rods. The three varieties of cones combine their effects to produce colour vision. At low light levels only the rods are triggered by the radiation and vision is then in black and white. The overall spectral sensitivities of the rods and cones differ (figures 1.1.3 and 1.1.5) with that of the rods peaking at about 510 nm and that of the cones at 550 nm. This shift in sensitivity is called the Purkinje effect. It can be a problem for double-star observers since it may cause a hot and bright star to have its magnitude underestimated in comparison with a cooler and fainter star, and vice versa. Upon entering a dark observatory from a brightly illuminated room, the rhodopsin in the rods slowly re-forms over about half an hour; their sensitivity therefore improves concurrently. Thus we have the well-known phenomenon of dark adaptation, whereby far more can be distinguished after a few minutes in the dark than can be detected initially. If sensitive vision is important for an observation, then, for optimum results, bright lights should be avoided for half an hour before it is due to be made. Most observatories are only faintly illuminated and usually with red light in order to try and minimize any loss of dark adaptation.

Astronomical observation is generally due to rod vision. Usually with a dark-adapted eye, between one and ten photons are required to trigger an individual rod. However, several rods must be triggered in order to result in a pulse being sent to the brain. This arises because many rods are connected to a single nerve fibre. Most cones are also multiply connected, although a few, particularly those in the *fovea centralis*, have a one-to-one relationship with nerve fibres. The total number of rods is about 10^8 with about 6×10^6 cones and about 10^6 nerve fibres, so that upwards of a hundred receptors can be connected to a single nerve. In addition there are many cross connections between groups of receptors. The cones are concentrated towards the *fovea centralis*, and this is the region of most acute vision. Hence the rods are more plentiful towards the periphery of the field of view, and so the phenomenon called averted vision, whereby a faint object only becomes visible when *not* looked at directly, arises. Its image is falling on to a region of the retina richer in rods when the eye is averted from its direct line of sight. The combination of differing receptor sensitivities, change in pigment concentration, and aperture adjustment by the iris means that the eye is usable over a range of illuminations differing by a factor of 10^9 or 10^{10}.

The Rayleigh limit (see later) of resolution of the eye is about $20''$ of arc when the iris has its maximum diameter of 5–7 mm, but for two separate images to be distinguished, they must be separated on the retina by at least one unexcited receptor cell, so that even for images on the *fovea centralis* the actual resolution is between $1'$ and $2'$ of arc. This is much better than elsewhere on the retina since the *fovea centralis* is populated almost exclusively by small tightly packed cones, many of which are singly connected to the nerve fibres. Its slightly depressed shape may also cause it to act as a diverging lens producing a slightly magnified image in that region. Away from the *fovea centralis* the multiple connection of the rods, which may be as high as a thousand receptors to a single nerve fibre, degrades the resolution far beyond this figure. Other aberrations and variations between individuals in their retinal structure means that the average resolution of the human eye lies between $5'$ and $10'$ of arc for point sources. Linear sources such as an illuminated grating can be resolved down to $1'$ of arc fairly commonly. The effect of the granularity of the retina is minimized in images by rapid oscillations of the eye through a few tens of seconds of arc with a frequency of a few hertz, so that several receptors are involved in the detection when averaged over a short time.

With areas of high contrast, the brighter area is generally seen as too large, a phenomenon that is known as irradiation. We may understand this as arising from stimulated responses of unexcited receptors due to their cross-connections with excited receptors. We may also understand the eye fatigue that occurs when staring fixedly at a source (for example when guiding on a star) as due to depletion of the sensitive pigment in those few cells covered by the image. Averting the eye very slightly will then focus the image on different cells so reducing the problem. Alternatively the eye can be rested momentarily by looking away from the eyepiece to allow the cells to recover.

The response of vision to changes in illumination is logarithmic (the Weber–Fechner law). That is to say, if two sources, A and B, are observed to differ in brightness by a certain amount, and a third source, C, appears midway in brightness between them, then the energy from C will differ from that from A by the same *factor* as that by which B differs from C.

In other words, if we use L to denote the perceived luminosity and E to denote the actual energy of the sources, we have

$$\tfrac{1}{2}(L_A + L_B) = L_C \qquad (1.1.1)$$

then

$$\frac{E_A}{E_C} = \frac{E_C}{E_B}. \qquad (1.1.2)$$

This phenomenon is the reason for the magnitude scale used by astronomers to measure stellar brightnesses (section 3.1), since that scale had its origins in the eye estimates of stellar luminosities by the ancient astronomers. The faintest stars visible to the dark-adapted naked eye from a good observing site on a good clear moonless night are of about magnitude six. This corresponds to the detection of about 3×10^{-15} W. Special circumstances or especially sensitive vision may enable this limit to be improved upon by some one to one and a half stellar magnitudes. Conversely the normal ageing processes in the eye mean that the retina of a 60-year-old person receives only about 30% of the amount of light seen by a person half that age. Eye diseases and problems, such as cataracts, may reduce this much farther. Thus observers should expect a reduction in their ability to perceive very faint objects as time goes by.

Semiconductors

The photomultiplier, CCD and several of the other detectors considered later derive their properties from the behaviour of semiconductors. Thus some discussion of the relevant aspects of these materials is a necessary prerequisite to a full understanding of detectors of this type.

Conduction in a solid may be understood by considering the electron energy levels. For a single isolated atom they are unperturbed and are the normal atomic energy levels. As two such isolated atoms approach each other their interaction causes the levels to split (figure 1.1.6). If further atoms approach, then the levels develop further splits, so that for N atoms in close proximity to each other, each original level is split into N sublevels (figure 1.1.7). Within a solid, therefore, each level becomes a pseudo-continuous band of permitted energies since the individual sublevels overlap each other. The energy level diagram for a solid thus has the appearance shown in figure 1.1.8. The innermost electrons remain bound to their nuclei, while the outermost electrons interact to bind the atoms together. They occupy an energy band called the valence band.

In order to conduct electricity through such a solid, the electrons must be able to move. From figure 1.1.8 we may see that free movement could occur within the valence and higher bands. However, if the original atomic level that became the valence band upon the formation of the solid was

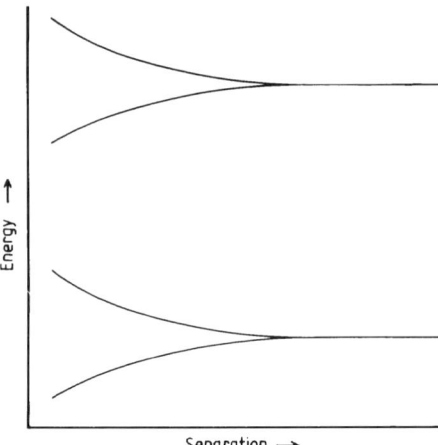

Figure 1.1.6. Schematic diagram of the splitting of two of the energy levels of an atom due to its proximity to another similar atom.

fully occupied, then all the sublevels within the valence band will still be fully occupied. If any given electron is to move under the influence of an electric potential, then its energy must increase. This it cannot do, since all the sublevels are occupied, and so there is no vacant level available for it at this higher energy. Thus the electron cannot move after all. Under these conditions we have an electrical insulator. If the material is to be a conductor, we can now see that there must be vacant sublevels that the conduction electron can enter. There are two ways in which such empty sublevels may

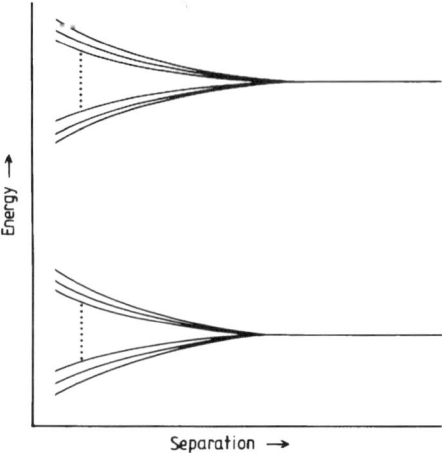

Figure 1.1.7. Schematic diagram of the splitting of two of the energy levels of an atom due to its proximity to many similar atoms.

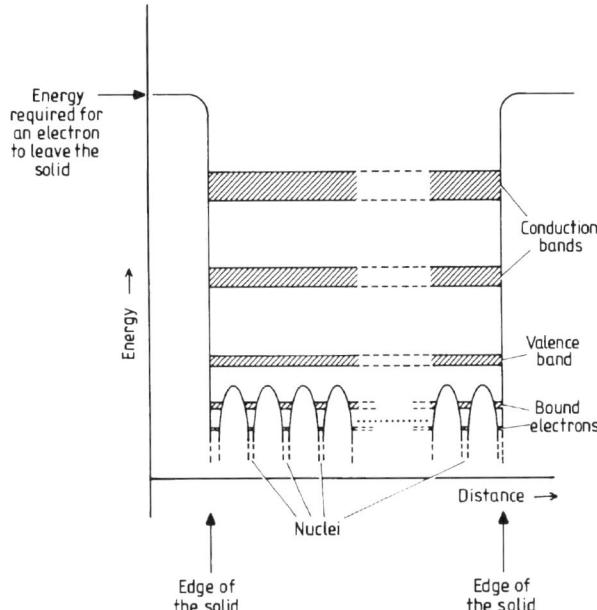

Figure 1.1.8. Schematic energy level diagram of a solid.

become available. Either the valence band is unfilled, for example when the original atomic level had only a single electron in an s subshell, or one of the higher energy bands becomes sufficiently broadened to overlap the valence band. In the latter case at a temperature of absolute zero, all the sublevels of both bands will be filled up to some energy that is called the Fermi level. Higher sublevels will be unoccupied. As the temperature rises some electrons will be excited to some of these higher sublevels, but will still leave room for conduction electrons.

A third type of behaviour occurs when the valence and higher bands do not actually overlap, but have only a small separation. Thermal excitation may then be sufficient to push a few electrons into some of the higher bands. An electric potential can now cause the electrons in either the valence or the higher band to move. The material is known as a semiconductor since its conductance is generally better than that of an insulator but considerably worse than that of a true conductor. The higher energy bands are usually known as the conduction bands. A pure substance will have equal numbers of electrons in its conduction bands and of spaces in its valence band. However, an imbalance can be induced by the presence of differing atomic species. If the valence band is full, and one atom is replaced by another that has a larger number of valence electrons (a donor atom), then the excess electron(s) usually occupy new levels in the gap between the valence and conduction bands. From there they may more easily be excited into the conduction band. The semiconductor is then an n-type since any current within it will largely be carried by the (negative) electrons in the conduction band. In the other case when an atom is replaced with one that has fewer valence electrons (an acceptor atom), spaces will become available within the valence band and the movement of electrons within this band will carry an electric current. Since the movement of an electron within the valence band is exactly equivalent to the movement of a positively charged hole in the opposite direction, this type of semiconductor is called p-type, and its currents are thought of as being transported by the positive holes in the valence band.

The photoelectric effect

The principle of the photoelectric effect is well known: the material absorbs a photon with a wavelength less than the limit for the material and an electron is emitted. The energy of the electron is a function of the energy of the photon, while the number of electrons depends upon the intensity of the illumination. In practice the situation is somewhat more complex, particularly when we are interested in specifying the properties of a *good* photoemitter.

The main requirements are that the material should absorb the required radiation efficiently, and that the mean free paths of the released electrons within the material should be greater than that of the photons. The relevance of these two requirements may be most simply understood from looking at the behaviour of metals in which neither condition is fulfilled. Since metals are conductors there are many vacant sublevels near their Fermi levels (see the earlier discussion). After absorption of a photon, an electron is moving rapidly and energetically within the metal. Collisions will occur with other electrons and since these electrons have other nearby sublevels that they may occupy, they can absorb some of the energy from our photoelectron. Thus the photoelectron is slowed by collisions until it may no longer have sufficient energy to escape from the metal even if it does reach the surface. For most metals the mean free path of a photon is about 10 nm and that of the released electron less than 1 nm, thus the eventually emitted number of electrons is very considerably reduced by collisional energy losses. Furthermore the high reflectivity of metals results in only a small fraction of the suitable photons being absorbed and so the actual number of electrons emitted is only a very small proportion of those potentially available.

A good photoemitter must thus have low energy loss mechanisms for its released electrons whilst they are within its confines. The loss mechanism in metals (collision) can be eliminated by the use of semiconductors or insulators. Then the photoelectron cannot lose significant amounts of energy to the valence electrons because there are no vacant levels for the latter to occupy. Neither can it lose much energy to the conduction electrons since there are very few of these around. In insulators and semiconductors, the

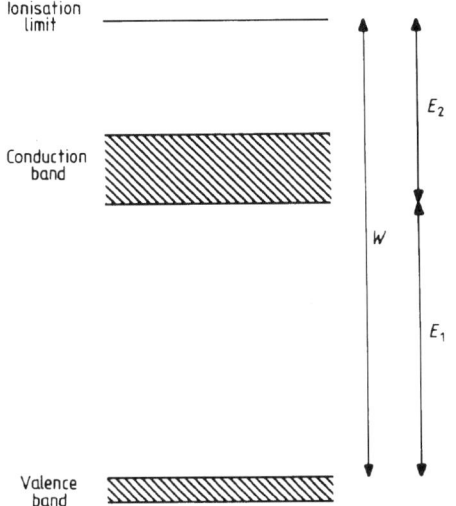

Figure 1.1.9. Schematic partial Grotrian diagram of a good photoemitter.

two important energy loss mechanisms are pair production and sound production. If the photoelectron is energetic enough, then it may collisionally excite other valence electrons into the conduction band thus producing *pairs* of electrons and holes. This process may be eliminated by requiring that E_1, the minimum energy to excite a valence electron into the conduction band of the material (figure 1.1.9), is larger than E_2, the excess energy available to the photoelectron. Sound waves or phonons are the vibrations of the atoms in the material and can be produced by collisions between the photoelectrons and the atoms especially at discontinuities in the crystal lattice etc. Only 1% or so of the electron's energy will be lost at each collision because the atom is so much more massive than the electron. However, the mean free path between such collisions is only 1 or 2 nm, so this energy loss mechanism becomes very significant when the photoelectron originates deep in the material. The losses may be reduced by cooling the material since this then reduces the available number of quantized vibration levels in the crystal and also increases the electron's mean free path.

  The minimum energy of a photon if it is to be able to produce photo-emission is known as the work function and is the difference between the ionization level and the top of the valence band (figure 1.1.9). Its value may, however, be increased by some or all of the energy loss mechanisms mentioned above. In practical photoemitters, pair production is particularly important at photon energies above the minimum and may reduce the expected flux considerably as the wavelength decreases. The work function is also strongly dependent upon the surface properties of the material; surface defects, oxidation products, impurities etc. can cause it to vary widely even among samples of the same substance. The work function may be reduced if the material is strongly p-type and is at an elevated temperature. Vacant levels in the valence band may then be populated by thermally excited electrons, bringing them nearer to the ionization level and so reducing the energy required to let them escape. Since most practical photoemitters are strongly p-type, this is an important process and confers sensitivity at longer wavelengths than the nominal cut-off point. The long-wave sensitivity, however, is variable since the degree to which the substance is p-type is strongly dependent upon the presence of impurities and so is very sensitive to small changes in the composition of the material.

A detector index

After the 'natural' detector formed by the eye, there are numerous types of 'artificial' optical detector. Before going on to look at some of them in more detail, however, it is necessary to place them in some sort of logical framework or the reader is likely to become more confused rather than more enlightened by this section. We may idealize any detector as simply a device wherein some measurable property changes in response to the effects of electromagnetic radiation. We may then classify the types of detector according to the property that is changing, and this is shown in table 1.1.1.

Other properties may be sensitive to radiation but fail to form the basis of a useful detector. For example, forests can catch fire, and some human

Table 1.1.1. Classification scheme for types of detector.

Sensitive parameter	Detector names	Class
Voltage	Photovoltaic cells	Quantum
	Thermocouples	Thermal
	Pyroelectric detectors	Thermal
Resistance	Blocked impurity band device (BIB)	Quantum
	Bolometer	Thermal
	Photoconductive cell	Quantum
	Phototransistor	Quantum
	Transition edge sensor (TES)	Thermal
Charge	Charge-coupled device (CCD)	Quantum
	Charge injection device (CID)	Quantum
Current	Superconducting tunnel junction (STJ)	Quantum
Electron excitation	Photographic emulsion	Quantum
Electron emission	Photomultiplier	Quantum
	Television	Quantum
	Image intensifier	Quantum
Chemical composition	Eye	Quantum

skins turn brown under the influence of solar radiation, but these are extravagant or slow ways of detecting the sun. Yet other properties may become the basis of detectors in the future. Such possibilities might include the initiation of stimulated radiation from excited atomic states as in the laser and the changeover from superconducting to normally conducting regimes in a material (see TES detectors).

Detector parameters

Before resuming discussion of the detector types listed earlier, we need to establish the definitions of some of the criteria used to assess and compare detectors. The most important of these are listed in table 1.1.2.

For purposes of comparison D^* is generally the most useful parameter. For photomultipliers in the visible region it is around 10^{15} to 10^{16}. Figures for the eye and for photographic emulsion are not directly obtainable, but values of 10^{12} to 10^{14} can perhaps be used for both to give an idea of their relative performances.

Cryostats

The noise level in many detectors can be reduced by cooling them to below ambient temperature. Indeed for some detectors, such as STJs and TESs (see below) cooling to very low temperatures is essential for their operation. Small CCDs produced for the amateur market are usually chilled by Peltier-effect coolers (see below), but almost all other approaches require the use of liquid nitrogen or liquid helium as the coolant. Since these materials are both liquids, they must be contained in such a way that the liquid does not spill out as the telescope moves. The container is called a cryostat, and while there are many different detailed designs, the basic requirements are much the same for all detectors. In addition to acting as a container for the coolant, the cryostat must ensure that the detector, and sometimes pre-amplifiers, filters etc., are cooled whatever the position of the telescope, that the coolant is retained for as long as possible, that the evaporated coolant can escape and that the detector and other cooled items do not ice up. These requirements invariably mean that the container is a Dewar (vacuum flask) and that the detector is behind a window within a vacuum or dry atmosphere. Sometimes the window is heated to avoid ice forming on its surface. Most cryostats are of the 'bath' design, and are essentially simply tanks containing the coolant with the detector attached to the outside of the tank, or linked to it by a thermally conducting rod. Such devices can only be half filled with the coolant if none is to overflow as they are tilted by the telescope's motion, though if the movement of the telescope is restricted, higher levels of filling may be possible. Hold times between refilling with coolant of a few days can currently be achieved. When operating at Nasmyth or Coudé foci,

Table 1.1.2. Some criteria for assessment and comparison of detectors.

QE (quantum efficiency)	Ratio of the actual number of photons that are detected to the number of incident photons.
DQE (detective quantum efficiency)	Square of the ratio of the output signal/noise ratio to the input signal/noise ratio.
τ (time constant)	This has various precise definitions. Probably the most widespread is that τ is the time required for the detector output to approach to within $(1 - e^{-1})$ of its final value after a change in the illumination, i.e. the time required for about 63% of the final change to have occurred.
Dark noise	The output from the detector when it is un-illuminated. It is usually measured as a root-mean-square voltage or current.
NEP (noise equivalent power or minimum detectable power)	The radiative flux as an input, that gives an output signal-to-noise ratio of unity. It can be defined for monochromatic or black body radiation, and is usually measured in watts.
D (detectivity)	Reciprocal of NEP. The signal-to-noise ratio for incident radiation of unit intensity.
D^* (normalized detectivity)	The detectivity normalized by multiplying by the square root of the detector area, and by the electrical bandwidth. It is usually pronounced 'dee star'.

$$D^* = \frac{(a\,\Delta f)^{1/2}}{\text{NEP}} \qquad (1.1.3)$$

	The units $\text{cm}\,\text{Hz}^{1/2}\,\text{W}^{-1}$ are commonly used and it then represents the signal-to-noise ratio when 1 W of radiation is incident on a detector with an area of $1\,\text{cm}^2$, and the electrical bandwidth is 1 Hz.
R (responsivity)	Detector output for unit intensity input. Units are usually volts per watt or amps per watt.
Dynamic range	Ratio of the saturation output to the dark signal. Sometimes only defined over the region of linear response.
Spectral response	The change in output signal as a function of changes in the wavelength of the input signal.
λ_m (peak wavelength)	The wavelength for which the detectivity is a maximum.
λ_c (cut-off wavelength)	There are various definitions. Among the commonest are: wavelength(s) at which the detectivity falls to zero; wavelength(s) at which the detectivity falls to 1% of its peak value; wavelength(s) at which D^* has fallen to half its peak value.

where the detector is not tilted, continuous flow cryostats can be used, where the coolant is supplied from a large external reservoir, and hold times of weeks or longer are then possible.

Bolometers, STJs and TESs require cooling to temperatures well below 1 K. Temperatures down to about 250 mK can be reached using liquid $^3_2$He. The $^3_2$He itself has to be cooled to below 2 K before it liquefies, and this is achieved by using $^4_2$He under reduced pressure. Temperatures down to a few mK require a dilution refrigerator. This uses a mix of $^3_2$He and $^4_2$He at a temperature lower than 900 mK. The two isotopes partially separate out under gravity, but the lower $^4_2$He layer still contains some $^3_2$He. The $^3_2$He is removed from the lower layer, distilling it off at 600 mK in a separate chamber. This forces some $^3_2$He from the upper layer to cross the boundary between the two layers to maintain the concentration. However, crossing the boundary requires energy, and this is drawn from the surroundings, so cooling them. SCUBA (see bolometers below) uses a dilution refrigeration system in order to operate at 100 mK. An alternative route to mK temperatures is the adiabatic demagnetization refrigerator. The ions in a paramagnetic salt are first aligned by a strong magnetic field. The heat generated in this process is transferred to liquid helium via a thermally conducting rod. The rod is then moved from contact with the salt, the magnetic field reduced and the salt cools adiabatically.

Charge-coupled devices (CCD)

Willard Boyle and George Smith invented CCDs in 1969 at the Bell telephone labs for use as a computer memory. The first application of CCDs within astronomy as optical detectors occurred in the late 1970s. Since then they have come to dominate completely the detection of optical radiation at professional observatories, and to be very widely used among amateur astronomers. Their popularity arises from their ability to integrate the detection over long intervals, their dynamic range ($>10^5$), high quantum efficiency and from the ease with which arrays can be formed to give two-dimensional imaging. In fact, CCDs can only be formed as an array; a single unit is of little use by itself.

The basic detection mechanism is related to the photoelectric effect. Light incident on a semiconductor (usually silicon) produces electron–hole pairs as we have already seen. These electrons are then trapped in potential wells produced by numerous small electrodes. There they accumulate until their total number is read out by charge coupling the detecting electrodes to a single read-out electrode.

An individual unit of a CCD is shown in figure 1.1.10. The electrode is insulated from the semiconductor by a thin oxide layer. In other words, the device is related to the metal oxide–silicon or MOS transistors. It is held at a small positive voltage that is sufficient to drive the positive holes in the p-type silicon away from its vicinity and to attract the electrons into a thin layer immediately beneath it. The electron–hole pairs produced in this depletion region by the incident radiation are thereby separated and the electrons accumulate in the storage region. Thus an electron charge is formed whose magnitude is a function of the intensity of the illuminating radiation. In effect, the unit is a radiation-driven capacitor.

Figure 1.1.10. Basic unit of a CCD.

Now if several such electrodes are formed on a single silicon chip and zones of very high p-type doping insulate the depletion regions from each other, then each will develop a charge that is proportional to its illuminating intensity. Thus we have a spatially digitized electric analogue of the original optical image (figure 1.1.11). All that remains is to retrieve this electron

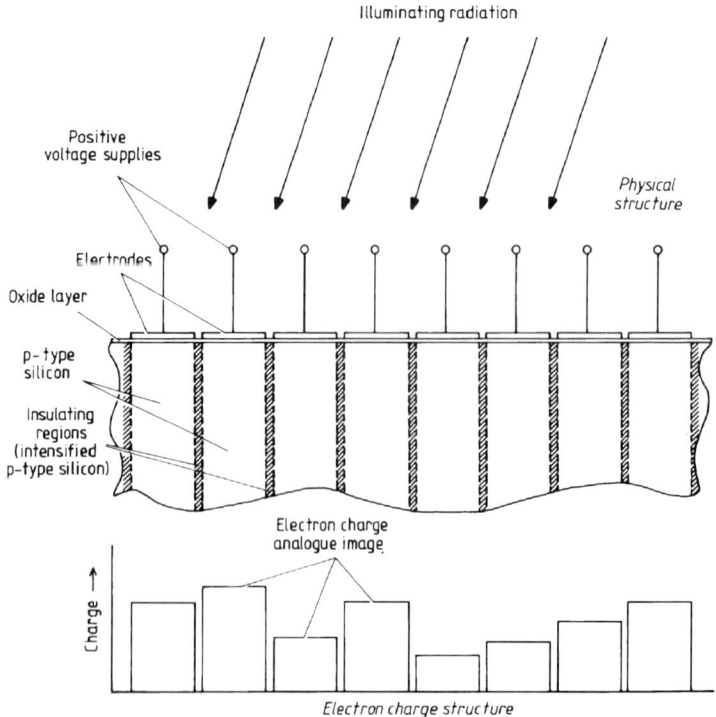

Figure 1.1.11. Array of CCD basic units.

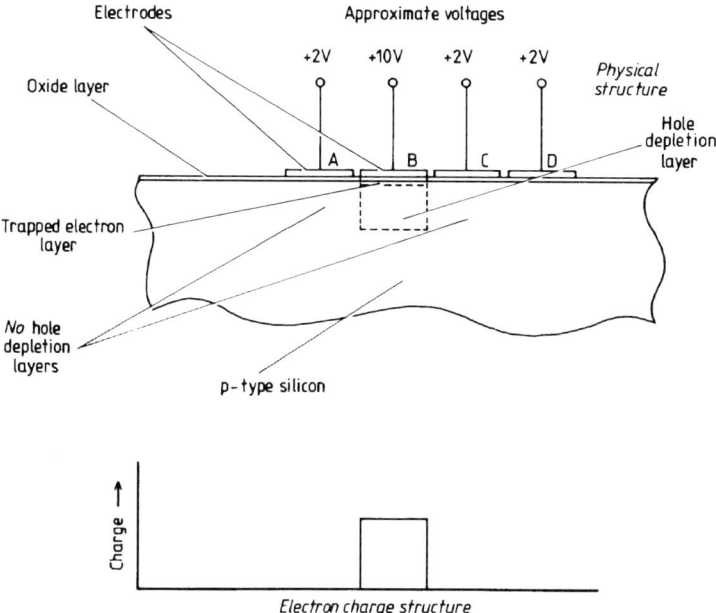

Figure 1.1.12. Active electron charge trapping in CCD.

image in some usable form. This is accomplished by charge coupling. Imagine an array of electrodes such as those we have already seen in figure 1.1.10, but without their insulating separating layers. Then if one such electrode acquired a charge, it would diffuse across to the nearby electrodes. However, if the voltage of the electrodes on either side of the one containing the charge were reduced, then their hole depletion regions would disappear and the charge would once again be contained within two p-type insulating regions (figure 1.1.12). This time, however, the insulating regions are not permanent but may be changed by varying the electrode voltage. Thus the stored electric charge may be moved physically through the structure of the device by sequentially changing the voltages of the electrodes. Hence in figure 1.1.12, if the voltage on electrode C is changed to about +10 V, then a second hole depletion zone will form adjacent to the first. The stored charge will diffuse across between the two regions until it is shared equally. Now if the voltage on electrode B is gradually reduced to +2 V, its hole depletion zone will gradually disappear and the remaining electrons will transfer across to be stored under electrode C. Thus by cycling the voltages of the electrodes as shown in figure 1.1.13, the electron charge is moved from electrode B to electrode C. With careful design the efficiency of this charge transfer (or coupling) may be made as high as 99.9999%. Furthermore we may obviously continue to move the charge through the structure to electrodes D, E, F etc. by continuing to cycle the voltages in an appropriate

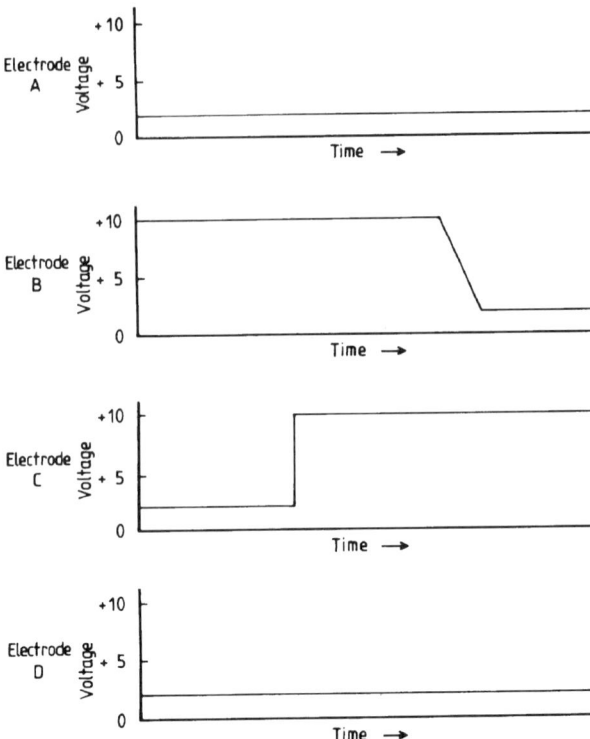

Figure 1.1.13. Voltage changes required to transfer charge from electrode B to electrode C in the array shown in figure 1.1.12.

fashion. Eventually the charge may be brought to an output electrode from whence its value may be determined by discharging it through an integrating current meter or some similar device. In the scheme outlined here, the system requires three separate voltage cycles to the electrodes in order to move the charge. Hence it is known as a three-phase CCD. Most CCDs used in astronomy are three-phase devices. Two-phase and virtual-phase CCDs are discussed below. Three separate circuits are formed, with each electrode connected to those three before and three after it (figure 1.1.14). The voltage supplies, alpha, beta and gamma (figure 1.1.14), follow the cycles shown in figure 1.1.15 in order to move charge packets from the left towards the right. Since only every third electrode holds a charge in this scheme the output follows the pattern shown schematically at the bottom of figure 1.1.15. The order of appearance of an output pulse at the output electrode is directly related to the position of its originating electrode in the array. Thus the original spatial charge pattern and hence the optical image may easily be inferred from the time-varying output.

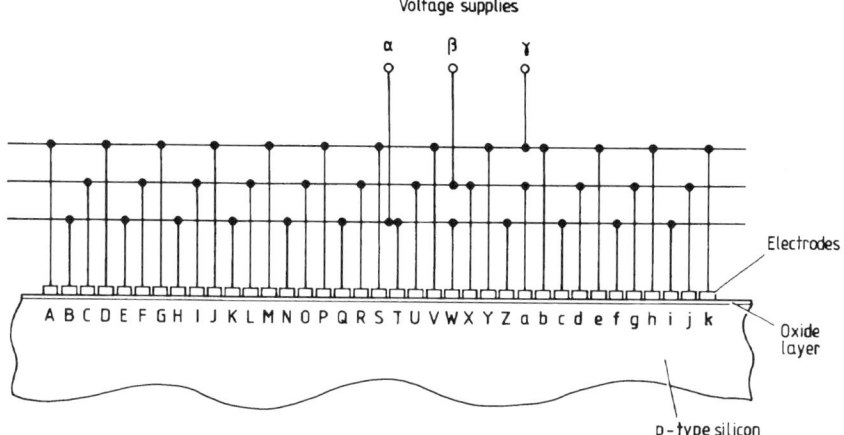

Figure 1.1.14. Connection diagram for a three-phase CCD.

The complete three-phase CCD is a combination of the detecting and charge transfer systems. Each pixel has three electrodes and is isolated from pixels in adjacent columns by insulating barriers (figure 1.1.16). During an exposure electrodes B are at their full voltage and the electrons from the whole area of the pixel accumulate beneath them. Electrodes A and C meanwhile are at a reduced voltage and so act to isolate each pixel from its neighbours along the column. When the exposure is completed, the voltages in the three electrode groups are cycled as shown in figure 1.1.15 until the first set of charges reaches the end of the column. At the end of the column a second set of electrodes running orthogonally to the columns (figure 1.1.16) receives the charges into the middle electrode for each column. That electrode is at the full voltage, and its neighbours are at reduced voltages, so that each charge package retains its identity. The voltages in the read-out row of electrodes are then cycled to move the charges to the output electrode where they appear as a series of pulses. When the first row of charges has been read out, the voltages on the column electrodes are cycled to bring the second row of charges to the read-out electrodes, and so on until the whole image has been retrieved. With the larger format CCDs, the read-out process can take some time. In order to improve observing efficiency, some devices therefore have a storage area between the imaging area and the read-out electrodes. This is simply a second CCD that is not exposed to radiation or half of the CCD is covered by a mask (frame transfer CCD). The image is transferred to the storage area in less than 0.1 ms, and while it is being read out from there, the next exposure can commence on the detecting part of the CCD. Even without a separate storage area, reading the top half of the pixels in a column upwards and the other half downwards can halve read-out times.

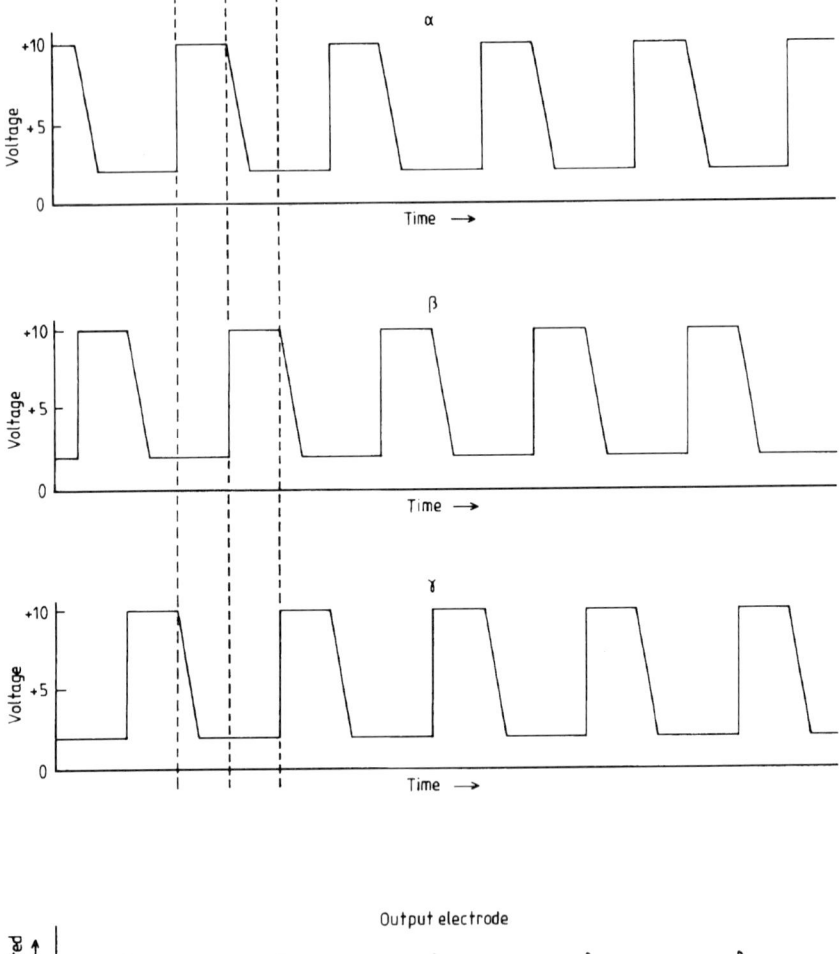

Figure 1.1.15. Voltage and output cycles for a three-phase CCD.

A two-phase CCD requires only a single clock, but needs double electrodes to provide directionality to the charge transfer (figure 1.1.17). The second electrode, buried in the oxide layer, provides a deeper well than that under the surface electrode and so the charge accumulates under the former. When voltages cycle between 2 V and 10 V (figure 1.1.18), the stored charge is attracted over to the nearer of the two neighbouring surface electrodes and then accumulates again under the buried electrode.

Figure 1.1.16. Schematic structure of a three-phase CCD.

Thus cycling the electrode voltages, which may be done from a single clock, causes the charge packets to move through the structure of the CCD, just as for the three-phase device.

A virtual-phase CCD requires just one set of electrodes. Additional wells with a fixed potential are produced by p and n implants directly into the silicon substrate. The active electrode can then be at a higher or lower potential as required to move the charge through the device. The active electrodes in a virtual phase CCD are physically separate from each other leaving parts of the substrate directly exposed to the incoming radiation. This enhances their sensitivity, especially at short wavelengths.

Non-tracking instruments such as the Carlsberg Meridian (section 5.1), liquid mirror and Hobby–Eberly telescopes (see below) can follow the

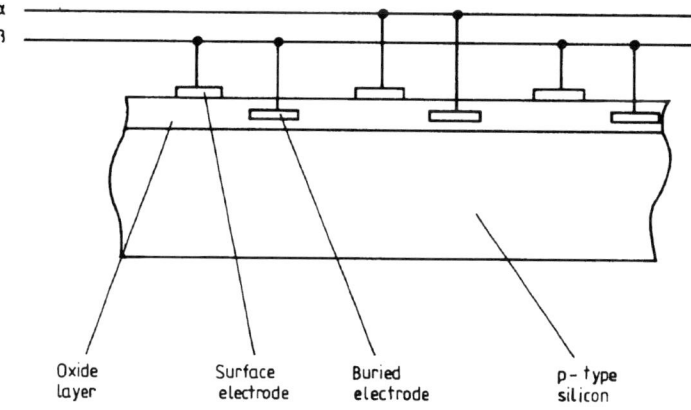

Figure 1.1.17. Physical structure of a two-phase CCD.

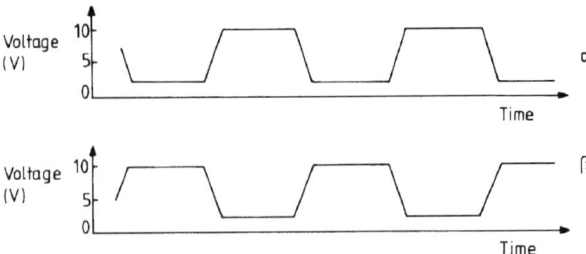

Figure 1.1.18. Voltage cycles for a two-phase CCD.

motion of objects in the sky by transferring the charges in their CCD detectors at the same speed as the image drifts across their focal planes (time delayed integration or TDI). To facilitate this, orthogonal transfer CCDs (OTCCDs) have recently been made. These can transfer the charge in up to eight directions (up/down, left/right, and at $45°$ between these directions). OTCCDs can also be used for active image motion compensation arising from scintillation, wind shake etc.

A combination of photomultiplier/image intensifier (sections 2.1 and 2.3) and CCD, known as an electron bombarded CCD or an intensified CCD (EBCCD or ICCD), has recently been developed. This places a negatively charged photocathode before the CCD. The photoelectron is accelerated by the voltage difference between the photocathode and the CCD, and hits the CCD at high energy, producing many electron–hole pairs in the CCD for each incident photon. This might seem to give the device a quantum efficiency of over 100% but it is in effect merely another type of amplifier; the signal-to-noise ratio remains that of the basic device (or worse), and so no additional information is obtained.

Interline transfer CCDs have an opaque column adjacent to each detecting column. The charge can be rapidly transferred into the opaque columns and read out from there more slowly while the next exposure is obtained using the detecting columns. This enables rapid exposures to be made, but half the detector is dead space. They are mainly used for digital video cameras, and rarely find astronomical applications.

The electrodes on or near the surface of a CCD can reflect some of the incident light, thereby reducing the quantum efficiency and changing the spectral response. To overcome this problem several approaches have been adopted. First, transparent polysilicon electrodes replace the metallic electrodes used in early devices. Second, the CCD may be illuminated from the back so that the radiation does not have to pass through the electrode structure at all. This, however, requires that the silicon forming the CCD be very thin so that the electrons produced by the incident radiation are collected efficiently. Nonetheless, even with thicknesses of only 10 to 20 μm, some additional cross talk (see below) may occur. More importantly, the process of thinning the CCD chips is risky and many devices may be damaged during the operation. Successfully thinned CCDs are therefore very expensive in order to cover the cost of the failures. They are also very fragile and can become warped, but they do have the advantage of being less affected by cosmic rays than thicker CCDs (see below). With a suitable anti-reflection coating, a back-illuminated CCD can now reach a quantum efficiency of 90% in the red and near infrared. Other methods of reducing reflection losses include using extremely thin electrodes, or an open electrode structure (as in a virtual-phase CCD) that leaves some of the silicon exposed directly to the radiation. At longer wavelengths, the thinned CCD may become semi-transparent. Not only does this reduce the efficiency because fewer photons are absorbed, but interference fringes may occur between the two faces of the chip. These can become very obtrusive and have to be removed as a part of the data reduction process.

The CCD as so far described suffers from loss of charge during transfer because of imperfections at the interface between the substrate and the insulating oxide layer. This affects the faintest images worst since the few electrons that have been accumulated are physically close to the interface. Artificially illuminating the CCD with a low-level uniform background during an exposure will ensure that even the faintest parts of the image have the surface states filled. This offset or 'fat zero' can then be removed later in signal processing. Alternatively, a positively charged layer of n-type silicon between the substrate and the insulating layer forces the charge packets away from the interface, producing a buried-channel CCD. Charge transfer efficiencies for CCDs now approach 99.9999%, leading to typical read-out noise levels of one to two electrons per pixel. Using a non-destructive read-out mechanism and repeating and averaging many read-out sequences can reduce the read-out noise further. To save time, only the lowest-intensity parts of the image are repeatedly read out, with the high-intensity parts being skipped. The devices have therefore become known as skipper CCDs.

The spectral sensitivity of CCDs ranges from 400 nm to 1100 nm, with a peak where the quantum efficiency can approach 90% near 750 nm. The quantum efficiency drops off further into the infrared as the silicon becomes more transparent. Short-wave sensitivity can be conferred by coating the device with an appropriate phosphor (see below) to convert the radiation to the sensitive region. The CCD regains sensitivity at very short wavelengths because x-rays are able to penetrate the device's surface electrode structure.

For integration times longer than a fraction of a second, it is usually necessary to cool the device in order to reduce its dark signal. Using liquid nitrogen for this purpose and operating at around $-100\,°C$, integration times of minutes to hours are easily attained. Small commercial CCDs produced for the amateur market, that nonetheless often find applications at professional observatories for guiding etc., often use Peltier coolers to get down to about 50 °C below the ambient temperature. Subtracting a dark frame from the image can further reduce the dark signal. The dark frame is in all respects (exposure length, temperature, etc.) identical to the main image, except that the camera shutter remains closed whilst it is obtained. It therefore just comprises the noise elements present within the detector and its electronics. Because the dark noise itself is noisy, it may be better to use the average of several dark frames obtained under identical conditions.

Two-dimensional arrays are made by stacking linear arrays. The operating principle is unchanged, and the correlation of the output with position within the image is only slightly more complex than before. The largest single CCD arrays currently produced are 2k × 4k (2048 × 4096) pixels in size. 10k × 10k CCDs are being envisaged, but this is probably approaching the practical limit because of the time taken to read out such large images. For an adaptive optics telescope (see later in this section) operating at $0.2''$ resolution, a 2k × 4k CCD covers only $200'' \times 400''$ of the sky if the resolution is to be preserved. Many applications require larger areas of the sky than this to be imaged, so several such CCDs must then be formed into a mosaic. However, the normal construction of a CCD with electrical connections on all four edges means that there will then be a gap of up to 10 mm between each device, resulting in large dead-space gaps in the eventual image. To minimize the dead space, 'three-edge-buttable' CCDs are now made. These have all the connections brought to one of the short edges, reducing the gaps between adjacent devices to about 0.2 mm on the other three edges. It is thus possible to form mosaics with two rows and as many columns as desired with a minimum of dead space. Currently the largest such mosaics use twelve 2k × 4k CCDs to give an 8k × 12k total area. A four-row by eight-column mosaic (OmegaCAM) is currently under construction for the 2.6 m VLT survey telescope. This will cover a 1° square field of view, but the two outer rows are separated from the central rows by wide dead spaces.

For typical noise levels and pixel capacities, astronomical CCDs have dynamic ranges of 100 000 to 500 000 (usable range in accurate brightness determination of up to 14.5^m). This therefore compares very favourably with the dynamic range of less than 1000 (brightness range of less than 7.5^m) available from a typical photographic image.

A major problem with CCDs used as astronomical detectors is the noise introduced by cosmic rays. A single cosmic ray particle passing through one of the pixels of the detector can cause a large number of ionizations. The resulting electrons accumulate in the storage region along with those produced by the photons. It is usually possible for the observer to recognize such events in the final image because of the intense 'spike' that is produced. Replacing the signal from the affected pixel by the average of the eight surrounding pixels improves the appearance of the image, but does not retrieve the original information. This correction often has to be done 'by hand' and is a time-consuming process. When two or more images of the same area are available, automatic removal of the cosmic ray spikes is possible with reasonable success rates.

Another serious defect of CCDs is the variation in background noise between pixels. This takes two forms. There may be a large-scale variation of 10–20% over the whole sensitive area, and there may be individual pixels with permanent high background levels (hot spots). The first problem has been much reduced by improved production techniques and may largely be eliminated during subsequent signal processing (flat fielding), if its effect can be determined by observing a uniform source. Commonly used sources for the flat field include the twilit sky and a white screen inside the telescope dome illuminated by a single distant light source. The flat field image is divided into the main image after dark frame subtraction from both images to reduce the large scale sensitivity variations. The effect of a single hot spot may also be reduced in signal processing by replacing its value with the mean of the four surrounding pixels. However, the hot spots are additionally often poor transferors of charge and there may be other bad pixels. All preceding pixels are then affected as their charge packets pass through the hot spot or bad pixel and a spurious line is introduced into the image. There is little that can be done to correct this last problem, other than to buy a new CCD. Even the 'good' pixels do not have 100% charge transfer efficiency, so that images containing very bright stars show a 'tail' to the star caused by the electrons lost to other pixels as the star's image is read out.

Yet another problem is that of cross talk or blooming. This occurs when an electron strays from its intended pixel to one nearby. It affects rear-illuminated CCDs because the electrons are produced at some distance from the electrodes, and is the reason why such CCDs have to be thinned. It can also occur for any CCD when the accumulating charge approaches the maximum capacity of the pixel. Then the mutual repulsion of the negatively charged electrons may force some over into adjacent pixels. The capacity of each pixel depends upon its physical size and is around half a million electrons for 25 μm sized pixels. Some CCDs have extra electrodes to enable excess charges to be bled away before they spread into nearby pixels. Such electrodes are known as drains, and their effect is often adjustable by means of an anti-blooming circuit. Anti-blooming should not be used if you intend making photometric measurements on the final image, but otherwise it can be very effective in improving the appearance of images containing both bright and faint objects.

The size of the pixels in a CCD can be too large to allow a telescope to operate at its limiting resolution. In such a situation, the images can be 'dithered' to provide improved resolution. Dithering consists simply of obtaining multiple images of the same field of view, with a shift of the position of the image on the CCD between each exposure by a fraction of the size of a pixel. The images are then combined to give a single image with sub-pixel resolution. A further improvement is given by 'drizzling' (also known as variable pixel linear reconstruction) in which the pixel size is first shrunk leaving gaps between the pixels. The images are then rotated before mapping on to a finer scale output grid.

One of the major advantages of a CCD over a photographic emulsion is the improvement in the sensitivity. However, widely different estimates of the degree of that improvement may be found in the literature, based upon different ways of assessing performance. At one extreme, there is the ability of a CCD to provide a recognizable image at very short exposures, because of its low noise. Based upon this measure, the CCD is perhaps 1000 times faster than photographic emulsion. At the other extreme, one may use the time taken to reach the mid-point of the dynamic range. Because of the much greater dynamic range of the CCD, this measure suggests CCDs are about five to ten times faster than photography. The most sensible measure of the increase in speed, however, is based upon the time required to reach the same signal-to-noise ratio in the images (i.e. to obtain the same information). This latter measure suggests that, in their most sensitive wavelength ranges (around 750 nm for CCDs and around 450 nm for photographic emulsion), CCDs are 20 to 50 times faster than photographic emulsion.

Photography

This is dealt with as a part of imaging in section 2.2. Here it is sufficient to point out that the basic mechanism involved in the formation of the latent image is pair production. The electrons excited into the conduction band are trapped at impurities, while the holes are removed chemically. Unsensitized emulsion is blue sensitive as a consequence of the minimum energy required to excite the valence electrons in silver bromide.

Photomultipliers

Electron multiplier phototubes (or photomultipliers as they are more commonly but less accurately known) were once the workhorses of optical photometry (chapter 3). They continue sometimes to be used when individual photons need to be detected as in the neutrino and cosmic ray Čerenkov detectors (sections 1.3 and 1.4) or when very rapid responses are required as in the observation of occultations (section 2.7). They may also be used on board spacecraft for ultraviolet measurements in the 10–300 nm region where CCDs are insensitive.

Photomultipliers detect photons through the photoelectric effect. The basic components and construction of the device are shown in figure 1.1.19.

Figure 1.1.19. Schematic arrangement for a photomultiplier.

The photoemitter is coated on to the cathode and this is at a negative potential of some 1000 V. Once a photoelectron has escaped from the photoemitter, it is accelerated by an electric potential until it strikes a second electron emitter. The primary electron's energy then goes into pair production and secondary electrons are emitted from the substance in a manner analogous to photo-electron emission. Since typically $1\,\text{eV}$ $(1.6 \times 10^{-19}\,\text{J})$ of energy is required for pair production and the primary's energy can reach $100\,\text{eV}$ or more by the time of impact, several secondary electron emissions result from a single primary electron. This multiplication of electrons cannot, however, be increased without bound, since if the primary's energy is too high it will penetrate too far into the material for the secondary electrons to escape. The upper limit, however, is a factor of one hundred or more and is not usually a restriction in practical devices in which multiplications by factors of ten or less are used. The properties of a good secondary electron emitter are the same as those of a good photoemitter except that its optical absorption and reflection are no longer important.

The secondary emitter is coated on to dynodes that are successively more positive than the cathode by $100\,\text{V}$ or so for each stage. The various electrodes are shaped and positioned so that the electrons are channelled towards the correct next electrode in the sequence after each interaction. The final signal pulse may contain 10^6 electrons for each incoming photon, and after arriving at the anode it may be further amplified and detected in any of the usual ways. This large intrinsic amplification of the photo-multiplier is one of the major advantages of the device.

The electrode may be opaque as illustrated in figure 1.1.19 so that the electrons emerge from the illuminated face, or it may be semitransparent, when they emerge from the opposite face. In the latter case some change to the spectral sensitivity occurs due to the absorption in the material, but there are the compensating advantages of higher quantum efficiency and convenience since the cathodes may be coated on the inside of the window. Semitransparent cathodes are usually only about 30 nm thick so that it is very easy for small thickness changes to occur and to alter their sensitivities. Their usual application is for long-wave radiation detection.

The micro-channel plate (figure 1.3.4) can also be used at optical wavelengths. With an array of anodes to collect the clouds of electrons emerging from the plate, it provides an imaging detector with a high degree of intrinsic amplification. Such devices are often referred to as MAMA (Multi-Anode Micro-channel Array) detectors.

The commonest photoelectric and secondary emitter materials in current use are listed in table 1.1.3.

Noise in the signal from a photomultiplier arises from many sources. The amplification of each pulse can vary by as much as a factor of ten

Table 1.1.3. Photoelectron emitting substances.

Substance	Long wavelength cut-off point (nm)
Sodium chloride (NaCl)	150
Potassium bromide (KBr)	155
Rubidium iodide (RbI)	185
Cuprous chloride (CuCl)	190
Caesium iodide (CsI)	200
Copper/beryllium (Cu/Be)	200
Copper iodide (CuI)	210
Rubidium telluride ($RbTe_2$)	300
Caesium telluride (Cs_2Te)	350
Caesium antimonide ($Cs_{2.9}Sb$)	600–700
Bi-alkali ((K_2Cs)Sb)	670
Tri-alkali ((Cs) Na_2KSb)	850
Gallium arsenide (GaAs (Cs))	1000
Silver/oxygen/caesium (Ag/Cs_2O)	1000–1100

Secondary electron emitting substances
Beryllium oxide (BeO (Cs))
Caesium antimonide (Cs_3Sb)
Gallium phosphide (GaP (Cs))
Magnesium oxide (MgO (Cs))
Potassium chloride (KCl)
Silver/magnesium (Ag/Mg)

through the sensitivity variations and changes in the number of secondary electrons lost between each stage. The final registration of the signal can be by analogue means and then the pulse strength variation is an important noise source. Alternatively individual pulses may be counted and then it is less significant. Indeed, when using pulse counting even the effects of other noise sources can be reduced by using a discriminator to eliminate the very large and very small pulses which do not originate in primary photon interactions. Unfortunately, however, pulse counting is limited in its usefulness to faint sources otherwise the individual pulses start to overlap. Electrons can be emitted from either the primary or secondary electron emitters through processes other than the required ones, and these electrons then contribute to the noise of the system. The most important such processes are thermionic emission and radioactivity. Cosmic rays and other sources of energetic particles and gamma rays contribute to the noise in several ways. Their direct interactions with the cathode or dynodes early in the chain can expel electrons that are then amplified in the normal manner. Alternatively electrons or ions may be produced from the residual gas or from other components of the structure of the photomultiplier. The most important interaction, however, is generally Čerenkov radiation produced in the window or directly in the cathode. Such Čerenkov pulses can be up to a hundred times more intense than a normal pulse. They can thus be easily discounted when using the photomultiplier in a pulse counting mode, but make a significant contribution to the signal when simply its overall strength is measured.

Superconducting tunnel junction detectors (STJs)

A possible replacement for the CCD in a few years is the superconducting tunnel junction detector (STJ). The STJ can operate from the ultraviolet to longwave infrared, and also in the x-ray region, can detect individual photons, has a very rapid response and provides intrinsic spectral resolution of perhaps 500 or 1000 in the visible. Its operating principle is based upon a Josephson junction. This has two superconducting layers separated by a very thin insulating layer. Electrons are able to tunnel across the junction because they have a wavelike behaviour as well as a particle-like behaviour, and so a current may flow across the junction despite the presence of the insulating layer. Within the superconductor, the lowest energy state for the electrons occurs when they link together to form Cooper pairs. The current flowing across the junction due to paired electrons can be suppressed by a magnetic field.

The STJ detector therefore comprises a Josephson junction based upon tantalum, hafnium, niobium etc., placed within a magnetic field to suppress the current and having an electric field applied across it. It is cooled to about a tenth of the critical temperature of the superconductor—normally under

1 K. A photon absorbed in the superconductor may split one of the Cooper pairs. This requires an energy of a milli-electron-volt or so compared with about an electron volt for pair production in a CCD. Potentially therefore the STJ can detect photons with wavelengths up to a millimetre. Shorter wavelength photons will split many Cooper pairs, with the number split being dependent upon the energy of the photon, hence the device's intrinsic spectral resolution. The free electrons are able to tunnel across the junction under the influence of the electric field, and produce a detectable burst of current. STJ detectors and arrays made from them are still very much under development at the time of writing, but have recently been tried on the William Herschel telescope. The S Cam used a 6×6 array of 25 μm square STJs and successfully observed the light curve of the Crab nebula pulsar and the colour changes in the rapidly variable binary, UZ For.

Other types of detector

Detectors other than those considered above may still find use in particular applications or have historical interest. Some of these devices are briefly surveyed below.

Photovoltaic cells

These are also known as photodiodes, photoconductors and barrier junction detectors. They rely upon the properties of a p–n junction in a semiconductor. The energy level diagram of such a junction is shown in figure 1.1.20. Electrons in the conduction band of the n-type material are at a higher energy than the holes in the valence band of the p-type material. Electrons therefore diffuse across the junction and produce a potential difference across it. Equilibrium occurs when the potential difference is sufficient to halt the electron flow. The two Fermi levels are then coincident, and the potential across the junction is equal to their original difference. The n-type material is positive and the p-type negative, and we have a simple p–n diode. Now if light of sufficiently short wavelength falls on to such a junction then it can generate electron–hole pairs in both the p- and the n-type materials. The electrons in the conduction band in the p region will be attracted towards the n region by the intrinsic potential difference across the junction, and they will be quite free to flow in that direction. The holes in the valence band of the p-type material will be opposed by the potential across the junction and so will not move. In the n-type region the electrons will be similarly trapped while the holes will be pushed across the junction. Thus a current is generated by the illuminating radiation and this may simply be monitored and used as a measure of the light intensity. For use as a radiation detector the p–n junction often has a region of undoped (or

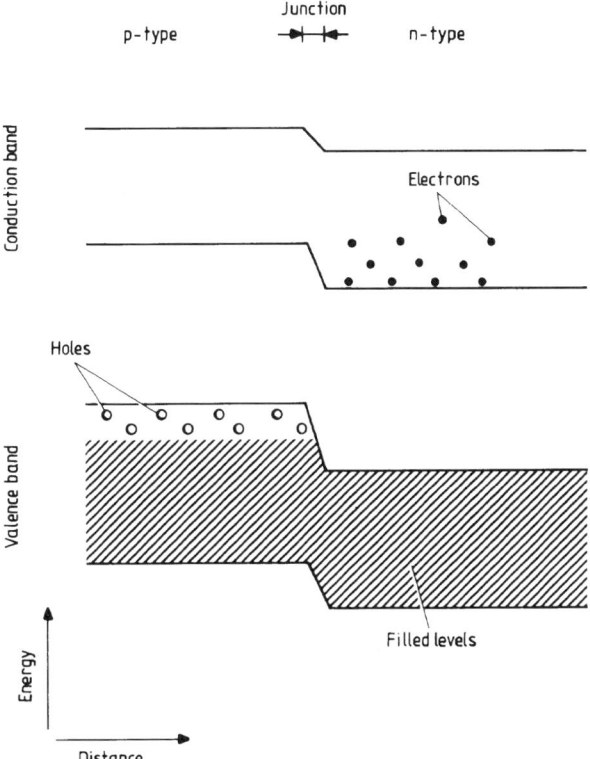

Figure 1.1.20. Schematic energy level diagram of a p–n junction at the instant of its formation.

intrinsic) material between the p and n regions in order to increase the size of the detecting area. These devices are then known as p–i–n junctions. Their operating principle does not differ from that of the simple p–n junction.

The response of a p–n junction to radiation is shown in figure 1.1.21. It can be operated under three different regimes. The simplest, labelled B in figure 1.1.21, has the junction short-circuited through a low-impedance meter. The current through the circuit is the measure of the illumination. In regime C the junction is connected to a high impedance so that the current is very close to zero, and it is the change in voltage that is the measure of the illumination. Finally in regime A the junction is back biased, i.e. an opposing voltage is applied to the junction. The voltage across a load resistor in series with the junction then measures the radiation intensity. In this mode the device is known as a photoconductor (see Infrared detectors below).

The construction of a typical photovoltaic cell is shown in figure 1.1.22. The material in most widespread use for the p and n semiconductors is silicon

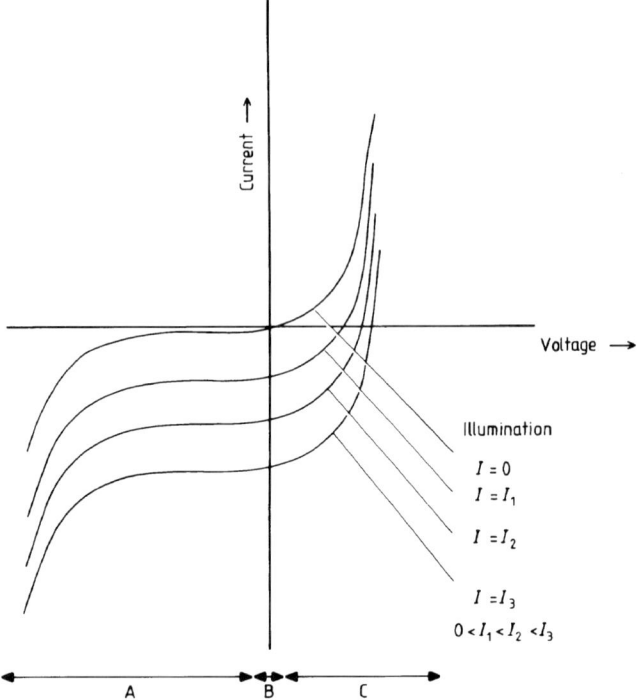

Figure 1.1.21. Schematic V/I curves for a p–n junction under different levels of illumination.

that has been doped with appropriate impurities. The solar power cells found on most artificial satellites are of this type. The silicon-based cells have a peak sensitivity near 900 nm and cut-off wavelengths near 400 and 1100 nm. Their quantum efficiency can be up to 50% near their peak sensitivity and D^* can be up to 10^{12}.

Figure 1.1.22. Cross section through a p–n junction photovoltaic detector.

Indium arsenide, indium selenide, indium antimonide, gallium arsenide and indium gallium arsenide can all be formed into photodiodes. They are particularly useful in the infrared where germanium doped with gallium out-performs bolometers (see below) for wavelengths up to 100 μm.

If a p–n junction is reverse biased to more than half its breakdown voltage an avalanche photodiode (APD) results. The original electron–hole pair produced by the absorption of a photon will be accelerated by the applied field sufficiently to cause further pair production through inelastic collisions. These secondary electrons and holes can in their turn produce further ionizations and so on (cf. Geiger and proportional counters). Eventually an avalanche of carriers is created, leading to an intrinsic gain in the signal by a factor of one hundred or more. The typical voltage used is 200 V, and the avalanche is quenched (because otherwise the current would continue to flow once started) by connecting a resistor of several hundred kΩ in series with the diode. Then as the current starts to flow the voltage across the diode is reduced and the breakdown is quenched. A major drawback for APDs is that the multiplication factor depends very sensitively upon the bias. The power supply typically has to be stable to a factor of ten better than the desired accuracy of the output. The devices also have to be cooled to reduce noise. Using, for example, gallium arsenide, gains of over a factor of 1000 are possible, enabling individual photons to be detected. APDs are used in the University of Hertfordshire's extra-solar planet polarimeter (section 5.2) but have found few other applications in astronomy since CCDs and now STJs are generally to be preferred.

Another type of junction photovoltaic cell uses a p-type semiconductor in contact with a metal. A potential difference develops over the junction as the electrons diffuse across from the metal. The sensitivity of devices of this form of construction, however, is very low, and they are slow in their response. Selenium is the most commonly used material and their major applications occur in the lightmeters of cameras.

Thermocouples

As is well known from school physics, two dissimilar metals in contact can develop a potential difference across their junction. This is called the Seebeck effect. It arises from the differing Fermi levels of the two metals. Electrons flow from the one with the higher level until the potential across the junction is sufficient to halt the transfer. Now the Fermi level at temperatures above absolute zero is the energy level that has a 50% chance of being occupied. As the temperature is increased, more and more of the electrons become thermally excited. However, since the holes will be densely packed near the top of the unexcited electrons while the excited electrons can be much more widely dispersed among the upper levels, the position of the Fermi level will change with temperature. The change in the Fermi level will not necessarily be identical for two different metals, and so at a junction the difference between the two Fermi levels will vary with temperature. Thus the Seebeck potential at the junction will change in order to continue to produce a balance. In a thermocouple, therefore, two dissimilar metals are joined into a circuit that incorporates a galvanometer. When one junction is at a different temperature from the other, their Seebeck potentials differ and a current flows around the circuit.

A practical thermocouple for radiation detection is made from two metals in the form of wires that are twisted together and blackened to improve their absorption. The other junction is kept in contact with something with a large thermal inertia so that it is kept at a constant temperature. The sensitive junction absorbs the radiation and so heats up slightly. The junctions are then at different temperatures so that a current flows as described above. A sensitive galvanometer can detect this current and it can be used to measure the intensity of the radiation. Several thermocouples are usually connected serially so that their potentials combine. Such an array is called a thermopile.

Practical thermocouples and thermopiles are usually made from antimony and bismuth or from nickel with various admixtures of copper, silicon, chromium, aluminium etc. They are useful wideband detectors, especially for infrared work. Their simplicity of operation and their robustness has led to many applications being found for them in satellite-borne instrumentation, in spite of their relatively low sensitivity (they have values of D^* only up to 10^9 to 10^{10}).

Phototransistors

These are of little direct use in astronomy because of their low sensitivity. They find wide application, however, within the apparatus used by astronomers, for example in conjunction with a light-emitting diode (LED) to generate a switching signal. They consist simply of a p–n–p or n–p–n transistor with the minority current carriers produced by the illumination instead of the normal emitter. Thus the current rises with increasing radiation and provides a measure of its intensity.

The photovoltaic cell discussed earlier when operated in mode A is acting as a phototransistor. It is then sometimes called a photodiode. Arrays of such diodes have found some astronomical applications in the past when allied to image intensifiers (section 2.3).

Charge injection devices (CID)

The detection mechanism of these devices is identical to that of the CCD. The difference between them occurs in their read-out system. With these devices two electrodes are used for each sensitive element and the charge can be switched between them by alternating their voltages. If the voltages of both elements are simultaneously reduced to zero, then the accumulated charge is deposited (injected) into the substrate. The resulting pulse can be detected and its magnitude is a measure of the original charge. Connecting all the first electrodes of each sensitive element serially along rows, while all the second electrodes are connected serially along columns, provides a two-dimensional read-out. Switching a given row and column to zero volts then causes the charge in the sensitive element at their intersection to be read out, while all other charges are either unaffected or jump to their alternative electrode. The connections for a CID device are more complex than those for a CCD, but its data are easier to retrieve separately, and can be read out non-destructively. The CID is now rare compared with the all-conquering CCD.

Television tubes

See section 2.3.

Image intensifiers

See section 2.3.

Infrared detectors

Many of the detectors just considered have some infrared sensitivity, especially out to 1 μm. However, at longer wavelengths, other types of detector are needed, though if the STJ fulfils its promise it may replace many of these devices in the future. The infrared region is conventionally divided into three: the near infrared (NIR), 0.7–5 μm, the mid infrared (MIR), 5–30 μm and the far infrared (FIR), 30–1000 μm. All infrared detectors need to be cooled, with the longer the operating wavelength, the colder the required temperature. Thus in the NIR, liquid nitrogen (77 K) generally suffices, in the MIR, liquid helium (4 K) is needed, while in the FIR, temperatures down to 100 mK are used. Currently there are two main types of infrared detector: the photoconductor for the near and mid infrared and somewhat into the far infrared and the bolometer for the far infrared.

The Earth's atmosphere is opaque over much of the infrared region, although there are narrow wavelength ranges (windows) where it becomes transparent to a greater or lesser degree. The windows can be enhanced by observing from high altitude, dry sites, or by flying telescopes on balloons or aircraft. Nonetheless, the sky background can still be sufficiently high that images have to be read out several hundred times a second so that they do not saturate. Much of the observing, however, has to be done from spacecraft. Conventional reflecting optics can be used for the telescope and the instruments, though the longer wavelength means that lower surface accuracies are adequate. Refractive optics, including achromatic lenses, can be used in the NIR, using materials such as barium, lithium and strontium fluoride, zinc sulphate or selenide and infrared-transmitting glasses.

Photoconductive cells

Photoconductive cells exhibit a change in conductivity with the intensity of their illumination. The mechanism for that change is the absorption of the radiation by the electrons in the valence band of a semiconductor and their consequent elevation to the conduction band. The conductivity therefore increases with increasing illumination, and is monitored by a small bias current. There is a cut-off point determined by the minimum energy required to excite a valence electron over the band gap. A very wide variety of materials may be used, with widely differing sensitivities, cut-off wavelengths, operating temperatures etc. The semiconductor may be intrinsic, such as silicon, germanium, mercury cadmium telluride, lead sulphide or indium antimonide. However, the band gaps in intrinsic semiconductors tend to be large, restricting their use to the near infrared. Doping of an intrinsic semiconductor produces an extrinsic semiconductor with the electrons or gaps from the doping atom occupying isolated levels within the band gap. These levels can be just above the top of the valence band, or close to the bottom of the conduction band so that much less energy is needed to excite electrons to or from them. Extrinsic semiconductors can therefore be made that are sensitive across most of the infrared. Doping is normally carried out during the melt stage of the formation of the material. However, this can lead to variable concentrations of the dopant, and so to variable responses for the detectors. For germanium doped with gallium (Ge(Ga)), the most widely used detector material at wavelengths longer than $50\,\mu m$, extremely uniform doping has recently been achieved by exposing pure germanium to a flux of thermal neutrons in a nuclear reactor. Some of the germanium nuclei absorb a neutron and become radioactive. The $^{70}_{32}Ge$ nucleus transmutes to $^{71}_{32}Ge$ that in turn decays to $^{71}_{31}Ga$ via β decay. Arsenic, an n-type dopant, is also produced from $^{74}_{32}Ge$ during the process, but only at 20% of the rate of production of the gallium. The process is known as neutron transmutation doping (NTD). The response of Ge(Ga) detectors may be changed by applying pressure or stressing the material along one of its crystal axes. The pressure is applied by a spring, and can change the detectivity range of the material (normally from ~40 to $\sim115\,\mu m$) to from ~80 to $\sim240\,\mu m$.

The InfraRed Astronomy Satellite (IRAS), launched in 1983, carried a cooled 0.57 m telescope and used small arrays of Ge(Ga) detectors to survey 98% of the sky at 12, 25, 60 and $100\,\mu m$. For the recently completed 2MASS (2 Micron All Sky Survey) project, three 256×256 mercury cadmium telluride arrays were used to observe at 1.25, 1.65 and $2.17\,\mu m$. The Space InfraRed Telescope Facility (SIRTF), due for launch in 2004, will carry an

NIR/MIR camera using indium antimonide arrays to observe at 3.6 and 4.5 μm and Si(As) BIB (see below) detectors for 5.8 and 8.0 μm. It will also carry a 32×32 array of unstressed Ge(Ga) detectors to observe at 70 μm, and a 2×20 array of stressed Ge(Ga) detectors to observe at 160 μm.

Ge(Ga) along with silicon doped with arsenic (Si(As)) or antimony (Si(Sb)) is also one of the materials used in the relatively recently developed blocked impurity band (BIB[1]) detectors. These use a thin layer of very heavily doped semiconductor to absorb the radiation. Such heavy doping would normally lead to high dark currents but a layer of undoped semiconductor blocks these. Ge(Ga) BIB detectors are sensitive out to about 180 μm, and are twice as sensitive as normal stressed Ge(Ga) photoconductors around the 140 μm region. SOFIA (Stratospheric Observatory For Infrared Astronomy) planned to start operating in 2004 will use a Boeing 747 aircraft to fly a 2.5 m telescope at high altitudes. Its MIR detector will use 256×256 Si(As) and Si(Sb) BIB arrays to observe over the 5–8, 17–25 and 25–40 μm ranges.

Photoconductive detectors that do not require the electrons to be excited all the way to the conduction band have been made using alternating layers of gallium arsenide (GaAs) and indium gallium arsenide phosphide (InGaAsP) or aluminium gallium arsenide (AlGaAs), each layer being only ten or so atoms thick. The detectors are known as QWIPs (Quantum Well Infrared Photodetectors). The lower energy required to excite the electron gives the devices a wavelength sensitivity ranging from 1 to 12 μm. The sensitivity region is quite narrow and can be tuned by changing the proportions of the elements. A 256×256 QWIP array has been used on the 5 m Hale telescope, and others in SETI,[2] but they have found few other astronomical applications so far.

Large (2048×2048) arrays can now be produced for some of the NIR detectors and 1024×1024 arrays for some MIR detectors, though at the long wave end of the MIR, arrays are 256×256 at maximum. The UK's 4 m infrared VISTA[3] telescope is planned to use 6144×6144 NIR mosaic arrays. In the FIR array sizes are still only up to 32×32. Unlike CCDs, infrared arrays are read out pixel by pixel. Although this complicates the connecting circuits, there are advantages; the pixels are read out non-destructively, and so can be read out several times and the results averaged to reduce the noise, there is no cross-talk (blooming) between the pixels, and one bad pixel does not affect any of the others. The sensitive material is usually bonded to a silicon substrate that contains the read-out electronics.

In the NIR, large format arrays have led to the abandonment of the once common practice of alternately observing the source and the background

[1] Also known as Impurity Band Conduction (IBC) detectors.
[2] Search for Extra-Terrestrial Intelligence.
[3] Visible and Infrared Survey Telescope for Astronomy.

Table 1.1.4. Some materials used for infrared photoconductors.

Material	Cut-off wavelength or wavelength range (μm)
Silicon (Si)	1.11
Germanium (Ge)	1.8
Gold-doped germanium (Ge(Au))	1–9
Mercury–cadmium–telluride (HgCdTe)	1–12
Gallium arsenide QWIPs (GaAs + InGaAsP or AlGaAs)	1–12
Lead sulphide (PbS)	3.5
Mercury-doped germanium (Ge(Hg))	4
Indium antimonide (InSb)	6.5
Copper-doped germanium (Ge(Cu))	6–30
Gallium-doped silicon (Si(Ga))	17
Arsenic-doped silicon BIB (Si(As))	23
Gallium-doped germanium (Ge(Ga))	~40 to ~115
Gallium-doped germanium stressed (Ge(Ga))	~80 to ~240
Boron-doped germanium (Ge(B))	120
Gallium-doped germanium BIB (Ge(Ga))	~180
Antimony-doped germanium (Ge(Sb))	130

(chopping), but at longer wavelengths the sky is sufficiently bright that chopping is still needed. Chopping may be via a rotating 'windmill' whose blades reflect the background (say) on to the detector, while the gaps between the blades allow the radiation from the source to fall on to the detector directly. Alternatively, some telescopes designed specifically for infrared work have secondary mirrors that can be oscillated to achieve this switching.

Details of some of the materials used for infrared photoconductors are listed in table 1.1.4. Those in current widespread use are in bold.

Bolometers

A bolometer is simply a device that changes its electrical resistivity in response to heating by illuminating radiation. At its simplest, two strips of the material are used as arms of a Wheatstone bridge. When one is heated by the radiation its resistance changes and so the balance of the bridge alters. Two strips of the material are used to counteract the effect of slower environmental temperature changes, since they will both vary in the same manner under that influence.

Bolometers used in astronomy are of two main varieties: room temperature thermistor bolometers and cooled semiconductor bolometers. The former have found in the past extensive application in rockets and spacecraft because of their simplicity and sturdiness. They consist of mixtures of manganese, nickel and cobalt oxides sintered together and coated with an appropriate absorber for the wavelength that is to be observed. They have a high negative coefficient of resistance. Values of D^* of up to 10^9 over a very wide range of wavelengths can be obtained.

Cooled semiconductor bolometers were once used as detectors over most of the infrared region. Photoconductive cells have now replaced them for near and mid-infrared work, but they are still used for the far infrared ($\sim 100\,\mu m$ to a few mm). Germanium doped with gallium (a p-type dopant) is widely used for the bolometer material with a metal-coated dielectric as the absorber. The bolometer is cooled to around 100 mK when in operation to reduce the thermal noise. Germanium doped with beryllium, silicon and silicon nitride are other possible bolometer materials.

SCUBA on the 15 m JCMT operates over the 350–450 μm and 750–850 μm bands using NTD germanium bolometer arrays with 91 and 37 pixels respectively. The individual detectors operate at a temperature of 100 mK and are fed by polished horns whose size matches the telescope's resolution. The current state of the art uses an absorber that is a mesh of metallized silicon nitride like a spider's web, with a much smaller bolometer bonded to its centre. This minimizes the noise produced by cosmic rays, but since the mesh size is much smaller than the operating wavelength, it still absorbs all the radiation. Arrays of bolometers up to 30×30 pixels in size can now be produced. The Herschel spacecraft, scheduled for launch in 2007 with a 3.5 m telescope, will carry a far infrared detector known as SPIRE. This will use spider-web bolometer arrays with feed horns to observe at 250, 350 and 500 μm. Each array will cover $4' \times 8'$ of the sky and contain 149, 88 and 43 detectors respectively.

A recent development that promises to result in much larger arrays is the transition edge sensor (TES). These detectors are thin films of a superconductor, such as tungsten, held at their transition temperature from the superconducting to the normally conducting state. There is a very strong dependence of the resistivity upon temperature in this region. The absorption of a photon increases the temperature slightly, and so increases the resistance. The resistance is monitored through a bias voltage. The replacement for SCUBA (SCUBA2), due in 2006, plans to use two 80×80 TES arrays. These should have twice the sensitivity of the current SCUBA detectors and be able to produce images up to 2000 times faster.

Other types of detector

For detection from the visual out to 30 μm or so, a solid-state photomultiplier can be used. This is closely related to the avalanche photodiode. It uses a layer of arsenic-doped silicon on a layer of undoped silicon. A potential difference is applied across the device. An incident photon produces an electron–hole pair in the doped silicon layer. The electron drifts under the potential difference towards the undoped layer. As it approaches the latter, the increasing potential difference accelerates it until it can ionize further atoms. An electrode collects the resulting avalanche of electrons on the far side of the undoped silicon layer. A gain of 10^4 or so is possible with this device. It has found little application in astronomy to date.

Platinum silicide acting as a Schottky diode can operate out to 5.6 µm. It is easy to fabricate into large arrays, but is of low sensitivity compared with photoconductors. Its main application is for terrestrial surveillance cameras.

Ultraviolet detectors

Some of the detectors that we have reviewed are intrinsically sensitive to short-wave radiation, although modification from their standard optical forms may be required. For example, photomultipliers will require suitable photo-emitters (table 1.1.3) and windows that are transparent to the required wavelengths. Lithium fluoride and sapphire are common materials for such windows. Thinned, rear illuminated CCDs have a moderate intrinsic sensitivity into the long wave ultraviolet region. EBCCDs (see CCDs above) with an appropriate ultraviolet photocathode can also be used. At shorter wavelengths micro-channel plates (section 1.3) take over. If the detector is sensitive to the visible region as well as the ultraviolet, then filters will be needed to exclude the usually much more intense longer wavelengths. Unfortunately the filters also absorb some of the ultraviolet radiation and can bring the overall quantum efficiency down to a few per cent. The term 'solar blind' is used for detectors or detector/filter combinations that are only ultraviolet sensitive.

Another common method of short-wave detection is to use a standard detector for the visual region, and to add a fluorescent or short glow phosphorescent material to convert the radiation to longer wavelengths. Sodium salicylate and tetraphenyl butadiene are the most popular such substances since their emission spectra are well matched to standard photocathodes. Sensitivity down to 60 nm can be achieved in this manner. Additional conversions to longer wavelengths can be added for matching to CCD and other solid-state detectors whose sensitivities peak in the red and infrared. Ruby (Al_2O_3) is suitable for this, and its emission peaks near 700 nm. In this way, the sensitivity of CCDs can be extended down to about 200 nm.

EBCCDs were used for some of the first generation instruments on board the HST. It now uses phosphor-coated CCDs for the WFPC2,[4] and microchannel plates for the STIS.[5]

---

[4] Wide Field Planetary Camera 2.
[5] Space Telescope Imaging Spectrograph.

Future possibilities

Many of the devices discussed above are in the early stages of their development and considerable improvement in their characteristics and performance may be expected over the next decade or so. Other possibilities such as laser action initiation and switching of superconducting states have already been mentioned. Another technique that is currently in the early stages of application to astronomy is optical or infrared heterodyning. A suitable laser beam is mixed with the light from the object. The slow beat frequency is at radio frequencies and may be detected using relatively normal radio techniques (section 1.2). Both the amplitude and phase information of the original signal can be preserved with this technique. Very high dispersion spectroscopy seems likely to be its main area of application.

Noise

In the absence of noise any detector would be capable of detecting any source, however, faint. Noise, however, is never absent and generally provides the major limitation on detector performance. A minimum signal-to-noise ratio of unity is required for reliable detection. However, most research work requires signal-to-noise ratios of at least 10, and preferably 100 or 1000. Noise sources in photomultipliers and CCDs have already been mentioned, and the noise for an un-illuminated detector (dark signal) is a part of the definitions of DQE, NEP, D^* and dynamic range (see earlier discussion). Now we must look at the nature of detector noise in more detail.

We may usefully separate noise sources into four main types: *intrinsic noise*, i.e. noise originating in the detector, *signal noise*, i.e. noise arising from the character of the incoming signal, particularly its quantum nature, *external noise* such as spurious signals from cosmic rays etc. and *processing noise*, arising from amplifiers etc. used to convert the signal from the detector into a usable form. We may generally assume processing noise to be negligible in any good detection system. Likewise, external noise sources should be reduced as far as possible by external measures. Thus an infrared detector should be in a cooled enclosure and be allied to a cooled telescope (e.g. the Infrared Astronomy Satellite, IRAS) to reduce thermal emission from its surroundings, or a photomultiplier used in a photon-counting mode should employ a discriminator to eliminate the large pulses arising from Čerenkov radiation from cosmic rays etc. Thus we are left with intrinsic and signal noise to consider further.

Intrinsic noise

Intrinsic noise in photomultipliers has already been discussed and arises from sources such as variation in photoemission efficiency over the photocathode, thermal electron emission from the photocathode and dynode chain etc. Noise in photographic emulsion, such as chemical fogging, is discussed later (section 2.2), while irradiation etc. in the eye has also been covered earlier.

In solid-state devices, intrinsic noise comes from four sources.

Thermal noise, also known as Johnson or Nyquist noise (see section 1.2) arises in any resistive material. It is due to the thermal motion of the charge carriers. These motions give rise to a current, whose mean value is zero, but which may have non-zero instantaneous values. The resulting fluctuating voltage is given by equation (1.2.5).

Shot noise occurs in junction devices and is due to variation in the diffusion rates in the neutral zone of the junction because of random thermal motions. The general form of the shot noise current is

$$i = (2ei\Delta f + 4eI_0\Delta f) \qquad (1.1.4)$$

where e is the charge on the electron, Δf is the measurement frequency bandwidth, I is the diode current and I_0 is the reverse bias or leakage current. When the detector is reverse biased, this equation simplifies to

$$i = (2eI_0\Delta f). \qquad (1.1.5)$$

Generation–recombination (g–r) noise is caused by fluctuations in the rate of generation and recombination of thermal charge carriers, which in turn leads to fluctuations in the device's resistivity. Generation–recombination noise has a flat spectrum up to the inverse of the mean carrier lifetime and then decreases roughly with the square of the frequency.

Flicker noise, or $1/f$ noise, occurs when the signal is modulated in time, either because of its intrinsic variations or because it is being 'chopped' (i.e. the source and background or comparison standard are alternately observed). The mechanism of flicker noise is unclear but its amplitude follows an f^{-n} power spectrum where f is the chopping frequency and n lies typically in the region 0.75 to 2.0. This noise source may obviously be minimized by increasing f. Furthermore, operating techniques such as phase-sensitive detection (section 3.2) are commonplace especially for infrared work, when the external noise may be many orders of magnitude larger than the desired signal. Modulation of the signal is therefore desirable or necessary on many occasions. However, the improvement of the detection by faster chopping is usually limited by the response time of the detector. Thus an optimum chopping frequency normally may be found for each type of detector. For example, this is about 100 Hz for bolometers and 1000 Hz or more for most photoconductive and photovoltaic cells.

The relative contributions of these noise sources are shown in figure 1.1.23.

Figure 1.1.23. Relative contributions of various noise sources (schematic).

Signal noise

Noise can be present in the signal for a variety of reasons. One obvious example is background noise. The source under observation will generally be superimposed upon a signal from the sky due to scattered terrestrial light sources, scattered starlight, diffuse galactic emission, zodiacal light, microwave background radiation etc. The usual practice is to reduce the importance of this noise by measuring the background and subtracting it from the main signal. Often the source and its background are observed in quick succession (chopping, see *flicker noise* above and section 3.2). Alternatively, there may only be measurements of the background at the beginning and end of an observing run. In either case, some noise will remain, due to fluctuations of the background signal about its mean level. This noise source also reduces as the resolution of the telescope improves. If the resolution is $1''$, then a point source has to have an energy equal to that coming from a square second of arc of the background sky in order to have a signal-to-noise ratio of unity. But if the resolution were to be $0.1''$, then the same point source would have a signal-to-noise ratio of 100 since it is only 'competing' with 0.01 square seconds of arc of the background. Since the light grasp of a telescope increases as the diameter squared, for *diffraction-limited* telescopes, the signal-to-noise ratio for point sources thus improves as D^4 (see discussion below on real-time atmospheric compensation).

Noise also arises from the quantum nature of light. At low signal levels photons arrive at the detector sporadically. A Poisson distribution gives the probability of arrival, and this has a standard deviation of $\sqrt{n}$ (where n is the mean number of photons per unit time). Thus the signal will fluctuate about its mean value. In order to reduce the fluctuations to less than $x\%$, the signal must be integrated for $10^4/(nx^2)$ times the unit time. At high photon densities, photons tend to cluster more than a Poisson distribution would suggest because they are subject to Bose–Einstein statistics. This latter noise source may dominate at radio wavelengths (section 1.2), but is not normally of importance over the optical region.

Digitization

Signals are digitized in two ways, signal strength and time. The effect of the first is obvious; there is an uncertainty (i.e. noise) in the measurement corresponding to plus or minus half the measurement resolution. The effect of sampling a time varying signal is more complex. The well-known sampling theorem (section 2.1) states that the highest frequency in a signal that can be determined is half the measurement frequency. Thus, if a signal is bandwidth limited to some frequency, f, then it may be completely determined by sampling at $2f$ or higher frequencies. In fact sampling at higher than twice the limiting (or Nyquist) frequency is a waste of effort; no further information will be gained. However, in a non-bandwidth-limited signal, or a signal containing components above the Nyquist frequency, errors or noise will be introduced into the measurements through the effect of those higher frequencies. In this latter effect, or 'aliasing' as it is known, the beat frequencies between the sampling rate and the higher frequency components of the signal appear as spurious low frequency components of the signal.

Telescopes

The *apologia* for the inclusion of telescopes in the section entitled Detectors (section 1.3) has already been given, but it is perhaps just worth repeating that for astronomers the use of a detector in isolation is almost unknown. Some device to restrict the angular acceptance zone and possibly to increase the radiation flux as well, is almost invariably a necessity. In the optical region this generally implies the use of a telescope, and so they are discussed here as a required ancillary to detectors.

The telescope may justly be regarded as the symbol of the astronomer, for notwithstanding the heroic work done with the naked eye in the past and the more recent advances in 'exotic' regions of the spectrum, our picture of the universe is still to a large extent based upon optical observations made through telescopes. The optical telescope has passed through four major phases of development, each of which has caused a quantum jump in astronomical knowledge. We are now at the start of the fifth phase when very large diffraction-limited telescopes can be formed from smaller mirrors or segments with active control of their alignment etc. We may hope for a similar rapid development of our knowledge as these instruments come on stream in the next few decades.

Optical theory

Before looking at the designs of telescopes and their properties it is necessary to cover some of the optical principles upon which they are based. It is assumed that the reader is familiar with the basics of the optics of lenses and mirrors: such terminology as focal ratio, magnification, optical axis etc., with laws such as Snell's law, the law of reflection and so on and with equations such as the lens and mirror formulae and the lens maker's formula. If it is required, however, any basic optics book or reasonably comprehensive physics book will suffice to provide this background.

We start by considering the resolution of a lens. That this is limited at all is due to the wave nature of light. As light passes through any opening it is diffracted and the wave fronts spread out in a shape given by the envelope of Huygens' secondary wavelets (figure 1.1.24). Huygens' secondary wavelets radiate spherically outwards from all points on a wave front with the velocity of light in the medium concerned. Three examples are shown in figure 1.1.24. Imaging the wave front after passage through a slit-shaped aperture produces an image whose structure is shown in figure 1.1.25. The variation in intensity is due to interference between waves originating from different parts of the aperture. The paths taken by such waves to arrive at a given point will differ and so the distances that they have travelled will also differ. The waves will be out of step with each other to a greater or

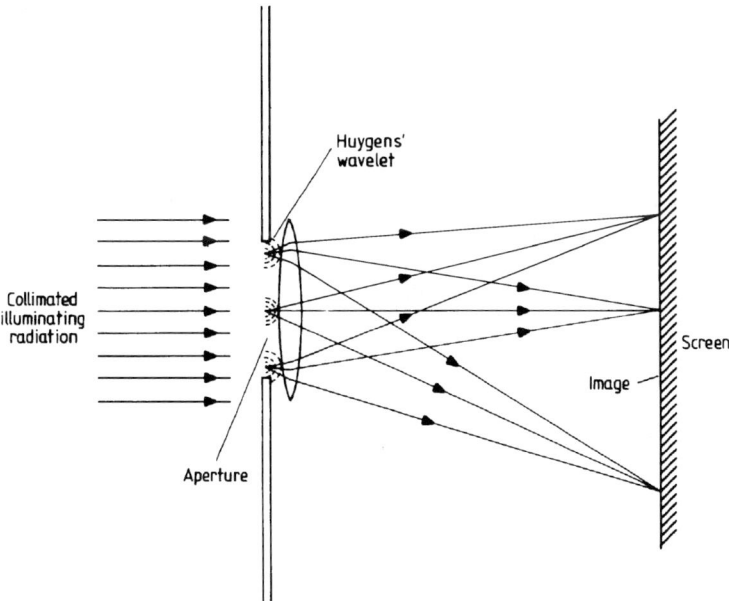

Figure 1.1.24. Fraunhofer diffraction at an aperture.

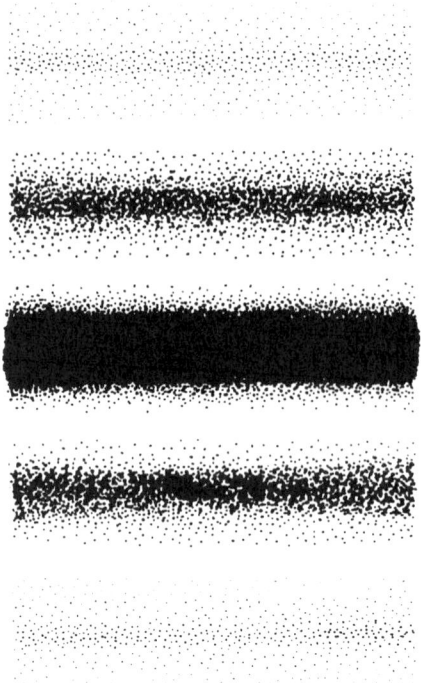

Figure 1.1.25. Image of a narrow slit (negative image).

lesser extent depending upon the magnitude of this path difference. When the path difference is a half wavelength or an integer number of wavelengths plus a half wavelength, then the waves will be 180° out of phase with each other and so will cancel out. When the path difference is a whole number of wavelengths, the waves will be in step and will reinforce each other. Other path differences will cause intermediate degrees of cancellation and reinforcement. The central maximum arises from the reinforcement of many waves, while the first minimum occurs when the path difference for waves originating at opposite edges of the aperture is one wavelength, then for every point in one half of the aperture there is a point in the other half such that their path difference is half a wavelength, and all the waves cancel out completely. The intensity at a point within the image of a narrow slit may be obtained from the result that the diffraction pattern of an aperture is the power spectrum of the Fourier transform of its shape, and is given by

$$I_\theta = I_0 \frac{\sin^2(\pi d \sin \theta/\lambda)}{(\pi d \sin \theta/\lambda)^2} \qquad (1.1.6)$$

where θ is the angle to the normal from the slit, d is the slit width and I_0 and I_θ are the intensities within the image on the normal and at an angle θ to the

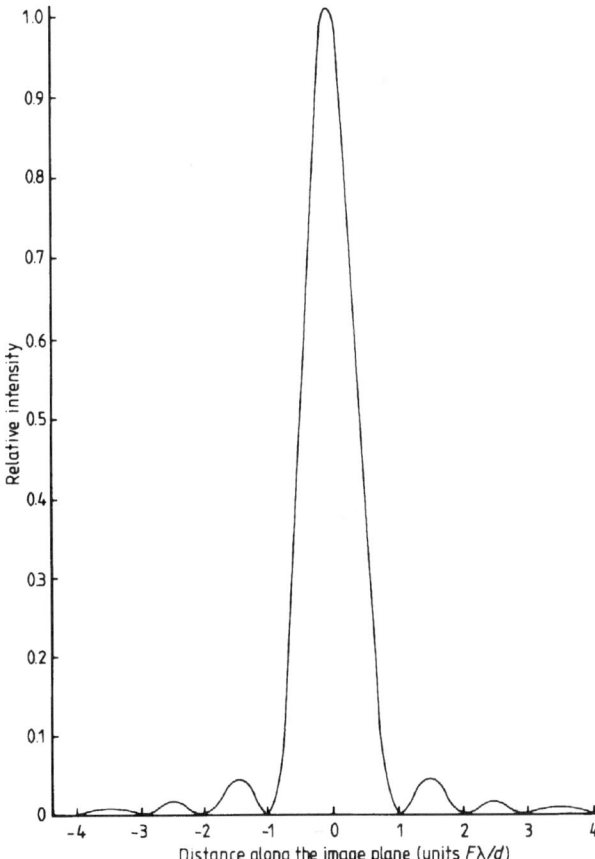

Figure 1.1.26. Cross-section through the image of a narrow slit.

normal from the slit respectively. With the image focused on to a screen at distance F from the lens, the image structure is as shown in figure 1.1.26, where d is assumed to be large when compared with the wavelength λ. For a rectangular aperture with dimensions $d \times l$, the image intensity is similarly

$$I(\theta, \phi) = I_0 \frac{\sin^2(\pi d \sin \theta/\lambda)}{(\pi d \sin \theta/\lambda)^2} \frac{\sin^2(\pi l \sin \phi/\lambda)}{(\pi l \sin \phi/\lambda)^2} \qquad (1.1.7)$$

where ϕ is the angle to the normal from the slit measured in the plane containing the side of length l. To obtain the image structure for a circular aperture, which is the case normally of interest in astronomy, we must integrate over the surface of the aperture. The image is then circular with concentric light and dark fringes. The central maximum is known as Airy's disc after the Astronomer Royal who first succeeded in completing the

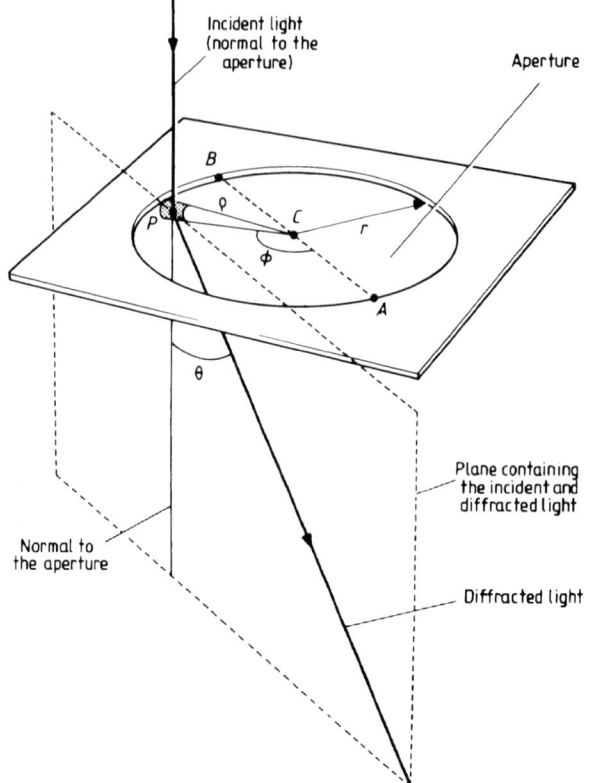

Figure 1.1.27. Diffraction by a circular aperture.

integration. The major difference from the previous case occurs in the position of the fringes.

Consider a circular aperture of radius r, illuminated by a beam of light normal to its plane (figure 1.1.27), and consider also the light which is diffracted at an angle θ to the normal to the aperture from a point P, whose cylindrical coordinates with respect to the centre of the aperture C, and the line AB which is drawn through C parallel to the plane containing the incident and diffracted rays, are (ϕ, ρ). The path difference, Δ, between diffracted rays from P and A, is then

$$\Delta = (r - \rho \cos \phi) \sin \theta \qquad (1.1.8)$$

and the phase difference is

$$\frac{2\pi\Delta}{\lambda} = \frac{2\pi}{\lambda}(r - \rho \cos \phi) \sin \theta. \qquad (1.1.9)$$

The elemental area at P is just

$$dA = \rho\, d\phi\, d\rho. \tag{1.1.10}$$

So the contribution to the electric vector of the radiation in the image plane by the elemental area around P for an angle θ to the normal is proportional to

$$\sin(\omega t + (2\pi/\lambda)(r - \rho\cos\phi)\sin\theta)\rho\, d\phi\, d\rho \tag{1.1.11}$$

where $\omega/2\pi$ is the frequency of the radiation, and the net effect is obtained by integrating over the aperture

$$\int_0^{2\pi}\int_0^r \sin\left[\omega t + \left(\frac{2\pi r\sin\theta}{\lambda}\right) - \left(\frac{2\pi\rho\cos\phi\sin\theta}{\lambda}\right)\right]\rho\, d\rho\, d\phi$$

$$= \sin\left(\omega t + \frac{2\pi r\sin\theta}{\lambda}\right)\int_0^{2\pi}\int_0^r \rho\cos\left(\frac{2\pi\rho\cos\phi\sin\theta}{\lambda}\right)d\rho\, d\phi$$

$$- \cos\left(\omega t + \frac{2\pi r\sin\theta}{\lambda}\right)\int_0^{2\pi}\int_0^r \rho\sin\left(\frac{2\pi\rho\cos\phi\sin\theta}{\lambda}\right)d\rho\, d\phi. \tag{1.1.12}$$

The second integral on the right-hand side is zero, which we may see by substituting

$$s = \frac{2\pi\rho\cos\phi\sin\theta}{\lambda} \tag{1.1.13}$$

so that

$$\cos\left(\omega t + \frac{2\pi r\sin\theta}{\lambda}\right)\int_0^{2\pi}\int_0^r \rho\sin\left(\frac{2\pi\rho\cos\phi\sin\theta}{\lambda}\right)d\rho\, d\phi \tag{1.1.14}$$

$$= \cos\left(\omega t + \frac{2\pi r\sin\theta}{\lambda}\right)\int_0^r \rho\int_{s=2\pi\rho\sin\theta/\lambda}^{s=2\pi\rho\sin\theta/\lambda}\frac{-\sin s}{[(4\pi^2\rho^2\sin^2\theta/\lambda^2) - s^2]^{1/2}}\, ds\, d\phi$$

$$= 0 \tag{1.1.15}$$

since the upper and lower limits of the integration with respect to s are identical. Thus we have the image intensity $I(\theta)$ in the direction θ to the normal

$$I(\theta) \propto \left[\sin\left(\omega t + \frac{2\pi r\sin\theta}{\lambda}\right)\int_0^{2\pi}\int_0^r \rho\cos\left(\frac{2\pi\rho\cos\phi\sin\theta}{\lambda}\right)d\rho\, d\phi\right]^2 \tag{1.1.16}$$

$$\propto \left(r^2\int_0^{2\pi}\frac{\sin(2m\cos\phi)}{2m\cos\phi}\, d\phi - \frac{1}{2}r^2\int_0^{2\pi}\frac{\sin^2(m\cos\phi)}{(m\cos\phi)^2}\, d\phi\right)^2 \tag{1.1.17}$$

where

$$m = \frac{\pi r\sin\theta}{\lambda}. \tag{1.1.18}$$

Now

$$\frac{\sin(2m\cos\phi)}{2m\cos\phi} = 1 - \frac{(2m\cos\phi)^2}{3!} + \frac{(2m\cos\phi)^4}{5!} - \cdots \qquad (1.1.19)$$

and

$$\frac{\sin^2(m\cos\phi)}{(m\cos\phi)^2} = 1 - \frac{2^3(m\cos\phi)^2}{4!} + \frac{2^5(m\cos\phi)^4}{6!} - \cdots \qquad (1.1.20)$$

so that

$$I(\theta) \propto \left(r^2 \sum_0^\infty (-1)^n \int_0^{2\pi} \frac{(2m\cos\phi)^{2n}}{(2n+1)!} \, d\phi \right. $$
$$\left. - \frac{1}{2} r^2 \sum_0^\infty (-1)^n \int_0^{2\pi} \frac{2^{2n+1}(m\cos\phi)^{2n}}{(2n+2)!} \, d\phi \right)^2. \qquad (1.1.21)$$

Now

$$\int_0^{2\pi} \cos^{2n}\phi \, d\phi = \frac{(2n)!}{2(n!)^2} \pi \qquad (1.1.22)$$

and so

$$I(\theta) \propto \pi^2 r^4 \left[\sum_0^\infty (-1)^n \frac{1}{n+1} \left(\frac{m^n}{n!} \right)^2 \right]^2 \qquad (1.1.23)$$

$$\propto \frac{\pi^2 r^4}{m^2} (J_1(2m))^2 \qquad (1.1.24)$$

where $J_1(2m)$ is the Bessel function of the first kind of order unity. Thus $I(\theta)$ is a maximum or zero accordingly as $J_1(2m)/m$ reaches an extremum or is zero. Its variation is shown in figure 1.1.28. The first few zeros occur for values of m of

$$m = 1.916, 3.508, 5.087, \ldots \qquad (1.1.25)$$

and so in the image, the dark fringes occur at values of θ given by

$$\sin\theta = \frac{1.916\lambda}{\pi r}, \frac{3.508\lambda}{\pi r}, \frac{5.087\lambda}{\pi r}, \ldots \qquad (1.1.26)$$

or for small θ

$$\theta \approx \frac{1.220\lambda}{d}, \frac{2.233\lambda}{d}, \frac{3.238\lambda}{d}, \ldots \qquad (1.1.27)$$

where d is now the diameter of the aperture. The image structure along a central cross section is therefore similar to that of figure 1.1.26, but with the minima expanded outwards to the points given in equation (1.1.27) and the fringes, of course, are circular.

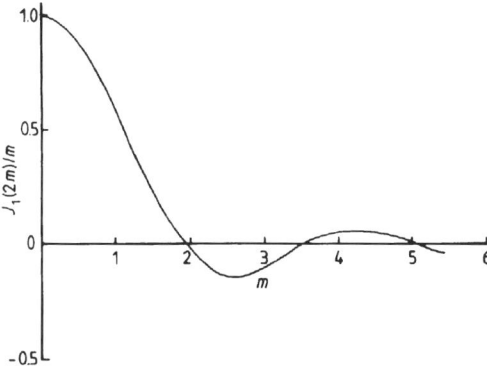

Figure 1.1.28. Variation of $J_1(2m)/m$ with m.

If we now consider two distant point sources separated by an angle α, then two such images will be produced and will be superimposed. There will not be any interference effects between the two images since the sources are mutually incoherent, and so their intensities will simply add together. The combined image will have an appearance akin to that shown in figure 1.1.29. When the centre of the Airy disc of one image is superimposed upon the first minimum of the other image (and vice versa), then we have Rayleigh's criterion for the resolution of a lens (or mirror). This is the normally accepted measure of the theoretical resolution of a lens. It is given in radians by (from equation (1.1.27))

$$\alpha = \frac{1.220\lambda}{d}. \qquad (1.1.28)$$

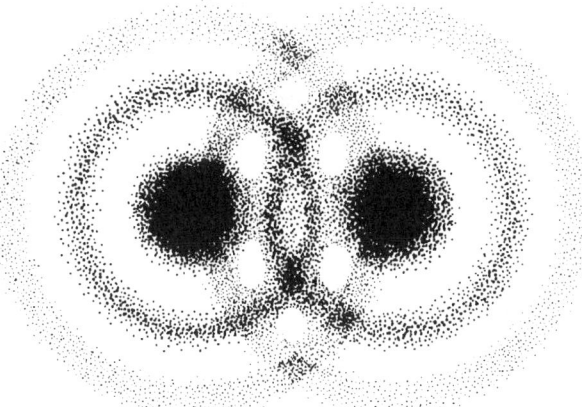

Figure 1.1.29. Image of two distant point sources through a circular aperture (negative image).

It is a convenient measure, but it is quite arbitrary. For sources of equal brightness the image will appear non-circular for separations of about one third of α, while for sources differing appreciably in brightness, the separation may need to be an order of magnitude larger than α for them to be resolved. For the eye, the effective wavelength is 510 nm for faint images so that the resolution of a telescope used visually is given by

$$R = \frac{0.128}{d} \qquad (1.1.29)$$

where d is the objective's diameter in metres and R is the angular resolution in seconds of arc. An empirical expression for the resolution of a telescope of $0.116''/d$, known as Dawes' limit, is often used by amateur astronomers since the slightly better value it gives for the resolution takes some account of the abilities of a skilled observer. To achieve either resolution in practice, the magnification must be sufficient for the angular separation of the images in the eyepiece to exceed the resolution of the eye (see the earlier discussion). Taking this to be an angle, β, we have the minimum magnification required to realize the Rayleigh limit of a telescope, M_m

$$M_m = \frac{\beta d}{1.220\lambda} \qquad (1.1.30)$$

so that for β equal to 3' of arc, which is about its average value,

$$M_m = 1300d \qquad (1.1.31)$$

where d is again measured in metres. Of course, most astronomical telescopes are actually limited in their resolution by the atmosphere. A 1 m telescope might reach its Rayleigh resolution on one night a year from a good observing site. On an average night, scintillation will spread stellar images to about 2'' of arc so that only telescopes smaller than about 0.07 m can regularly attain their diffraction limit. Since telescopes are rarely used visually for serious work such high magnifications as are implied by equation (1.1.31) are hardly ever encountered today. However, some of William Herschel's eyepieces still exist and these, if used on his 1.2 m telescope, would have given magnifications of up to 8000 times.

The theoretical considerations of resolution that we have just seen are only applicable if the lens or mirror is of sufficient optical quality that the image is not already degraded beyond this limit. There are many effects that will blur the image and these are known as aberrations. With one exception they can all affect the images produced by either lenses or mirrors. The universal or monochromatic aberrations are known as the Seidel aberrations after Ludwig von Seidel who first analysed them. The exception is chromatic aberration and the related second-order effects of transverse chromatic aberration and secondary colour, and these affect only lenses.

Table 1.1.5.

	Refractive index at the specified wavelengths (nm)				
Glass type	361	486	589	656	768
Crown	1.539	1.523	1.517	1.514	1.511
High dispersion crown	1.546	1.527	1.520	1.517	1.514
Light flint	1.614	1.585	1.575	1.571	1.567
Dense flint	1.705	1.664	1.650	1.644	1.638

Chromatic aberration arises through the change in the refractive index of glass or other optical material with the wavelength of the illuminating radiation. Table 1.1.5 shows some typical values of the refractive index of some commonly used optical glasses.

The degree to which the refractive index varies with wavelength is called the dispersion, and is measured by the constringence, ν

$$\nu = \frac{\mu_{589} - 1}{\mu_{486} - \mu_{656}} \tag{1.1.32}$$

where μ_λ is the refractive index at wavelength λ. The three wavelengths that are chosen for the definition of ν are those of strong Fraunhofer lines: 486 nm, the F line (Hβ); 589 nm, the D lines (Na); and 656 nm, the C line (Hα). Thus for the glasses listed earlier, the constringence varies from 57 for the crown glass to 33 for the dense flint (note that the *higher* the value of the constringence, the *less* the rays of different wavelengths diverge from each other). The effect of the dispersion upon an image is to string it out into a series of different coloured images along the optical axis (figure 1.1.30). Looking at this sequence of images with an eyepiece, then at a particular point along the optical axis, the observed image will consist of a

Figure 1.1.30. Chromatic aberration.

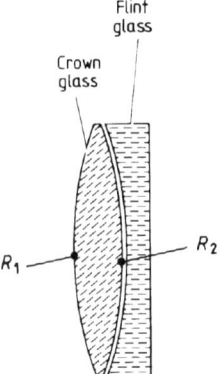

Figure 1.1.31. An achromatic doublet.

sharp image in the light of one wavelength surrounded by blurred images of varying sizes in the light of all the remaining wavelengths. To the eye, the best image occurs when yellow light is focused since it is less sensitive to the red and blue light. The image size at this point is called the circle of least confusion. The spread of colours along the optical axis is called the longitudinal chromatic aberration, while that along the image plane containing the circle of least confusion is called the transverse chromatic aberration.

Two lenses of different glasses may be combined to reduce the effect of chromatic aberration. Commonly in astronomical refractors, a biconvex crown glass lens is allied to a planoconcave flint glass lens to produce an achromatic doublet. In the infrared, achromats can be formed using barium, lithium and strontium fluorides, zinc sulphate or selenide and infrared-transmitting glasses. In the sub-millimetre region (i.e. wavelengths of several hundred microns) crystal quartz and germanium can be used. The lenses are either cemented together or separated by only a small distance (figure 1.1.31). Despite its name there is still some chromatic aberration remaining in this design of lens since it can only bring two wavelengths to a common focus. If the radii of the curved surfaces are all equal, then the condition for two given wavelengths, λ_1 and λ_2, to have coincident images is

$$2\Delta\mu_C = \Delta\mu_F \qquad (1.1.33)$$

where $\Delta\mu_C$ and $\Delta\mu_F$ are the differences between the refractive indices at λ_1 and λ_2 for the crown glass and the flint glass respectively. More flexibility in design can be attained if the two surfaces of the converging lens have differing radii. The condition for achromatism is then

$$\frac{|R_1| + |R_2|}{|R_1|} \Delta\mu_C = \Delta\mu_F \qquad (1.1.34)$$

where R_2 is the radius of the surface of the crown glass lens that is in contact with the flint lens (NB: the radius of the flint lens surface is almost invariably R_2 as well, in order to facilitate alignment and cementing) and R_1 is the radius of the other surface of the crown glass lens. By careful selection of λ_1 and λ_2 an achromatic doublet can be constructed to give tolerable images. For example, by achromatizing at 486 and 656 nm, the longitudinal chromatic aberration is reduced when compared with a simple lens of the same focal length by a factor of about 30. Nevertheless, since chromatic aberration varies as the square of the diameter of the objective and inversely with its focal length, refractors larger than about 0.25 m still have obtrusively coloured images. More seriously, if filters are used then the focal position will vary with the filter. Similarly the scale on photographic plates will alter with the wavelengths of their sensitive regions. Further lenses may be added to produce apochromats that have three corrected wavelengths and super-apochromats with four corrected wavelengths. But such designs are impossibly expensive for telescope objectives of any size, although eyepieces and camera lenses may have eight or ten components and achieve very high levels of correction.

A common and severe aberration of both lenses and mirrors is spherical aberration. In this effect, annuli of the lens or mirror that are of different radii have different focal lengths, illustrated in figure 1.1.32 for a spherical mirror. For rays parallel to the optical axis it can be eliminated completely by

Figure 1.1.32. Spherical aberration.

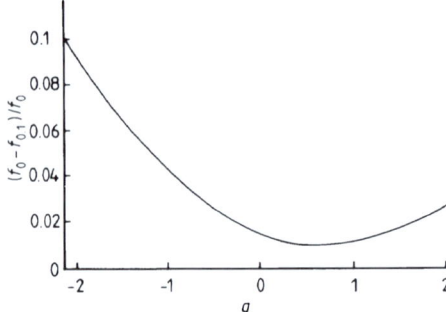

Figure 1.1.33. Spherical aberration in thin lenses. f_x is the focal length for rays parallel to the optical axis, and at a distance x times the paraxial focal length away from it.

deepening the sphere to a paraboloidal surface for the mirror. It cannot be eliminated from a simple lens without using aspheric surfaces, but for a given focal length it may be minimized. The shape of a simple lens is measured by the shape factor, q,

$$q = \frac{R_2 + R_1}{R_2 - R_1} \qquad (1.1.35)$$

where R_1 is the radius of the first surface of the lens and R_2 is the radius of the second surface of the lens. The spherical aberration of a thin lens then varies with q as shown in figure 1.1.33, with a minimum at $q = +0.6$. The lens is then biconvex with the radius of the surface nearer to the image three times the radius of the other surface. Judicious choice of surface radii in an achromatic doublet can lead to some correction of spherical aberration while still retaining the colour correction. Spherical aberration in lenses may also be reduced by using high-refractive-index glass since the curvatures required for the lens' surfaces are lessened, but this is likely to increase chromatic aberration. Spherical aberration increases as the cube of the aperture.

The deepening of a spherical mirror to a paraboloidal one in order to correct for spherical aberration unfortunately introduces a new aberration called coma. This also afflicts mirrors of other shapes and lenses. It causes the images for objects away from the optical axis to consist of a series of circles that correspond to the various annular zones of the lens or mirror and which are progressively shifted towards or away from the optical axis (figure 1.1.34). The severity of coma is proportional to the square of the aperture. It is zero in a system that obeys Abbe's sine condition

$$\frac{\sin \theta}{\sin \phi} = \frac{\theta_p}{\phi_p} = \text{constant} \qquad (1.1.36)$$

where the angles are defined in figure 1.1.35. A doublet lens can be simultaneously corrected for chromatic and spherical aberrations, and coma within

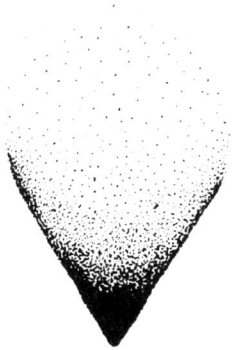

Figure 1.1.34. Shape of the image of a point source due to coma.

acceptable limits, if the two lenses can be separated. Such a system is called an aplanatic lens. A parabolic mirror can be corrected for coma by adding thin correcting lenses before or after the mirror, as discussed in more detail later in this section under the heading 'telescope designs'. The severity of the coma at a given angular distance from the optical axis is inversely proportional to the square of the focal ratio. Hence using as large a focal ratio as possible can also reduce its effect. In Newtonian reflectors a focal ratio of f8 or larger gives acceptable coma for most purposes. At f3, coma will limit the useful field of view to about $1'$ of arc so that prime focus imaging almost always requires the use of a correcting lens to give reasonable fields of view.

Astigmatism is an effect whereby the focal length differs for rays in the plane containing an off-axis object and the optical axis (the tangential plane), in comparison with rays in the plane at right angles to this (the sagittal plane). It decreases more slowly with focal ratio than coma so that it may

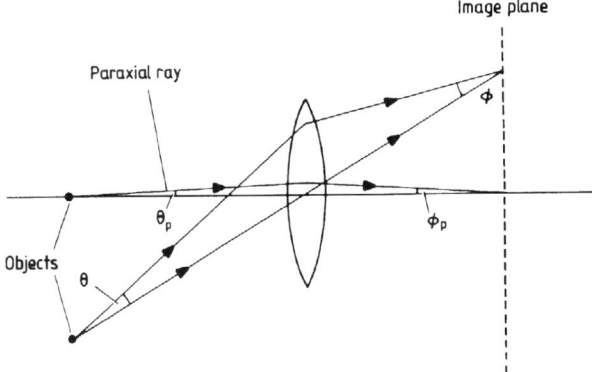

Figure 1.1.35. Parameters for Abbe's sine condition.

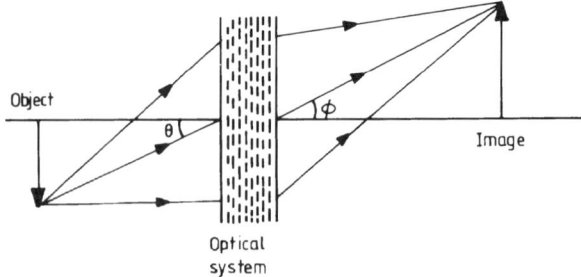

Figure 1.1.36. Terminology for distortion.

become the dominant effect for large focal ratios. It is possible to correct astigmatism, but only at the expense of introducing yet another aberration, field curvature. This is simply that the surface containing the sharply focused images is no longer a flat plane but is curved. A system in which a flat image plane is retained and astigmatism is corrected for at least two radii is termed an anastigmatic system.

The final aberration is distortion, and this is a variation in the magnification over the image plane. An optical system will be free of distortion only if the condition

$$\frac{\tan \theta}{\tan \phi} = \text{constant} \tag{1.1.37}$$

holds for all values of θ (see figure 1.1.36 for the definition of the angles). Failure of this condition to hold results in pincushion or barrel distortion (figure 1.1.37) accordingly as the magnification increases or decreases with distance from the optical axis. A simple lens is very little affected by distortion and it can frequently be reduced in more complex systems by the judicious placing of stops within the system.

A fault of optical instruments as a whole is vignetting. This is not an attribute of a lens or mirror and so is not included among the aberrations. It arises from the uneven illumination of the image plane, usually due to obstruction of the light path by parts of the instrument. Normally it can be avoided by careful design, but it may become important if stops are used in a system to reduce other aberrations.

This long catalogue of faults of optical systems may well have led the reader to wonder if the Rayleigh limit can ever be reached in practice. However, optical designers have a wide variety of variables to play with: refractive indices, dispersion, focal length, mirror surfaces in the form of various conic sections, spacings of the elements, number of elements and so on, so that it is usually possible to produce a system which will give an adequate image for a particular purpose. Other criteria such as cost, weight, production difficulties etc. may well prevent the realization of the

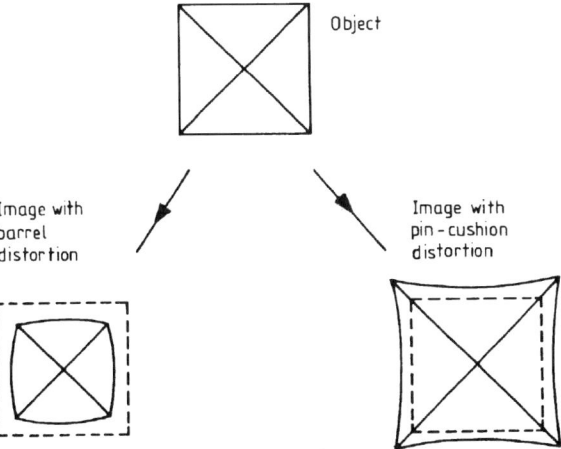

Figure 1.1.37. Distortion.

system in practice even though it is theoretically possible. Multi-purpose systems can usually only be designed to give lower quality results than single-purpose systems. Thus the practical telescope designs that are discussed later in this section are optimized for objects at infinity and if they are used for nearby objects their image quality will deteriorate. The approach to designing an optical system is largely an empirical one. There are two important methods in current use. The older approach requires an analytical expression for each of the aberrations. For example, the third-order approximation for spherical aberration of a lens is

$$\frac{1}{f_x} - \frac{1}{f_p} = \frac{x^2}{8f^3\mu(\mu-1)}\left[\frac{\mu+2}{\mu-1}q^2 + 4(\mu+1)\left(\frac{2f}{v}-1\right)q\right.$$

$$\left. + (3\mu+2)(\mu-1)\left(\frac{2f}{v}-1\right)^2 + \frac{\mu^2}{\mu-1}\right] \qquad (1.1.38)$$

where f_x is the focal distance for rays passing through the lens at a distance x from the optical axis, f_p is the focal distance for paraxial rays from the object (paraxial rays are rays which are always close to the optical axis and which are inclined to it by only small angles), f is the focal length for paraxial rays which are initially parallel to the optical axis and v is the object distance. For precise work it may be necessary to involve fifth-order approximations of the aberrations, and so the calculations rapidly become very cumbersome. The alternative and more modern approach is via ray tracing. The concepts involved are much simpler since the method consists simply of accurately following the path of a selected ray from the object through the system and finding its arrival point on the image plane. Only the basic formulae

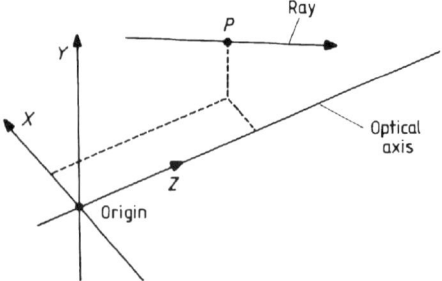

Figure 1.1.38. Ray tracing coordinate system.

are required: Snell's law for lenses

$$\sin i = \frac{\mu_1}{\mu_2} \sin r \qquad (1.1.39)$$

and the law of reflection for mirrors

$$i = r \qquad (1.1.40)$$

where i is the angle of incidence, r is the angle of refraction or reflection as appropriate and μ_1 and μ_2 are the refractive indices of the materials on either side of the interface.

The calculation of i and r for a general ray requires knowledge of the ray's position and direction in space, and the specification of this is rather more cumbersome. Consider first of all a ray passing through a point P, which is within an optical system (figure 1.1.38). We may completely describe the ray by the coordinates of the point P, together with the angles that the ray makes with the coordinate system axes (figure 1.1.39). We may without any loss of generality set the length of the ray, l, equal to unity, and therefore write

$$\gamma = \cos\theta \qquad (1.1.41)$$

$$\delta = \cos\phi \qquad (1.1.42)$$

$$\varepsilon = \cos\psi. \qquad (1.1.43)$$

The angular direction of the ray is thus specified by the vector

$$\mathbf{v} = (\gamma, \delta, \varepsilon). \qquad (1.1.44)$$

The quantities γ, δ and ε are commonly referred to as the direction cosines of the ray. If we now consider the point P as a part of an optical surface, then we require the angle of the ray with the normal to that surface in order to obtain i, and thereafter, r. Now we may similarly specify the normal at P by its

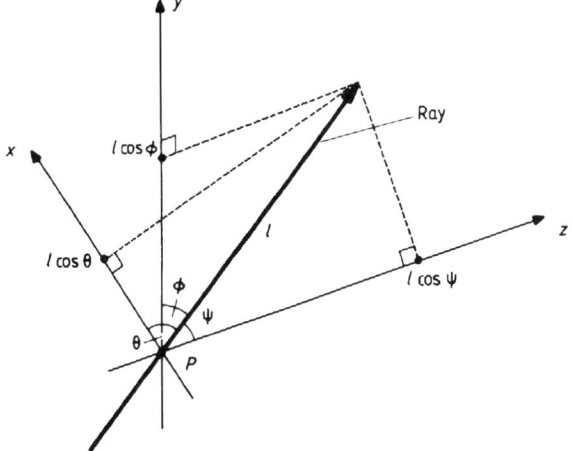

Figure 1.1.39. Ray tracing angular coordinate system.

direction cosines, forming the vector $\mathbf{v}'$,

$$\mathbf{v}' = (\gamma', \delta', \varepsilon') \tag{1.1.45}$$

and we then have

$$\cos i = \frac{\mathbf{v} \cdot \mathbf{v}'}{\|\mathbf{v}\| \, \|\mathbf{v}'\|} \tag{1.1.46}$$

or

$$i = \cos^{-1}(\gamma\gamma' + \delta\delta' + \varepsilon\varepsilon'). \tag{1.1.47}$$

The value of r can now be obtained from equations (1.1.39) or (1.1.40). We may again specify its track by direction cosines (the vector $\mathbf{v}''$)

$$\mathbf{v}'' = (\gamma'', \delta'', \varepsilon'') \tag{1.1.48}$$

and we may find the values of the components as follows. Three simultaneous equations can be obtained: the first from the angle between the incident ray and the refracted or reflected ray, the second from the angle between the normal and the reflected or refracted ray, and the third by requiring the incident ray, the normal, and the refracted or reflected ray to be coplanar, and these are

$$\gamma\gamma'' + \delta\delta'' + \varepsilon\varepsilon'' = \cos 2r \qquad \text{(reflection)} \tag{1.1.49}$$

$$= \cos(i - r) \qquad \text{(refraction)} \tag{1.1.50}$$

$$\gamma'\gamma'' + \delta'\delta'' + \varepsilon'\varepsilon'' = \cos r \tag{1.1.51}$$

$$(\varepsilon\delta' - \varepsilon'\delta)\gamma'' + (\gamma\varepsilon' - \gamma'\varepsilon)\delta'' + (\delta\gamma' - \delta'\gamma)\varepsilon'' = \cos r. \tag{1.1.52}$$

After a considerable amount of manipulation one obtains

$$\varepsilon'' = \frac{\{[(\varepsilon\delta' - \varepsilon'\delta)\delta' - (\gamma\varepsilon' - \gamma'\varepsilon)\gamma'](\gamma'\cos\alpha - \gamma\cos r) \atop -(\gamma'\delta - \gamma\delta')(\varepsilon\delta' - \varepsilon'\delta)\cos r\}}{\{[(\varepsilon\delta' - \varepsilon'\delta)\delta' - (\gamma\varepsilon' - \gamma'\varepsilon)\gamma'](\gamma'\varepsilon - \gamma\varepsilon') \atop -[(\varepsilon\delta' - \varepsilon'\delta)\varepsilon' - (\delta\gamma' - \delta'\gamma)\gamma'](\gamma'\delta - \gamma\delta')\}}$$

(1.1.53)

$$\delta'' = \frac{\gamma'\cos\alpha - \gamma\cos r - (\gamma'\varepsilon - \gamma\varepsilon')\varepsilon''}{(\gamma'\delta - \gamma\delta')}$$

(1.1.54)

$$\gamma = \frac{\cos\alpha - \delta\delta'' - \varepsilon\varepsilon''}{\gamma}$$

(1.1.55)

where α is equal to $2r$ for reflection, and $(i - r)$ for refraction.

The direction cosines of the normal to the surface are easy to obtain when its centre of curvature is on the optical axis

$$\gamma' = \frac{x_1}{R}$$

(1.1.56)

$$\delta' = \frac{y_1}{R}$$

(1.1.57)

$$\varepsilon' = \frac{(z_1 - z_R)}{R}$$

(1.1.58)

where (x_1, y_1, z_1) is the position of P and $(0, 0, z_R)$ is the position of the centre of curvature. If the next surface is at a distance s from the surface under consideration, then the ray will arrive on it at a point (x_2, y_2, z_2) still with the direction cosines γ'', δ'' and ε'', where

$$x_2 = \frac{\gamma''s}{\varepsilon''} + x_1$$

(1.1.59)

$$y_2 = \frac{\delta''s}{\varepsilon''} + y_1$$

(1.1.60)

$$z_2 = s + z_1.$$

(1.1.61)

We may now repeat the calculation again for this surface, and so on. Ray tracing has the advantage that all the aberrations are automatically included by it, but it has the disadvantage that many rays have to be followed in order to build up the structure of the image of a point source at any given place on the image plane, and many images have to be calculated in order to assess the overall performance of the system. Ray tracing, however, is eminently suitable for programming on to a computer, while the analytical approach is not, since it requires frequent value judgements. The approach of a designer, however, to a design problem is similar whichever method is used to assess the system. The initial prototype is set up purely on the basis of what the designer's experience suggests may fulfil the major specifications such as cost, size, weight, resolution etc. A general rule of thumb is that the number of optical surfaces needed will be at least as many as the number of aberrations to be corrected. The performance is then assessed either analytically or by ray tracing. In most cases it will not be good enough, so a slight alteration is made with the intention of improving the performance, and it is reassessed. This process continues until the original prototype has been optimized for its purpose. If the optimum solution is within the specifications then there is no further problem. If it is outside the specifications, even after optimization, then the whole procedure is repeated starting from a different prototype. The performances of some of the optical systems favoured by astronomers are considered in the next subsection.

Even after a design has been perfected, there remains the not inconsiderable task of physically producing the optical components to within the accuracies specified by the design. The manufacturing steps for both lenses and mirrors are broadly similar, although the details may vary. The surface is roughly shaped by moulding or by diamond milling. It is then matched to another surface formed in the same material whose shape is its inverse, called the tool. The two surfaces are ground together with coarse carborundum or other grinding powder between them until the required surface begins to approach its specifications. The pits left behind by this coarse grinding stage are removed by a second grinding stage in which finer powder is used. The pits left by this stage are then removed in turn by a third stage using still finer powder, and so on. As many as eight or ten such stages may be necessary. When the grinding pits are reduced to a micron or so in size, the surface may be polished. This process employs a softer powder such as iron oxide or cerium oxide that is embedded in a soft matrix such as pitch. Once the surface has been polished it can be tested for the accuracy of its fit to its specifications. Since in general it will not be within the specifications after the initial polishing, a final stage, which is termed figuring, is necessary. This is simply additional polishing to adjust the surface's shape until it is correct. The magnitude of the changes involved during this stage is only about a micron or two, so if the alteration needed is larger than this then it may be necessary to return to a previous stage in the grinding to obtain a better approximation. There are a number of tests that can determine the shape of the mirror's surface to a precision of $\pm 50\,\mathrm{nm}$ or better, such as the Foucault, Ronchi, Hartmann and Null tests. The details of these tests are beyond the scope of this book but may be found in books on optics and telescope making.

Small mirrors can be produced in this way by hand, and many an amateur astronomer has acquired his or her telescope at very low cost by making their own optics. Larger mirrors require machines that move the tool (now smaller than the mirror) in an epicyclic fashion. The motion of the tool is similar to that of the planets under the Ptolemaic model of the solar system, and so such machines have become known as planetary polishers. The epicyclic motion can be produced by a mechanical arrangement, but commercial production of large mirrors now relies on computer-controlled planetary polishers. The deepening of the surface to a parabolic, hyperbolic or other shape is accomplished by preferential polishing and sometimes by the use of a flexible tool whose shape can be adjusted. The mirror segments for large instruments like the 10 m Keck telescopes are clearly small, off-axis parts of the total hyperbolic shape. These, and also the complex shapes required for Schmidt telescope corrector lenses, have been produced using stressed polishing. The blank is deformed by carefully designed forces from a warping harness, and then polished to a spherical shape (or flat for Schmidt corrector plates). When the deforming forces are released, the blank springs into the required non-spherical asymmetric shape. For the Keck and Hobby–Eberly mirror segments the final polishing was undertaken using ion beams. The ion beam is an accelerated stream of ions, such as argon, that wears away the surface an atom at a time. Since it applies almost no force to the mirror, the process can be used to correct defects left by normal polishing techniques such as deformations at the edge and print-through of the honeycomb back on lightweight mirrors.

Recently, the requirements for non-axisymmetric mirrors for segmented mirror telescopes (see later discussion) and for glancing incidence x-ray telescopes (section 1.3) have led to the development of numerically controlled diamond milling machines which can produce the required shaped and polished surface directly to an accuracy of 10 nm or better.

The defects in an image that are due to surface imperfections on a mirror will not exceed the Rayleigh limit if those imperfections are less than about one eighth of the wavelength of the radiation for which the mirror is intended. Thus we have the commonly quoted $\lambda/8$ requirement for the maximum allowable deviation of a surface from its specifications. Sometimes the resulting deviations of the wavefront are specified rather than those of the optical surface. This is twice the value for the surface, i.e. a limit of $\lambda/4$. The restriction on lens surfaces is about twice as large as those of mirror surfaces since the ray deflection is distributed over the two lens faces. However, as we have seen, the Rayleigh limit is an arbitrary one, and for some purposes the fit must be several times better than this limit. This is particularly important when viewing extended objects such as planets, and improvements in the contrast can continue to be obtained by figuring surfaces to $\lambda/20$ or better.

The surface must normally receive its reflecting coating after its production. The vast majority of astronomical mirror surfaces have a thin layer of aluminium evaporated on to them by heating aluminium wires suspended over the mirror inside a vacuum chamber. Other materials such as silicon carbide are occasionally used, especially for ultraviolet work, since the reflectivity of aluminium falls off below 300 nm. The initial reflectivity of an aluminium coating in the visual region is around 90%. This, however, can fall to 75% or less within a few months as the coating ages. Mirrors therefore have to be re-aluminized at regular intervals. The intervals between re-aluminizing can be lengthened by gently cleaning the mirror every month or so. The currently favoured methods of cleaning are rinsing with de-ionized water and/or drifting carbon dioxide snow across the mirror surface. Mirrors coated with a suitably protected silver layer can achieve 99.5% reflectivity in the visible, and are becoming more and more required since some modern telescope designs can have four or five reflections. With ancillary instrumentation the total number of reflections can then reach 10 or more.

Lens surfaces also normally receive a coating after their manufacture, but in this case the purpose is to reduce reflection. Uncoated lenses reflect about 5% of the incident light from each surface so that a system containing, say, five uncoated lenses could lose 40% of its available light through this process. To reduce the reflection losses a thin layer of material covers the lens, for which

$$\mu' = \sqrt{\mu} \qquad (1.1.62)$$

where μ' is the refractive index of the coating and μ is the refractive index of the lens material. The thickness, t, should be

$$t = \frac{\lambda}{4\mu'}. \qquad (1.1.63)$$

This gives almost total elimination of reflection at the selected wavelength, but some will still remain at other wavelengths. Lithium fluoride and silicon dioxide are commonly used materials for the coatings. Recently it has become possible to produce anti-reflection coatings that are effective simultaneously over a number of different wavelength regions through the use of several interfering layers.

Mirrors only require the glass as a support for their reflecting film. Thus there is no requirement for it to be transparent; on the other hand it is essential for its thermal expansion coefficient to be low. Today's large telescope mirrors are therefore made from materials other than glass. The coefficient of thermal expansion of glass is about $9 \times 10^{-6} \, \mathrm{K}^{-1}$, that of Pyrex about $3 \times 10^{-6} \, \mathrm{K}^{-1}$, and for fused quartz it is about $4 \times 10^{-7} \, \mathrm{K}^{-1}$. Among the largest telescopes almost the only glass mirror is the one for the 2.5 m Hooker telescope on Mount Wilson. After its construction Pyrex became the favoured material, until the last three decades, when quartz or artificial materials with a similar low coefficient of expansion such as 'CerVit' and 'Zerodur' etc., but which are easier to manufacture, have been used. These low expansion materials have a ceramic glass structure that is partially crystallized. The crystals are around 50 nm across and contract as the temperature rises, while the surrounding matrix of amorphous material expands. An appropriate ratio of crystals to matrix provides very small expansion ($10^{-7} \, \mathrm{K}^{-1}$) over a wide range of temperatures. Ultra-low expansion (ULE) fused silica has an even smaller coefficient of expansion

$(3 \times 10^{-8} \, \text{K}^{-1})$. A radically different alternative is to use a material with a very high thermal conductivity, such as silicon carbide, graphite epoxy, steel, beryllium or aluminium. The mirror is then always at a uniform temperature and its surface has no stress distortion. Most metallic mirrors, however, have a relatively coarse crystalline surface that cannot be polished adequately. They must therefore be coated with a thin layer of nickel before polishing, and there is always some risk of this being worn through if the polishing process extends for too long.

Provided that they are mounted properly, small solid mirrors can be made sufficiently thick that, once ground and polished, their shape is maintained simply by their mechanical rigidity. However, the thickness required for such rigidity scales as the cube of the size so that the weight of a solid rigid mirror scales as D^5. Increasing solid mirror blanks to larger sizes thus rapidly becomes expensive and impractical. Most mirrors larger than about 0.5 to 1 m therefore have to have their weight reduced in some way. There are two main approaches to reducing the mirror weight: thin mirrors and honeycomb mirrors. The thin mirrors are also sub-divided into monolithic and segmented mirrors. In both cases, active support of the mirror is required in order to retain its correct optical shape. The 8.2 m mirrors of ESO's VLT are an example of thin monolithic mirrors: they are 178 mm thick Zerodur, with a weight of 23 tonnes each, but need 150 actuators to maintain their shape. The 10 m Keck telescopes use thin, segmented Zerodur mirrors. There are 36 individual hexagonal segments in each main mirror, with each segment 1.8 m across and about 70 mm thick, giving a total mirror weight of just 14.4 tonnes (compared with 14.8 tonnes for the 5 m Hale telescope mirror).

Honeycomb mirrors are essentially thick solid blanks that have had a lot of the material behind the reflecting surface removed, leaving only thin struts to support that surface. The struts often surround hexagonal cells of removed material giving the appearance of honeycomb, but other cell shapes such as square, circular or triangular can be used. The mould for a honeycomb mirror blank has shaped cores protruding from its base that produce the empty cells, with the channels between the cores filling with the molten material to produce the supporting ribs. Once the blank has solidified and been annealed it may be ground and polished as though it were solid. However, if the original surface of the blank is flat, then considerable amounts of material will remain near the edge after grinding, adding to the final weight of the mirror. Roger Angel of the University of Arizona has therefore pioneered the technique of spin-casting honeycomb mirror blanks. The whole furnace containing the mould and molten material is rotated so that the surface forms a parabola (see liquid mirrors below) close to the finally required shape. The surface shape is preserved as the furnace cools and the material solidifies. Potentially honeycomb mirrors up to 8 m in diameter may be produced by this method. However it may have been produced, thinning the ribs and the under-side of the surface may further reduce the weight of the blank. The thinning is accomplished by milling and/or etching with hydrogen fluoride. In this way the 1 m diameter Zerodur secondary mirrors produced for the 8 m Gemini telescopes have supporting ribs just 3 mm wide, and weigh less than 50 kg.

Whatever approach is taken to reducing the mirror's weight, it will be aided if the material used for the blank has an intrinsic high stiffness (resistance to bending). Beryllium has already been used to produce the 1.1 m secondary mirrors for the VLT, whose weights are only 42 kg. Silicon carbide and graphite epoxy are other high stiffness materials that may be used in the future.

A quite different approach to producing mirrors, that perhaps surprisingly gives low weights, is to use a rotating bath of mercury. Isaac Newton was the first to realize that the surface of a steadily rotating liquid would take up a paraboloidal shape under the combined influences of gravity and centrifugal acceleration. If that liquid reflects light, like mercury, gallium, gallium–indium alloy or an oil suffused with reflecting particles, then it can act as a telescope's primary mirror. Of course the mirror has to remain accurately horizontal, so that it always points towards the zenith, but with suitable, perhaps active, correcting secondary optics the detector can be moved around the image plane to give a field of view currently tens of minutes of arc wide, and conceivably eventually up to $8°$ across. Moving the electrons across the CCD detector at the same speed as the image movement enables short time exposures to be obtained (time delayed integration, TDI). The Earth's rotation enables such a zenith telescope to sample up to 7% of the sky for less than 7% of the cost of an equivalent fully steerable telescope. The 'bath' containing the mercury is in fact a lightweight structure whose surface is parabolic to within 0.1 mm. Only a thin layer of mercury is thus needed to produce the accurate mirror surface, and so the mirror overall is a lightweight one despite the high density of mercury. The bath is rotated smoothly around an accurately vertical axis at a constant angular velocity using a large air bearing. Mercury is toxic, so that suitable precautions have to be taken to protect the operators. Also its reflectivity is less than 80% in the visible, but since all the other optics in the instrument can be conventional mirrors or lenses, this penalty is not too serious. Until recently NASA operated a 3 m liquid mirror telescope in New Mexico to track debris orbiting the Earth (NASA Orbital Debris Observatory or NODO), and a 2.7 m liquid mirror telescope has operated since 1995 in British Columbia. The Large Zenith Telescope (LZT), also in British Columbia, is due to start regular observing shortly using a 6 m diameter liquid mirror.

LAMA,[6] a proposal for an array of eighteen 10 m liquid mirror telescopes, would give the equivalent collecting power of a single 42 m

[6] Large Aperture Mirror Array.

dish. In space a liquid mirror of potentially almost any size could be formed from a ferromagnetic liquid confined by electromagnetic forces.

An approach to producing mirrors whose shape can be rapidly adjusted is to use a thin membrane with a reflecting coating. The membrane forms one side of a pressure chamber, and its shape can be altered by changing the gas pressure inside the chamber. $\lambda/2$ surface quality or better can currently be achieved over mirror diameters up to 5 mm, and such mirrors have found application in optical aperture synthesis systems (section 2.5).

Telescope designs

Background

Most serious work with telescopes uses equipment placed directly at the focus of the telescope. But for visual work such as finding and guiding on objects, an eyepiece is necessary. Often it matters little whether the image produced by the eyepiece is of a high quality or not. Ideally, however, the eyepiece should not degrade the image noticeably more than the main optical system. There are an extremely large number of eyepiece designs, whose individual properties can vary widely. For example, one of the earliest eyepiece designs of reasonable quality is the Kellner. This combines an achromat and a simple lens and typically has a field of view of 40° to 50°. The Plössl uses two achromats and has a slightly wider field of view. More recently, the Erfle design employs six or seven components and gives fields of view of 60° to 70°, while the current state-of-the-art is represented by designs such as the Nagler with eight or more components and fields of view up to 85°. Details of these and other designs may generally be found in books on general astronomy or on optics or from the manufacturers. For small telescopes used visually, a single low magnification wide angle eyepiece may be worth purchasing for the magnificent views of large objects like the Orion Nebula that it will provide. There is little point in the higher power eyepieces being wide angle ones (which are very expensive), since these will normally be used to look at angularly small objects. The use of a Barlow lens provides an adjustable magnification for any eyepiece. The Barlow lens is an achromatic negative lens placed just before the telescope's focus. It increases the telescope's effective focal length, and so the magnification.

For our purposes, only four aspects of eyepieces are of concern: light losses, eye relief, exit pupil and angular field of view. Light loss occurs through vignetting when parts of the light beam fail to be intercepted by the eyepiece optics, or are obstructed by a part of the structure of the eyepiece, and also through reflection, scattering and absorption by the optical components. The first of these can generally be avoided by careful eyepiece design and selection, while the latter effects can be minimized by antireflection coatings and by keeping the eyepieces clean.

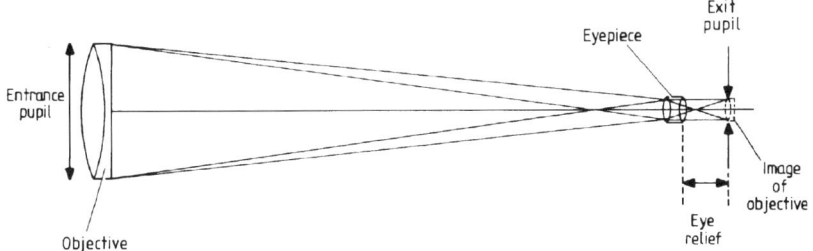

Figure 1.1.40. Exit pupil and eye relief.

The exit pupil is the image of the objective produced by the eyepiece (figure 1.1.40). All the rays from the object pass through the exit pupil, so that it must be smaller than the pupil of the human eye if all of the light gathered by the objective is to be utilized. Its diameter, E, is given by

$$E = \frac{F_e D}{F_o} \tag{1.1.64}$$

where D is the objective's diameter, F_e is the focal length of the eyepiece and F_o is the focal length of the objective. Since magnification is given by

$$M = \frac{F_o}{F_e} \tag{1.1.65}$$

and the diameter of the pupil of the dark-adapted eye is 6 or 7 mm, we must therefore have

$$M \geq \sim 170D \tag{1.1.66}$$

where D is in metres, if the whole of the light from the telescope is to pass into the eye. In practice, for small telescopes being used visually, the maximum useable magnification is around $\times 2000D$ (i.e. $\times 300$ for a 0.15 m telescope etc.).

The eye relief is the distance from the final lens of the eyepiece to the exit pupil. It should be about 6–10 mm for comfortable viewing. If you wear spectacles, however, then the eye relief may need to be up to 20 mm.

The angular field of view is defined by the acceptance angle of the eyepiece, θ'. Usually this is about 40–60°, but it may be up to 90° for wide-angle eyepieces. The angular diameter of the area of sky which is visible when the eye is positioned at the exit pupil, and which is known as the angular field of view, θ, is then just

$$\theta = \frac{\theta'}{M}. \tag{1.1.67}$$

The brightness of an image viewed through a telescope is generally expected to be greater than when it is viewed directly. However, this is not the case for extended objects. The naked eye brightness of a source is proportional to the eye's pupil diameter squared, while that of the image in a telescope is proportional to the objective diameter squared. If the eye looks at that image, then its angular size is increased by the telescope's magnification. Hence the increased brightness of the image is spread over a greater area. Thus we have

$$R = \frac{\text{brightness through a telescope}}{\text{brightness to the naked eye}} = \frac{D^2}{M^2 P^2} \tag{1.1.68}$$

where D is the objective's diameter, P is the diameter of the pupil of the eye and M is the magnification. But from equations (1.1.64) and (1.1.65) we have

$$R = 1 \tag{1.1.69}$$

when the exit pupil diameter is equal to the diameter of the eye's pupil, and

$$R < 1 \tag{1.1.70}$$

when it is smaller than the eye's pupil.

If the magnification is less than $\times 170D$, then the exit pupil is larger than the pupil of the eye and some of the light gathered by the telescope will be lost. Since the telescope will generally have other light losses due to scattering, imperfect reflection and absorption, the surface brightness of an extended source is always fainter when viewed through a telescope than when viewed directly with the naked eye. This result is in fact a simple consequence of the second law of thermodynamics; if it were not the case, then one could have a net energy flow from a cooler to a hotter object. The apparent increase in image brightness when using a telescope arises partly from the increased angular size of the object so that the image on the retina must fall to some extent on to the regions containing more rods, even when looked at directly, and partly from the increased contrast resulting from the exclusion of extraneous light by the optical system. Even with the naked eye, looking through a long cardboard tube enables one to see faint extended sources such as M31 more easily.

The analysis that we have just seen does not apply to images that are physically smaller than the detecting element. For this situation, the image brightness is proportional to D^2. Again, however, there is an upper limit to the increase in brightness that is imposed when the energy density at the image is equal to that at the source. This limit is never approached in practice since it would require 4π steradians for the angular field of view. Thus stars may be seen through a telescope that are fainter by a factor called the light grasp than those visible to the naked eye. The light grasp is simply given by D^2/P^2. Taking $+6^m$ as the magnitude of the faintest star visible to the naked eye (see section 3.1), the faintest star that may be seen through a telescope has a magnitude, m_l, which is the limiting magnitude for that telescope

$$m_l = 17 + 5\log_{10} D \tag{1.1.71}$$

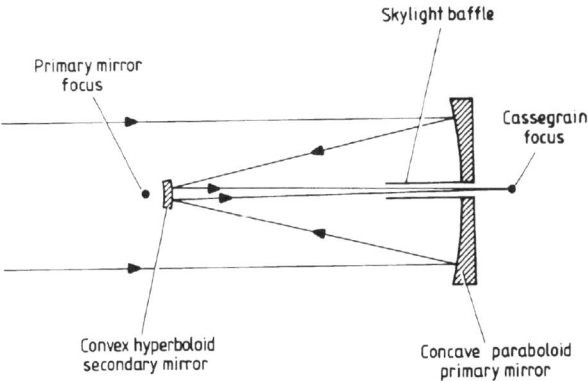

Figure 1.1.41. Cassegrain telescope optical system.

where D is in metres. If the stellar image is magnified to the point where it spreads over more than one detecting element, then we must return to the analysis for the extended sources. For an average eye, this upper limit to the magnification is given by

$$M \approx 850D \qquad (1.1.72)$$

where D is again in metres.

Designs

Probably the commonest format for large telescopes is the Cassegrain system, although most large telescopes can usually be used in several alternative different modes by interchanging their secondary mirrors. The Cassegrain system is based upon a paraboloidal primary mirror and a convex hyperboloidal secondary mirror (figure 1.1.41). The nearer focus of the conic section which forms the surface of the secondary is coincident with the focus of the primary, and the Cassegrain focus is then at the more distant focus of the secondary mirror's surface. The major advantage of the Cassegrain system lies in its telephoto characteristic; the secondary mirror serves to expand the beam from the primary mirror so that the effective focal length of the whole system is several times that of the primary mirror. A compact and hence rigid and relatively cheap mounting can thus be used to hold the optical components while retaining the advantages of long focal length and large image scale. The Cassegrain design is afflicted with coma and spherical aberration to about the same degree as an equivalent Newtonian telescope, or indeed to just a single parabolic mirror with a focal length equal to the effective focal length of the Cassegrain. The beam expanding effect of the secondary mirror means that Cassegrain telescopes normally work at focal ratios between 12 and 30, even though their primary mirror may be f3 or f4. Thus the images remain tolerable over a

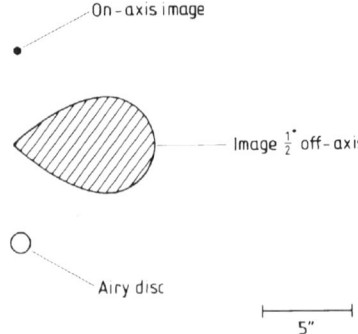

Figure 1.1.42. Images in a 0.25 m f4/f16 Cassegrain telescope. (The images were obtained by ray tracing. Since this does not allow for the effects of diffraction, the on-axis image in practice will be the same as the Airy disc, and the 0.5° off-axis image will be blurred even further.)

field of view that may be several tenths of a degree across (figure 1.1.42). Astigmatism and field curvature are stronger than in an equivalent Newtonian system, however. Focusing of the final image in a Cassegrain system is customarily accomplished by moving the secondary mirror along the optical axis. The amplification due to the secondary means that it only has to be moved a short distance away from its optimum position in order to move the focal plane considerably. The movement away from the optimum position, however, introduces severe spherical aberration. For the 0.25 m f4/f16 system whose images are shown in figure 1.1.42, the secondary mirror can only be moved by 6 mm either side of the optimum position before even the on-axis images without diffraction broadening become comparable in size with the Airy disc. Critical work with Cassegrain telescopes should therefore always be undertaken with the secondary mirror at or very near to its optimum position.

A very great improvement to the quality of the images may be obtained if the Cassegrain design is altered slightly to the Ritchey–Chrétien system. The optical arrangement is identical with that shown in figure 1.1.41 except that the primary mirror is deepened to a hyperboloid and a stronger hyperboloid is used for the secondary. With such a design both coma and spherical aberration can be corrected and we have an aplanatic system. The improvement in the images can be seen by comparing figure 1.1.43 with figure 1.1.42. It should be noted, however, that the improvement is in fact considerably more spectacular since we have a 0.5 m Ritchey–Chrétien and a 0.25 m Cassegrain with the same effective focal lengths. A 0.5 m Cassegrain telescope would have its off-axis image twice the size of that shown in figure 1.1.42 and its Airy disc half the size shown there.

Another variant on the Cassegrain system is the Dall–Kirkham telescope that has a concave ellipsoidal main mirror and spherical convex

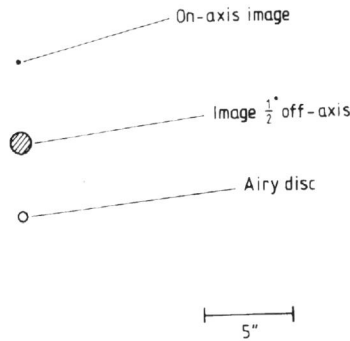

Figure 1.1.43. Ray tracing images in a 0.5 m f3/f8 Ritchey–Chrétien telescope.

secondary mirror. The design is popular with amateur telescope makers since the curves on the mirrors are easier to produce. However, it suffers badly from aberrations and its field of sharp focus is only about a third that of an equivalent conventional Cassegrain design.

Alternatively, a Cassegrain or Ritchey–Chrétien system can be improved by the addition of correctors just before the focus. The correctors are low or zero-power lenses whose aberrations oppose those of the main system. There are numerous successful designs for correctors although many of them require aspheric surfaces and/or the use of exotic materials such as fused quartz. Images can be reduced to less than the size of the seeing disc over fields of view of up to 1°, or sometimes over even larger angles. The 3.9 m Anglo-Australian Telescope (AAT), for example, uses correcting lenses to provide an unvignetted 2° field of view (two-degree field or 2dF), over which 400 optical fibres can be positioned to feed a spectroscope. Reflective corrective optics are also possible and a four-mirror system including aspheric active optics is proposed for the OWL project (see below). The corrective optics may be combined with a focal reducer to enable the increased field of view to be covered by the detector array. A focal reducer is the inverse of a Barlow lens and is a positive lens, usually an apochromatic triplet, placed just before the focal point of the telescope that decreases the effective focal length and so gives a smaller image scale.

Another telescope design that is again very closely related to the Cassegrain is termed the Coudé system. It is in effect a very long focal length Cassegrain or Ritchey–Chrétien whose light beam is folded and guided by additional flat mirrors to give a focus whose position is fixed in space irrespective of the telescope position. One way of accomplishing this is shown in figure 1.1.44. After reflection from the secondary, the light is reflected down the hollow declination axis by a diagonal flat mirror, and then down the hollow polar axis by a second diagonal. The light beam

Figure 1.1.44. Coudé system for a modified English mounting.

then always emerges from the end of the polar axis, whatever part of the sky the telescope may be inspecting. Designs with similar properties can be devised for most other types of mounting although additional flat mirrors may be needed in some cases. With alt–az mountings (which have motions in altitude and azimuth) the light beam can be directed along the altitude axis to one of the two Nasmyth foci on the side of the mounting. These foci still rotate as the telescope changes its azimuth, but this poses far fewer problems than the changing altitude and attitude of a conventional Cassegrain focus. On large modern telescopes, platforms of considerable size are often constructed at the Nasmyth foci allowing large ancillary instruments to be used. The fixed focus of the Coudé and Nasmyth systems is a very great advantage when bulky items of equipment, such as high dispersion spectrographs, are to be used, since these can be permanently mounted in a nearby separate laboratory and the light brought to them, rather than have to have the equipment mounted on the telescope. The design also has several disadvantages: the field of view rotates as the telescope tracks an object across the sky, and is very tiny due to the large effective focal ratio (f25 to f40) which is generally required to bring the focus through the axes, and finally the additional reflections will cause loss of light.

The simplest of all designs for a telescope is a mirror used at its prime focus. That is, the primary mirror is used directly to produce the images and the detector is placed at the top end of the telescope. The largest telescopes have a platform or cage which replaces the secondary mirror and which is large enough for the observer to ride in while he or she operates and guides the telescope from the prime focus position. With smaller instruments too much light would be blocked, so they must be guided using a separate guide telescope or use a separate detector monitoring another star (the guide star) within the main telescope's field of view. The image quality at the prime focus is usually poor even a few tens of seconds of arc away from the optical axis, because the primary mirror's focal ratio may be as short as f3 or less in order to reduce the length of the instrument to a minimum. Thus, correcting lenses are invariably essential to give acceptable images and reasonably large fields of view. These are similar to those used for correcting Cassegrain telescopes and are placed immediately before the prime focus.

A system that is almost identical to the use of the telescope at prime focus and which was the first design to be made into a working reflecting telescope, is that due to Newton and hence is called the Newtonian telescope. A secondary mirror is used which is a flat diagonal placed just before the prime focus. This reflects the light beam to the side of the telescope from where access to it is relatively easy (figure 1.1.45). The simplicity and cheapness of the design make it very popular as a small telescope for the amateur market, but it is rarely encountered in telescopes larger than about 1 m. There are several reasons for its lack of popularity for large telescope designs, the main ones being that it has no advantage over the prime focus position for large telescopes since the equipment used to detect the image blocks out no more light than the secondary mirror, the secondary mirror introduces additional light losses and the position of the equipment high on the side of the telescope tube causes difficulties of access and counterbalancing. The images in a Newtonian system and at prime focus are very

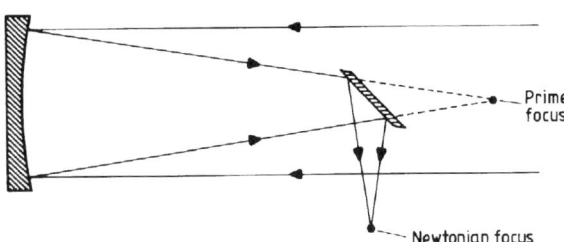

Figure 1.1.45. Newtonian telescope optical system.

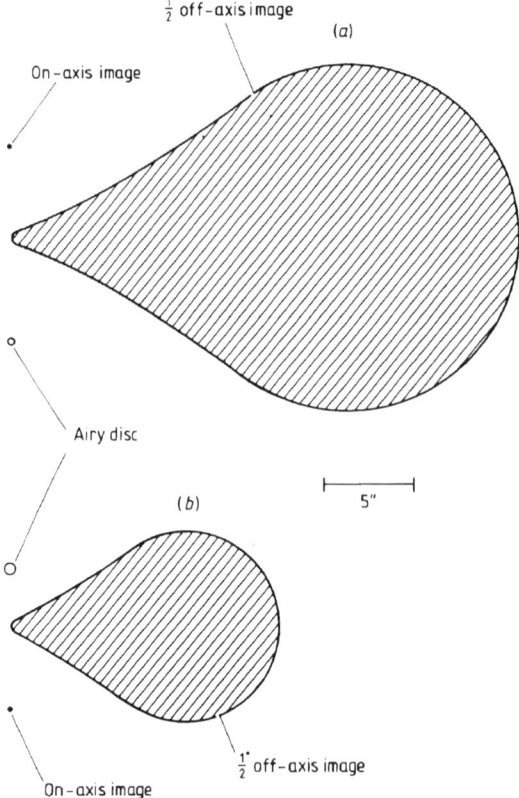

$\frac{1}{2}^\bullet$off-axis image

(a)

On-axis image

Airy disc

(b)

5"

(b)

$\frac{1}{2}^\bullet$ off-axis image

On-axis image

Figure 1.1.46. Images in Newtonian telescopes. (a) 1 m f4 telescope; (b) 0.5 m f8 telescope.

similar and are of poor quality away from the optical axis, as shown in figure 1.1.46.

A very great variety of other catoptric (reflecting objective) telescope designs abound, but most have few or no advantages over the two groups of designs which are discussed above. Some find specialist applications; for example, the Gregorian design is similar to the Cassegrain except that the secondary mirror is a concave ellipsoid and is placed after the prime focus. It was used for the Stratoscope II telescope, and the two 8.4 m instruments of the Large Binocular Telescope on Mount Graham include Gregorian secondary mirrors for infrared work. A few such specialized designs may also be built as small telescopes for amateur use because of minor advantages in their production processes, or for space-saving reasons as in the folded Schiefspieglers, but most such designs will be encountered very rarely.

Of the dioptric (refracting objective) telescopes, only the basic refractor using an achromatic doublet, or very occasionally a triplet, as its objective is

Figure 1.1.47. The astronomical refractor.

in any general use (figure 1.1.47). Apart from the large refractors that were built towards the end of the 19th century, most refractors are now found as small instruments for the amateur market or as the guide telescopes of larger instruments. The enclosed tube and relatively firmly mounted optics of a refractor means that they need little adjustment once aligned with the main telescope.

The one remaining class of optical telescopes is the catadioptric group, of which the Schmidt camera is probably the best known. A catadioptric system uses both lenses and mirrors in its primary light gathering section. Very high degrees of correction of the aberrations can be achieved because of the wide range of variable parameters that become available to the designer in such systems. The Schmidt camera uses a spherical primary mirror so that coma is eliminated by not introducing it into the system in the first place! The resulting spherical aberration is eliminated by a thin correcting lens at the mirror's radius of curvature (figure 1.1.48). The only major remaining aberration is field curvature and the effect of this is eliminated by moulding the photographic plate to the shape of the image surface or through the use of additional correcting lenses (field flatteners) for CCDs and other array detectors. The correcting lens can introduce

Figure 1.1.48. The Schmidt camera optical system.

small amounts of coma and chromatic aberration, but is usually so thin that these aberrations are negligible. Diffraction-limited performance over fields of view of several degrees with focal ratios as fast as f1.5 or f2 is possible. As larger CCDs are becoming available, they are being used on Schmidt cameras. They are still smaller than photographic plates, and so fields of view are restricted to only a degree or so. However, the high detecting efficiency, the linear response, and the machine-readable output of the CCD means that they may be preferred to the photographic plate for many purposes.

The need to use a lens in the Schmidt design has limited the largest such instruments to an entrance aperture 1.35 m in diameter (the Tautenburg Schmidt camera). The Large sky Area Multi-Object fibre Spectroscopic Telescope (LAMOST) currently under construction at the Xinglong Station of the Chinese National Astronomical Observatory, however, will be of 4 m diameter. It will achieve this by using a horizontally fixed primary mirror fed by a coelostat. The coelostat mirror will be actively controlled and also act as the corrector. With up to 4000 optical fibres, the instrument will be one of the most rapid spectroscopic survey tools in the world when it starts operating in 2005.

The Schmidt design suffers from having a tube length at least twice its focal length. A number of alternative designs to provide high quality images over fields of view of 5° to 10° have therefore been proposed, though none with any significant sizes have been built. For example, the Willstrop three-mirror telescope (figure 1.1.49) is completely achromatic and can potentially be made as large as the largest conventional optical telescope. The primary mirror is close to parabolic: the corrector (secondary mirror) is close to spherical with a turned-down edge, while the tertiary is also close to spherical in shape. It has a 5° field and a tube length one third to one half that of the equivalent Schmidt camera; but its focal surface is curved like that of the Schmidt, and all except the central 2° of the field of view suffers from vignetting. A 0.5 m version of the design has been successfully built and operated at Cambridge.

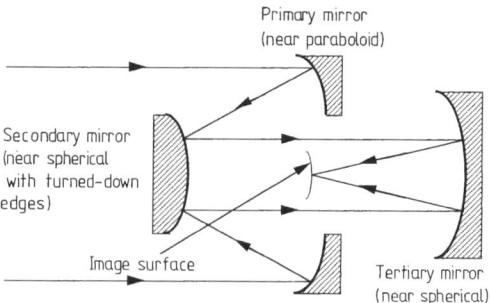

Figure 1.1.49. The Willstrop three-mirror camera optical system.

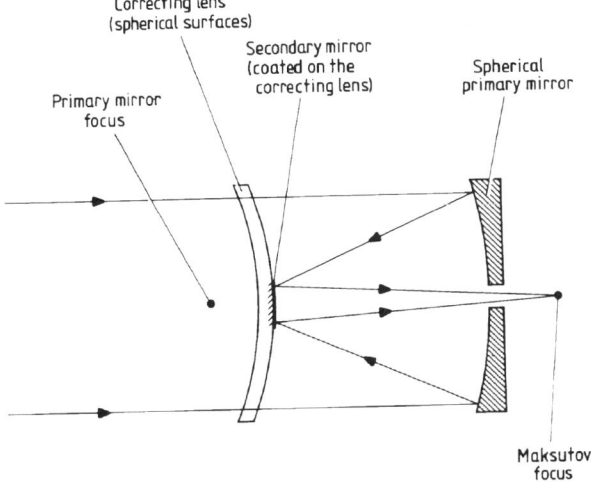

Figure 1.1.50. The Maksutov optical system.

The Schmidt camera cannot be used visually since its focus is inaccessible. There are several modifications of its design, however, that produce an external focus while retaining most of the beneficial properties of the Schmidt. One of the best of these is the Maksutov (figure 1.1.50) that originally also had an inaccessible focus, but which is now a kind of Schmidt–Cassegrain hybrid. All the optical surfaces are spherical, and spherical aberration, astigmatism and coma are almost eliminated, while the chromatic aberration is almost negligible. A similar system is the Schmidt–Cassegrain telescope itself. This uses a thin correcting lens like the Schmidt camera and has a separate secondary mirror. The focus is accessible at the rear of the primary mirror as in the Maksutov system. Schmidt–Cassegrain telescopes are now available commercially in sizes up to 0.4 m diameter. They are produced in large numbers by several firms, at very reasonable costs, for the amateur and education markets. They are also finding increasing use in specific applications, such as site testing, for professional astronomy.

Although it is not a telescope, there is one further system that deserves mention and that is the coelostat. This comprises two flat mirrors that are driven so that a beam of light from any part of the sky is routed into a fixed direction. They are particularly used in conjunction with solar telescopes whose extremely long focal lengths make them impossible to move. One mirror of the coelostat is mounted on a polar axis and driven at half sidereal rate. The second mirror is mounted and driven to reflect the light into the fixed telescope. Various arrangements and adjustments of the mirrors enable differing declinations to be observed.

No major further direct developments in the optical design of large tele-scopes seem likely to lead to improved resolution, since such instruments are already limited by the atmosphere. Better resolution therefore requires that the telescope be lifted above the Earth's atmosphere, or that the distortions that it introduces be overcome by a more subtle approach. Telescopes mounted on rockets, satellites and balloons are discussed later in this section, while the four ways of improving the resolution of Earth-based telescopes (interferometry, speckle interferometry, occultations and real-time compensation) are discussed in sections 2.5, 2.6, 2.7 and later in this section, respectively. A fairly common technique for improving resolution, deconvolution, is discussed in section 2.1. This requires that the precise nature of the degradations of the image be known so that they can be removed. Since the atmosphere's effects are changing in a fairly random manner on a timescale of milliseconds, it is not an appropriate technique for application here, although it should be noted that one approach to speckle interferometry is in fact related to this method and deconvolution may be needed on the compensated image.

A technique termed apodization (literally 'removal of the feet', see also sections 2.5 and 4.1) gives an apparent increase in resolution. It is actually, however, a method of enabling the telescope to achieve its normal resolution, and sometimes even reduces that resolution slightly. For optical telescopes, apodization is usually accomplished through the use of masks over the entrance aperture of the telescope. These have the effect of changing the dif-fraction pattern (see earlier discussion about the Airy disk, and also section 2.1). The nominal resolution given by the size of the central disk may then worsen, but the outer fringes disappear, or become of a more convenient shape for the purpose in mind. A couple of examples will illustrate the process. When searching for a faint object next to a bright one, such as Sirius B next to Sirius A, or for an extra-solar planet next to a star, a square aperture with fuzzy edges, or a four-armed star aperture, will produce fringes in the form of a cross, and suppress the normal circular fringes. If the aperture is rotated so that the faint object lies in one of the regions between the arms of the cross, then it will have a much higher signal-to-noise ratio than if superimposed upon a bright fringe from a circular aperture, and so be more easily detected. Secondly, for observing low-contrast features on planets, a mask that is a neutral density filter varying in opacity in a Gaussian fashion from clear at the centre to nearly opaque at the edge of the objective may be used. This has the effect of doubling the size of the Airy disk, but also of eliminating the outer fringes completely. The resulting image will be significantly clearer since the low contrast features on extended objects are normally swamped by the outer fringes of the PSF (section 2.1). A professionally made neutral density filter of this type would be extremely expensive, but a good substitute may be made using a number of masks formed out of a mesh. Each mask is an annulus with an outer diameter equal to that of the objective and an inner diameter that varies appropriately in size. When they are all superimposed on the objective, and provided that the meshes are *not* aligned, then they will obstruct an increasingly large proportion of the incoming light towards the edge of the objective. The overall effect of the mesh masks will thus be similar to the Gaussian neutral density filter.

Although resolution may not be improved in any straightforward manner (but see discussion on real-time atmospheric compensation below), the position is less hopeless for the betterment of the other main function of a telescope, which is light-gathering. The upper limit to the size of an individual mirror is probably being approached with the 8 m telescopes currently in existence. A diameter of 10–15 m for a metal-on-glass mirror would probably be about the ultimate possible with present-day techniques and their foreseeable developments such as a segmented honeycomb construction for the mirror blank or a thin mirror on active supports that compensate for its flexure. The ESO's VLT for example has four main telescopes each with thin actively supported 8.2 m monolithic mirrors. Currently under construction, the Large Binocular Telescope (LBT) on Mount Graham, Arizona, will have two 8.4 m diameter monolithic honeycomb mirrors on a single mounting. The mirrors will have very short focal ratios (f1.14) and the telescopes will be of Gregorian design. An additional practical constraint on mirror size is that imposed by needing to transport the mirror from the manufacturer to the telescope's observing site. Even with 8 m mirrors, roads may need widening, obstructions levelling, bends have their curvature reduced etc. and this can add very significantly to the cost of the project.

Greater light-gathering power may therefore be acquired only by the use of several mirrors, by the use of a single mirror formed from separate segments, by aperture synthesis (section 2.5) or perhaps in the future by using liquid mirrors (see above). Several telescopes are currently operating or are under construction using single large mirrors formed from several independent segments, for example the two 10 m Keck telescopes that are each constructed from thirty-six 1.8 m hexagonal segments (see figure 1.1.51) and the Hobby–Eberly spectroscopic telescope that has ninety-one 1.0 m spherical segments forming an 11 m × 10 m mirror. The latter uses correcting optics to produce an adequate image and is on a fixed mounting (see later discussion). This type of design requires that each segment of the mirror be independently mounted, and that their positions be continuously monitored and adjusted to keep them within a tenth of a wavelength of their correct positions (active optics—see Real-time atmospheric compensation below). The Gran Telescopio Canarias (GTC), sited at La Palma in the Canary Isles, is a Spanish-led project to build a 10.4 m segmented mirror telescope. It will have 36 segments that can be actively positioned and deformed to maintain their shape. It is currently scheduled for first light in 2004.

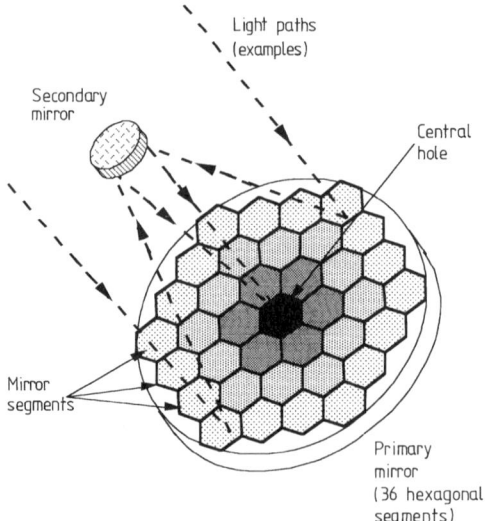

Figure 1.1.51. The optical arrangement of the Keck I and II 10 m telescopes.

Alternatively, the secondary mirror may be segmented as well, and those segments moved to compensate for the variations in the primary mirror segments. The monitoring and control of the segments would be very similar to those required for the remaining option: the multi-mirror telescope. The first such telescope was built at Mount Hopkins about three decades ago. This telescope was based upon six fairly conventional 1.8 m Cassegrain systems, with mirrors of honeycomb construction, and arranged in a hexagonal array 6.9 m across. They were supported on a single mounting and fed a common focus. The system was equivalent to a 4.4 m telescope in area, but only cost about one third as much. Since the cost of a telescope varies as about the cube of its diameter, far larger savings would be possible with the use of more mirrors. The Mount Hopkins telescope, however, experienced problems, especially with the mirror position monitoring system. It has therefore now been refitted with a single 6.5 m mirror.

The accuracy and stability of surfaces, alignments etc. required for both the multi-mirror and the segmented mirror systems are a function of the operating wavelength. For infrared and short microwave work, far less technically demanding limits are therefore possible and their telescopes are correspondingly easier to construct. Specialist infrared telescopes do, however, have other requirements that add to their complexity. First, it is often necessary to chop between the source and the background in order to subtract the latter. Some purpose-built infrared telescopes such as the 3.8 m UKIRT[7] oscillate the secondary mirror through a small angle to

---

[7] United Kingdom Infrared Telescope.

achieve this (a mass equal to that of the chopping mirror may need to be moved in the opposite direction to avoid vibrating the telescope). The UKIRT secondary mirror can also be moved to correct for flexure of the telescope tube and mounting and for buffeting by the wind. For NIR observations, the telescope must be sited above as much of the atmospheric water vapour as possible, limiting the sites to places like Mauna Kea on Hawaii, the Chilean altiplano or the Antarctic plateau. For MIR and some FIR work it may be necessary also to cool the telescope, or parts of it, to reduce its thermal emission, and such telescopes have to be lifted above most of the atmosphere by balloon or aircraft or flown on a spacecraft. The Stratospheric Observatory For Infrared Astronomy (SOFIA) for example, scheduled to start observing in 2005, will be a 2.5 m telescope flown on board a Boeing 747 aircraft at altitudes up to 13 700 m. It will operate at 250 K and be able to cover the range from 0.3 to 1600 μm.

The atmosphere is completely opaque to radiation shorter than about 320 nm. Telescopes designed for ultraviolet observations therefore have to be launched on rockets or carried on board spacecraft. Except in the extreme ultraviolet (EUV: ∼6 to ∼90 nm), where glancing optics are needed (see section 1.3), conventional telescope designs are used. Aluminium suffices as the mirror coating down to 100 nm. At shorter wavelengths, silicon carbide may be used.

The limits may also be relaxed if the system is required as a 'light bucket' rather than for imaging. Limits of ten times the operating wavelength may still give acceptable results when a photometer or spectroscope is used on the telescope, provided that close double stars of similar magnitudes are avoided.

Telescopes in space

The most direct way to improve the resolution of a telescope that is limited by atmospheric degradation is to lift it above the atmosphere, or at least above the lower, denser parts of the atmosphere. The three main methods of accomplishing this are to place the telescope on an aircraft, a balloon-borne platform or an artificial satellite. The Kuiper Airborne Observatory (KAO, a 0.9 m telescope on board a C141 aircraft that operated until 1995) and SOFIA (see above) are examples of the first, but aircraft are limited to maximum altitudes of about 15 km. Balloon-borne telescopes, of which the 0.9 m Stratoscope II is the best known but far from the only example, have the advantages of relative cheapness, and that their images may be recorded by fairly conventional means. Their disadvantages include low maximum altitude (40 km) and short flight duration (a few days).

Many small telescopes have already been launched on satellites. The telescope aboard the International Ultraviolet Explorer (IUE) spacecraft was a fairly typical example. This spacecraft used a 0.45 m Ritchey–Chrétien telescope that fed images to a pair of ultraviolet spectrometers. In the infrared, the entire telescope and its ancillary instrumentation may be cooled. The Infrared Space Observatory (ISO) spacecraft, for example, had a 0.6 m telescope cooled to 3 K with the detectors held at 1.8 K, using 2300 litres of liquid helium. The use of unconventional telescope designs, however, is not uncommon for this type of work, since they may have minor advantages in their weight or volume which become vital when the restrictions imposed by the balloon's lift or launcher's power are taken into account. For example, off-axis systems, Gregorian telescopes (in Strato-scope II), and the Wynne camera, which is a sort of inverse Cassegrain system, have all been used recently. Glancing incidence systems, an extreme form of the off-axis format, are of great importance for short-wave ultraviolet and x-ray observations and are discussed in section 1.3. The Hubble space telescope is a 2.4 m f2.3/f24 Ritchey–Chrétien telescope designed to operate from 115 nm to 1 mm, with a variety of instrumentation at its focus. With its corrective optics to compensate for the deficiencies of the primary mirror, and up-dated ancillary instruments, it is now regularly pro-viding diffraction-limited images. Its replacement scheduled for launch in 2010, the James Webb Space Telescope (JWST), will have a 6.5 m mirror. The mirror will have 36 segments and be folded for launch and during the spacecraft's journey to its working position at the Earth–Sun L2[8] point. It will be able to observe from 600 nm to 10 μm or longer wavelengths and will operate at a temperature of 50 K. Prior to the JWST, ESA's Herschel[9] spacecraft is due for launch in 2007 and will carry a 12-segment 3.5 m silicon carbide mirror. It will also operate at the L2 point, and will observe from 80 to 670 μm. The telescope will be passively cooled to 80 K, with active cooling for the instrumentation down to less than 3 K.

Suggestions for the future include arrays of orbiting optical telescopes with separations of 100 km or more, capable of directly imaging Earth-sized planets around other stars.

Mountings

The functions of a telescope mounting are simple—to hold the optical components in their correct mutual alignment, and to direct the optical axis towards the object to be observed. The problems in successfully accom-plishing this to within the accuracy and stability required by astronomers are so great, however, that the cost of the mounting can be the major item in funding a telescope. We may consider the functions of a mounting under three separate aspects: first supporting the optical components, second

---

[8] The outer Lagrangian point, 1.5 million km farther from the Sun than the Earth on the Sun–Earth line.
[9] Previously named FIRST—Far InfraRed and Sub-millimetre Telescope.

preserving their correct spatial relationships and third acquiring and holding the object of interest in the field of view.

Mounting the optical components is largely a question of supporting them in a manner that ensures their surfaces maintain their correct shapes. Lenses may only be supported at their edges, and it is the impossibility of doing this adequately for large lenses that limits their practicable sizes to a little over a metre. There is little difficulty with small lenses and mirrors since most types of mount may grip them firmly enough to hold them in place without at the same time straining them. However, large mirrors require very careful mounting. They are usually held on a number of mounting points to distribute their weight, and these mounting points, especially those around the edges, may need to be active so that their support changes to compensate for the different directions of the mirror's weight as the telescope moves. The active support can be arranged by systems of pivoted weights, especially on older telescopes or more normally nowadays by computer control of the supports. As discussed earlier some recently built telescopes and many of those planned for the future have active supports for the primary mirror that deliberately stress the mirror so that it retains its correct shape whatever the orientation of the telescope or the temperature of the mirror. Segmented mirror telescopes additionally need the supports to maintain the segments in their correct mutual alignments. On the 10 m Keck telescopes, for example, the active supports adjust the mirror segment's positions twice a second to maintain their places correctly to within ± 4 nm.

The optical components are held in their relative positions by the telescope tube. With most large telescopes, the 'tube' is in fact an open structure, but the name is still retained. For small instruments the tube may be made sufficiently rigid that its flexure is negligible. But this becomes impossible as the size increases. The solution then is to design the flexure so that it is identical for both the primary and secondary mirrors. The optical components then remain aligned on the optical axis, but are no longer symmetrically arranged within the mounting. The commonest structure in use that allows this equal degree of flexure and also maintains the parallelism of the optical components is the Serrurier truss (figure 1.1.52). However, this design is not without its own problems. The lower trusses may need to be excessively thin or the upper ones excessively thick in order to provide the required amount of flexure. Even so, many of the large reflecting telescopes built in the past four decades have used Serrurier truss designs for their tubes. More recently computer-aided design has enabled other truss systems to be developed. This, however, is mostly for economic reasons, and the design objectives of such supports remains unchanged.

A widely used means of mounting the telescope tube so that it may be pointed at an object and then moved to follow the object's motion across

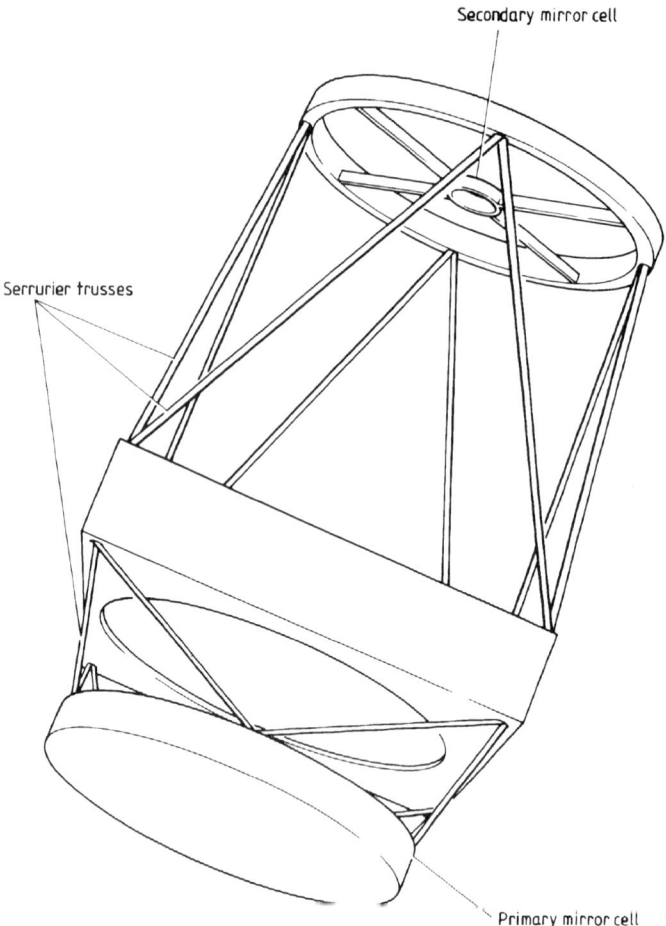

Secondary mirror cell

Serrurier trusses

Primary mirror cell

Figure 1.1.52. Telescope tube construction based on Serrurier trusses.

the sky is the equatorial mounting. This is a two-axis mounting, with one axis, the polar axis, aligned parallel with the Earth's rotational axis, and the other, the declination axis, perpendicular to the polar axis. The design has the enormous advantage that only a single constant velocity motor is required to rotate the mounting around the polar axis in order to track an object. It is also very convenient in that angular read-outs on the two axes give the hour angle or right ascension and declination directly.

A large variety of different arrangements for the basic equatorial mounting exist, but these will not be reviewed here. Books on telescopes and most general astronomy books list their details by the legion, and should be checked if the reader requires further information.

The alt–az mounting (which has motions in <u>al</u>titude and <u>az</u>imuth) is the other main two-axis mounting system. Structurally it is a much simpler form than the equatorial and has therefore long been adopted for most large radio telescope dishes. Its drawbacks are that the field of view rotates with the telescope motion, and that it needs driving continuously in both axes and with variable speeds in order to track an object. For the past two decades most new large optical telescopes have used alt–az mountings, and the reduction in price and increase in capacity of small computers means that smaller telescopes even down to a few centimetres in size frequently have such a mounting as an option.

Telescopes up to 1 m in diameter constructed for or by amateur astronomers are often mounted on a Dobsonian alt–az mounting. This may be simply and cheaply constructed from sheets of plywood or similar material. The telescope is supported by two circles, which rest in two semi-circular cut-outs in a three-sided cradle shaped like an inverted 'Π'. These allow the telescope to move in altitude. The cradle then pivots at its centre to give motion in azimuth. Telescopes on such mountings are usually un-driven, and are moved by hand to find and follow objects in the sky. Small telescopes on any alt–az design of mounting can be given a tracking motion for a short interval of time by placing them on a platform that is itself equatorially driven. There are several designs for such equatorial platforms using sloping planes or inclined bearings, and details may be found in modern books aimed at the amateur observer.

Large telescopes, whether on equatorial or alt–az mountings, almost universally use hydrostatic bearings for the main moving parts. These bearings use a thin layer of pressurized oil between the moving and static parts, and the oil is continuously circulated by a small pump. The two solid surfaces are thus never in direct contact, and movement around the bearing is very smooth and with low friction.

Several telescopes have fixed positions. These include liquid-mirror telescopes that always point near the zenith (see above), and the 11 m Hobby–Eberly and SALT telescopes (see above). Tracking is then accomplished by moving the detector and any correcting optics to follow the motion of the image, or for short exposures, by moving the charges along the pixels of a CCD detector (TDI; see above). Zenith-pointing telescopes are generally limited to a few degrees either side of the zenith, but the Hobby–Eberly telescope can point to objects with declinations ranging from $-11°$ to $+71°$. It accomplishes this because it points to a fixed zenith angle of $35°$, but can be rotated in azimuth between observations. There are also a few instruments mounted on altitude–altitude mountings. These use a universal joint (such as a ball and socket) and tip the instrument in two orthogonal directions to enable it to point anywhere in the sky.

With any type of mounting, acquisition and tracking accuracies of a second of arc or better are required, but these may not be achieved, especially for the smaller Earth-based telescopes. Thus the observer may have to search for his or her object of interest over an area tens to hundreds of seconds of arc across after initial acquisition, and then guide, either himself or herself, or with an automatic system, to ensure that the telescope tracks the object sufficiently closely for his or her purposes. With balloon-borne and space telescopes such direct intervention is difficult or impossible. However, the reduction of external disturbances and possibly the loss of weight mean that the initial pointing accuracy is higher. For space telescopes tracking is easy since the telescope will simply remain pointing in the required direction once it is fixed, apart from minor perturbations such as light pressure, the solar wind, gravitational anomalies and internal disturbances from the spacecraft. Automatic control systems can therefore usually be relied upon to operate the telescope. Earlier spacecraft such as the IUE, however, had facilities for transmitting a picture of the field of view after the initial acquisition to the observer on the ground, followed by corrections to the position in order to set on to the correct object.

With terrestrial telescopes the guiding may be undertaken via a second telescope that is attached to the main telescope and aligned with it, or a small fraction of the light from the main telescope may be diverted for the purpose. The latter method may be accomplished by beam splitters, dichroic mirrors, or by intercepting a fraction of the main beam with a small mirror, depending upon the technique being used to study the main image. The whole telescope can then be moved using its slow-motion controls to keep the image fixed, or additional optical components can be incorporated into the light path whose movement counteracts the image drift, or the detector can be moved around the image plane to follow the image motion or the charges can be moved within a CCD chip (TDI). Since far less mass needs to be moved with the latter methods, their response times can be much faster than that of the first method. The natural development of the second method leads on to the active surface control which is discussed later in this section and which can go some way towards eliminating the effects of scintillation.

If guiding is attempted by the observer then there is little further to add, except to advise a plentiful supply of black coffee in order to avoid going to sleep through the boredom of the operation! Automatic guiding has two major problem areas. The first is sensing the image movement and generating the error signal, while the second is the design of the control system to compensate for this movement. Detecting the image movement has been attempted in two main ways—CCDs and quadrant detectors.

With a CCD guide, the guide image is read out at frequent intervals, and any movement of the object being tracked is detected via software that also generates the correction signals. The guide CCD is adjacent to the main detector (which may or may not be a CCD itself), and thus it guides on objects other than those being imaged. This is termed off-set guiding and often has advantages over guiding directly on the object of interest. The off-set guide star can be brighter than the object of interest, and if the latter is an extended object, then guiding on it directly will be difficult. However, for comets, asteroids and other moving objects, the object itself must be followed. In some small CCD systems aimed at the amateur market, it is the main image that is read out at frequent intervals and the resulting multiple images are then shifted into mutual alignment before being added together. This enables poor tracking to be corrected provided that the image shift is small during any single exposure.

A quadrant detector is one whose detecting area is divided into four independent sectors, the signal from each of which can be separately accessed. If the image is centred on the intersection of the sectors, then their signals will all be equal. If the image moves, then at least two of the signals will become unbalanced and the required error signal can then be generated. A variant of this system uses a pyramidal 'prism'. The image is normally placed on the vertex, and so is divided into four segments each of which can then be separately detected. When guiding on stars by any of these automatic methods it may be advantageous to increase the image size slightly by operating slightly out of focus. Scintillation 'jitter' then becomes less important.

The problem of the design of the control system for the telescope to compensate for the image movement should be handed to a competent control engineer. If it is to work adequately, then proper damping, feedback, rates of motion and so on must be calculated so that the correct response to an error signal occurs without 'hunting' or excessive delays.

Real-time atmospheric compensation

The resolution of ground-based telescopes of more than a fraction of a metre in diameter is limited by the turbulence in the atmosphere. The maximum diameter of a telescope before it becomes seriously affected by atmospheric turbulence is given by Fried's coherence length, r_0,

$$r_0 \approx 0.114 \left(\frac{\lambda \cos z}{550} \right)^{0.6} \text{ m} \qquad (1.1.73)$$

where λ is the operating wavelength in nm and z is the zenith angle. Fried's coherence length, r_0, is the distance over which the phase difference is one radian. Thus, for visual work, telescopes of more than about 11.5 cm diameter will always have their images degraded by atmospheric turbulence.

In an effort to reduce the effects of this turbulence (the 'seeing', 'scintillation' or 'twinkling'), many large telescopes are sited at high altitudes, or placed on board high-flying aircraft or balloons. The ultimate, though very expensive, solution of course, is to orbit the telescope beyond the Earth's atmosphere out in space.

An alternative approach to obtaining diffraction-limited performance for large telescopes, that is relatively inexpensive and widely applicable, is to correct the distortions in the incoming light beam produced by the atmosphere. This atmospheric compensation is achieved through the use of adaptive optics. In such systems, one or more of the optical components can be changed rapidly and in such a manner that the undesired distortions in the light beam are reduced or eliminated. Although a relatively recent development in its application to large telescopes, such as the VLT, Gemini, Keck, William Herschel and Subaru telescopes, adaptive optics is actually a very ancient technique familiar to us all. That is because the eye operates via an adaptive optic system in order to keep objects in focus, with the lens being stretched or compressed by the ciliary muscle (figure 1.1.1).

The efficiency of an adaptive optics system is measured by the Strehl ratio. This quantity is the ratio of the intensity at the centre of the corrected image to that at the centre of a perfect diffraction-limited image of the same source. The normalized Strehl ratio is also used. This is the Strehl ratio of the corrected image divided by that for the uncorrected image. Strehl ratios up to 0.6 are currently being achieved, and may reach 0.8 in the near future.

Adaptive optics is not a complete substitute for spacecraft-based telescopes, however, because in the visual and near infrared, the correction only extends over a very small area (the isoplanatic patch, see below). Thus, for example, if applied to producing improved visual images of Jupiter, only a small percentage of the planet's surface would be seen sharply. Also, of course, ground-based telescopes are still limited in their wavelength coverage by atmospheric absorption.

There is often some confusion in the literature between adaptive optics and active optics. However, the most widespread usage of the two terms is that an adaptive optics system is a fast closed-loop system, and an active optics system a more slowly operating open- or closed-loop system. The division is made at a response time of a few seconds. Thus the tracking of a star by the telescope drive system can be considered as an active optics system that is open-loop if no guiding is used, and closed-loop if guiding is used. Large thin mirror optical telescopes and radio telescopes may suffer distortion due to buffeting by the wind at a frequency of 0.1 Hz or so; they may also distort under gravitational loadings or thermal stresses, and have residual errors in their surfaces from the manufacturing process. Correction of all of these effects would also be classified under active optics. There is additionally the term active support that refers to the mountings used for the optical components in either an adaptive or an active optics system.

An atmospheric compensation system contains three main components, a sampling system, a wave front sensor and a correcting system. We will look at each of these in turn.

Sampling system

The sampling system provides the sensor with the distorted wave front or an accurate simulacrum thereof. For astronomical adaptive optics systems, a beam splitter is commonly used. This is just a partially reflecting mirror that typically diverts about 10% of the radiation to the sensor, while allowing the other 90% to continue on to form the image. A dichroic mirror can also be used which allows all the light at the desired wavelength to pass into the image while diverting light of a different wavelength to the sensor. However, atmospheric effects change with wavelength, and so this latter approach may not give an accurate reproduction of the distortions unless the operating and sampling wavelengths are close together.

Since astronomers go to great lengths to gather photons as efficiently as possible, the loss of even 10% to the sensor is to be regretted. Many adaptive optics systems therefore use a guide star rather than the object of interest to determine the wave front distortions. This becomes essential when the object of interest is a large extended object, since most sensors need to operate on point or near-point images. The guide star must be very close in the sky to the object of interest, or its wave front will have undergone different atmospheric distortion (figure 1.1.53). For solar work small sunspots or pores can be used as the guide object. The region of the sky over which images have been similarly affected by the atmosphere is called the isoplanatic area or patch. It is defined by the distance over which the Strehl ratio improvement due to the adaptive optics halves. In the visible it is about 15″ across. The size of the isoplanatic patch scales as $\lambda^{1.2}$, so it is larger in the infrared, reaching 80″ at 2.2 µm (K band—see section 3.1).

The small size of the isoplanatic area means that few objects have suitable guide stars. Less than 1% of the sky can be covered using real stars as guides, even in the infrared. Recently therefore, artificial guide stars have been produced. This is accomplished using a powerful laser[10] pointed skywards. The laser is tuned to one of the sodium D line frequencies and excites the free sodium atoms in the atmosphere at a height of about 90 km. The glowing atoms appear as a starlike patch that can be placed as near in the sky to the object of interest as required. Guide stars at lower altitudes and at other wavelengths can be produced through back scattering by air molecules of a laser beam. The laser light can be sent out through the main telescope or more usually using an auxiliary telescope mounted on the main telescope. Laser-produced guide stars have two problems, however,

[10] Continuous wave lasers with powers of up to 10 W may be used. These radiate sufficient energy to cause skin burns and retinal damage. Care therefore has to be taken to ensure that the beam does not intercept an aircraft. The VLT, for example, uses a pair of cameras and an automatic detection system to close a shutter over the laser beam should an aircraft approach.

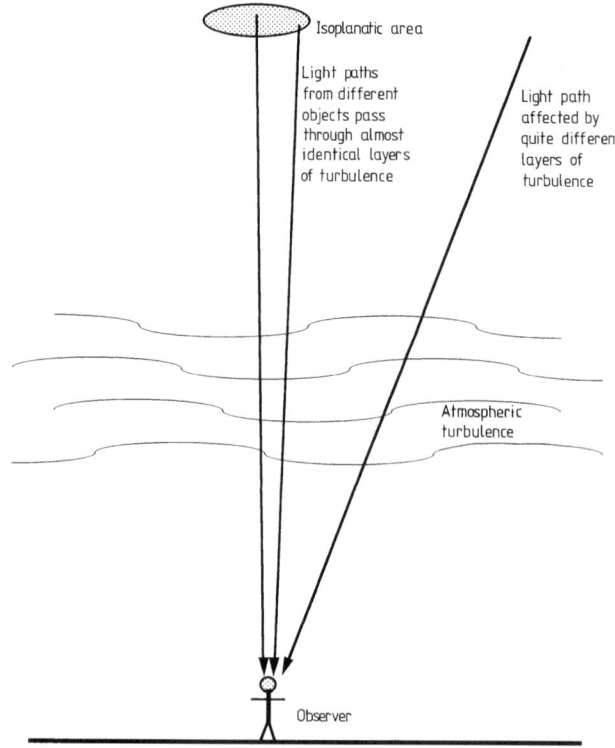

Figure 1.1.53. The isoplanatic area.

that limit their usefulness. First, for larger telescopes, the relatively low height of the guide star means that the light path from it to the telescope differs significantly from that for the object being observed (the cone problem—figure 1.1.54). At 1 μm, the gain in Strehl ratio is halved through this effect, and in the visible it results in almost no improvement at all to the images. Second, the outgoing laser beam is affected by atmospheric turbulence, and therefore the guide star moves with respect to the object. This limits the correction of the overall inclination (usually known as the tip–tilt) of the wave front resulting in a blurred image on longer exposures. A real star can, however, be used to determine the tip–tilt of the wave front separately. The wave front sensor simply needs to detect the motion of the star's image and, since the whole telescope aperture can be utilized for this purpose, very faint stars can be observed. So most objects have a suitable star sufficiently nearby to act as a tip–tilt reference.

An adaptive optics system using two or more guide stars, such as that proposed for the Gemini telescopes and known as MCAO (Multi-Conjugate Adaptive Optics), eliminates the cone effect and produces an isoplanatic patch up to 120″ across. The two guide stars are separated by a small

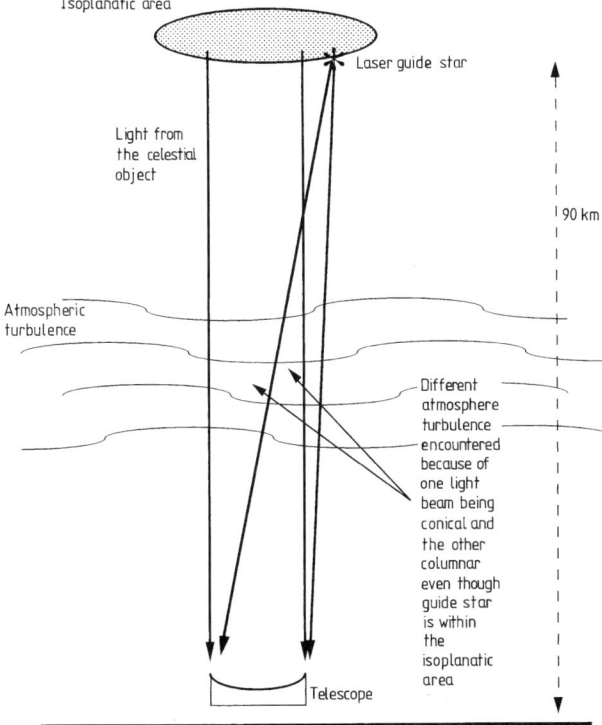

Figure 1.1.54. Light paths from a celestial object and a laser guide star to the telescope.

angle and detected by two wave front sensors. This enables the atmospheric turbulence to be modelled as a function of altitude. Two or three subsidiary deformable mirrors then correct the wave front distortion.

Wave front sensing

The wave front sensor detects the residual and changing distortions in the wave front provided by the sampler after reflection from the correcting mirror. The Hartmann (also known as the Hartmann–Shack or the Shack–Hartmann) sensor is in widespread use in astronomical adaptive optics systems. This uses a two-dimensional array of small lenses (figure 1.1.55). Each lens produces an image that is sensed by an array detector. In the absence of wave front distortion, each image will be centred on each detector. Distortion will displace the images from the centres of the detectors, and the degree of displacement and its direction is used to generate the error signal. An alternative sensor is based upon the shearing interferometer (figure 1.1.56). This is a standard interferometer but with the mirrors marginally turned so that the two beams are slightly displaced with respect to each other when they are recombined. The deformations of the fringes in the

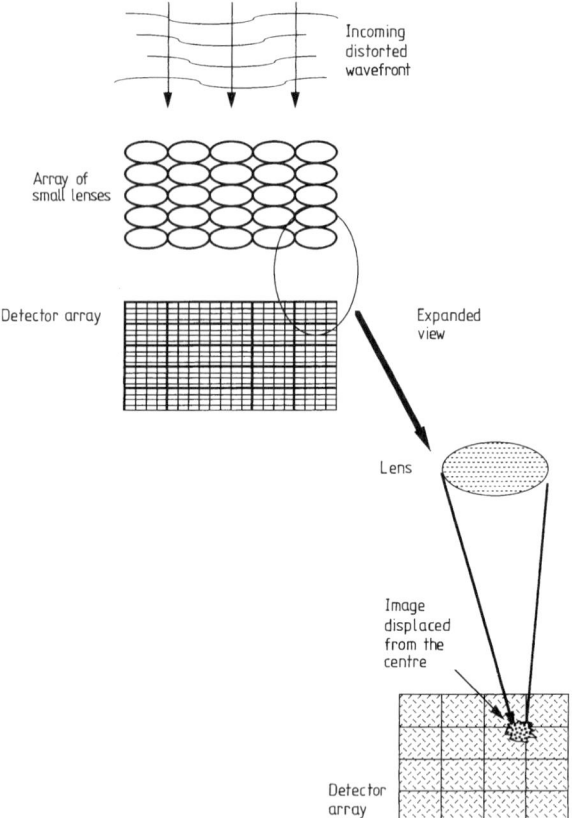

Figure 1.1.55. The Hartmann sensor.

overlap region then provide the slopes of the distortions in the incoming wave front. The shearing interferometer was widely used initially for adaptive optics systems, but has now largely been replaced by the Hartman sensor. A new sensor has been developed recently, known as the curvature sensor. It detects the wavefront distortions by comparing the illumination variations across slightly defocused images just inside and outside the focal point. A vibrating mirror is used to change the focus at kilohertz frequencies.

Wave front correction

In most astronomical adaptive optics systems, the correction of the wave front is achieved by distorting a subsidiary mirror. Since the atmosphere changes on a timescale of 10 ms or so, the sampling, sensing and correction have to occur in 1 ms or less. In the simplest systems only the overall tip and tilt of the wave front introduced by the atmosphere is corrected. That is accomplished by suitably inclining a plane or segmented mirror placed in

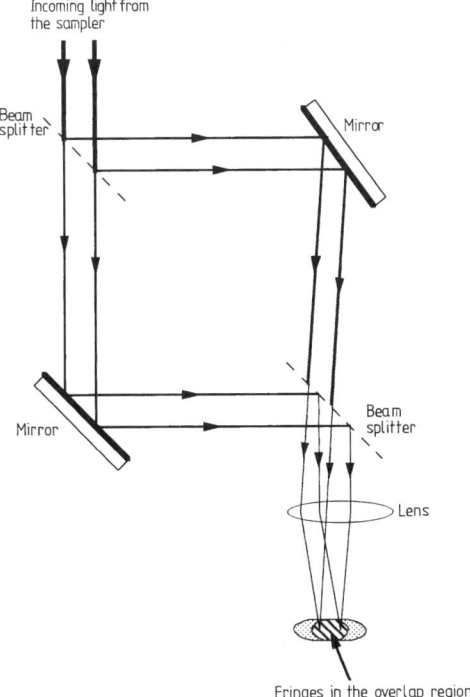

Figure 1.1.56. The shearing interferometer.

the light beam from the telescope in the opposite direction (figure 1.1.57). Tip–tilt correction systems for small telescopes are now available commercially at about 20% of the cost of a 0.3 m Schmidt–Cassegrain telescope and improve image sharpness by about a factor of two. An equivalent procedure, since the overall tilt of the wave front causes the image to move, is 'shift and add'. Multiple short exposure images are shifted until their brightest points are aligned, and then added together. Even this simple correction, however, can result in a considerable improvement of the images.

More sophisticated approaches provide better corrections; either just of the relative displacements within the distorted wave front, or of both displacement and fine scale tilt. In some systems the overall tilt of the wave front is corrected by a separate system using a flat mirror whose angle can be changed. Displacement correction would typically use a thin mirror capable of being distorted by up to 100 piezo-electric or other actuators placed underneath it. The error signal from the sensor is used to distort the mirror in the opposite manner to the distortions in the incoming wave front. The reflected wave front is therefore almost flat.

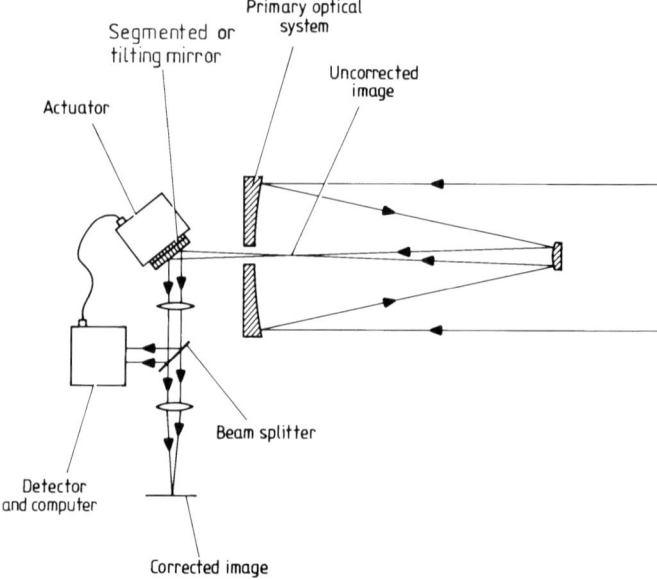

Figure 1.1.57. Schematic optical system for real-time atmospheric compensation.

The non-correction of the fine scale tilt, however, does leave some small imperfections in the reflected wave front. Nonetheless, currently operating systems using this approach can achieve diffraction-limited performance in the near infrared for telescopes of 3 or 4 m diameter (i.e. about $0.2''$ at $2\,\mu m$ wavelength). A recent proposal to improve the correcting mirrors is for a liquid mirror based upon reflective particles floating on a thin layer of oil. The oil contains magnetic grains and its surface can be shaped by the use of small electromagnets, but this has yet to be applied to astronomical image correction. At visual wavelengths, reductions of the uncorrected image size by about a factor of ten are currently being reached. Further improvements can sometimes be achieved by applying blind or myopic deconvolution (section 2.1) to the corrected images after they have been obtained.

Correction of fine scale tilt within the distorted wave front as well as displacement is now being investigated in the laboratory. It requires an array of small mirrors rather than a single 'bendy' mirror. Each mirror in the array is mounted on four actuators so that it can be tilted in any direction as well as being moved linearly. At the time of writing it is not clear whether such systems will be applied to large astronomical telescopes, because at visual wavelengths few of them have optical surfaces good enough to take advantage of the improvements such a system would bring. Plans for future 50 m and 100 m telescopes, however, include adaptive secondary or tertiary mirrors up to 8 m in diameter, requiring up to 500 000 actuators.

Future developments

As a general guide to the future we may look to the past. Figure 1.1.58 shows the way in which the collecting area of optical telescopes has increased with time. Simple extrapolation of the trends shown there suggests collecting areas of 800 m^2 by the year 2050 and of 4500 m^2 by the year 2100 (diameters of 32 m and 75 m respectively for filled circular apertures). Alternative historical analyses suggest a doubling of aperture every 30 years, leading to 100 m optical telescopes by the year 2100. However, there seem to be no fundamental technical differences between constructing 10 m segmented mirrors like those of the Keck telescopes and 30 m or even 100 m segmented mirrors. It is just a case of doing the same thing more times and finding

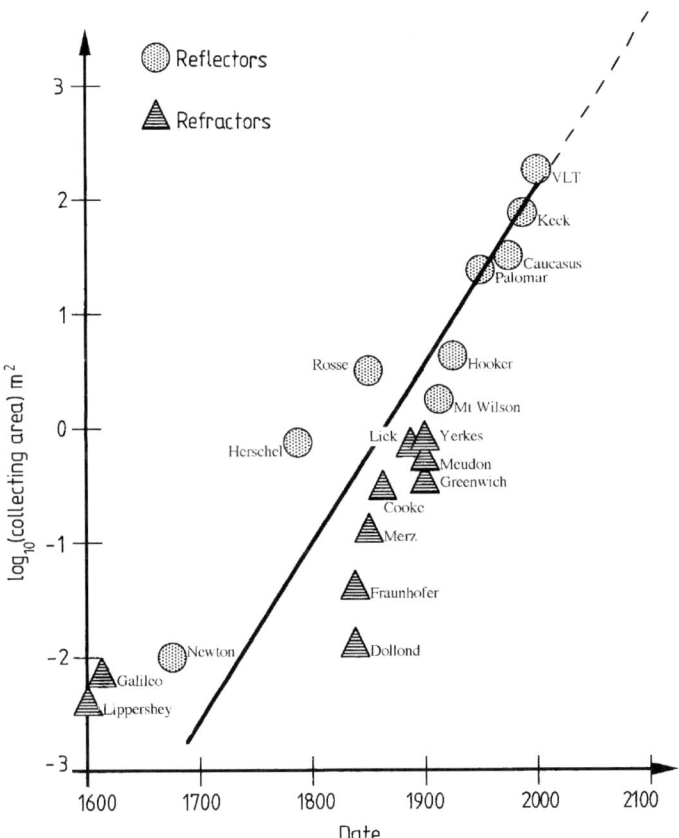

Figure 1.1.58. Optical telescope collecting area as a function of time.

the increased funding required. Traditionally the cost of a telescope is expected to rise as $D^{2.6}$ or thereabouts. This would suggest a cost of £10 000 million to £30 000 million for a 100 m class telescope. However, the use of lightweight mirrors, numerous small identical segments, short focal ratios etc. means that not only is the cost of the optics reduced, but the mounting and enclosure are much cheaper to produce as well. Other expensive items of ancillary equipment such as re-aluminizing vacuum chambers can be reduced in size, and so on. Thus current estimates are around £200–300 million for a 30 m telescope and £600–800 million for a 100 m telescope. These are still large sums, but not beyond the realms of possibility. The JWST, for example, is expected to cost around £500 million. The Apollo space programme in today's money cost around £30 000 million, and the HST has so far cost some £1500 million.

There are several proposals for telescopes significantly larger than 10 m. Apart from the VLT and the Kecks none of them have yet got beyond an initial planning stage, and some are sure to fall by the wayside, but it seems likely that in the next two, perhaps three, decades one or more 30–100 m class telescopes will be built. The VLT is currently starting to act as an aperture synthesis system with a sensitivity equal to that of a 16 m telescope and an un-filled aperture diameter of 100 m. The Keck telescopes will soon be able to act as an aperture synthesis system with a sensitivity equal to that of a 14 m telescope and an un-filled aperture diameter of 85 m. LAMA, based upon liquid mirrors with a 42 m equivalent aperture, has already been mentioned. The 20 m VLOT[11] proposal is currently being investigated by Canada. There are two USA-based proposals for 30 m telescopes, GSMT and CELT.[12] A European consortium, including the UK, is considering a 50 m telescope, Euro50. Perhaps the most ambitious concept is an ESO proposal for OWL,[13] a 100 m telescope. This would have a segmented secondary mirror as well as the primary and costs would be kept down by using mirror segments with spherical surfaces so that they could be mass-produced. Several smaller active mirrors would then provide correction for aberrations and atmospheric effects. With MCAO correction of the images (which would require 500 000 active elements), such a telescope would have a resolution of 1 mas[14] in the visible region. The limiting magnitude for such a telescope might be 35^m in the visible, not just because of its increased light grasp, but because a 1 mas stellar image only has to equal the brightness of a 1 mas^2 area of the background sky in order to be detectable. This would enable the telescope to observe Jupiter-sized planets directly out to a distance of about 100 pc. Telescopes

[11] Very Large Optical Telescope.
[12] Giant Segmented Mirror Telescope and California Extremely Large Telescope.
[13] OverWhelmingly Large telescope.
[14] Milli-arc second.

with diameters of several tens of metres have attracted, perhaps appropriately, the acronym GODs (Giant Optical Devices)!

The multi-mirror concept may be extended or made easier by the development of fibre optics. Fused silica fibres can now be produced whose light losses and focal ratio degradation are acceptable for lengths of tens of metres. Thus mirrors on independent mounts with a total area of hundreds of square metres which all feed a common focus will become technically feasible within a decade.

A technique that is in occasional use today but which may well gain in popularity is daytime observation of stars. Suitable filters and large focal ratios (that diminish the sky background brightness, but not that of the star, see the earlier discussion) enable useful observations to be made of stars as faint as the seventh magnitude. The technique is particularly applicable in the infrared where the scattered solar radiation is minimal. If diffraction-limited 100 m telescopes are built then with 10 nm bandwidths they would be able to observe stars (or other sub-mas sources, but not extended objects) down to a magnitude of 23^m in the visible during the day.

Lenses seem unlikely to make a comeback as primary light-gathering components. However, they are still used extensively in eyepieces and in ancillary equipment. Three major developments seem likely to affect these applications, principally by making lens systems simpler and/or cheaper. The first is the use of plastic to form the lens. High quality lenses then can be cheaply produced in quantity by moulding. The second advance is related to the first and it is the use of aspherical surfaces. Such surfaces have been used when they were essential in the past, but their production was expensive and time-consuming. The investment in an aspherical mould, however, would be small if that mould were then to produce thousands of lenses. Hence cheap aspherical lenses could become available. The final possibility is for lenses whose refractive index varies across their diameter or with depth. Several techniques exist for introducing such variability to the refractive index of a material, such as the diffusion of silver into glass and the growing of crystals from a melt whose composition varies with time. Lenses made from liquid crystal are now being made in small sizes. The refractive index in this material depends upon an externally applied electric field, and can thus be changed rapidly. Such lenses may find application to wavefront correction in adaptive optics systems in due course. Thus very highly corrected and relatively cheap lens systems may become available within a few years through the use of some or all of these techniques.

Space telescopes seem likely to follow the developments of terrestrially based telescopes towards increasing sizes, though if terrestrial diffraction-limited telescopes with diameters of several tens of metres or more become available, the space telescopes will only have advantages for spectral regions outside the atmospheric windows. Other developments may be along the lines of LISA (section 1.6) with two or more separately orbiting telescopes forming very high resolution interferometers.

Observing domes, enclosures and sites

Any permanently mounted optical telescope requires protection from the elements. Traditionally this has taken the form of a hemispherical dome with an aperture through which the telescope can observe. The dome can be rotated to enable the telescope to observe any part of the sky, and the aperture may be closed to protect the telescope during the day and during inclement weather. Many recently built large telescopes have used cylindrical or other shapes for the moving parts of the enclosure for economic reasons. However, such structures are still clearly related to the conventional hemisphere. The dome is an expensive item and it can amount to a third of the cost of the entire observatory, including the telescope. Domes and other enclosures also cause problems through heating-up during the day so inducing convection currents at night and through the generation of eddies as the wind blows across the aperture. Future very large telescopes may therefore operate without any enclosure at all, just having a movable shelter to protect them when not in use. This will expose the telescopes to wind buffeting, but active optics can now compensate for that at any wind speed at which it is safe to operate the telescope.

The selection of a site for a major telescope is at least as important for the future usefulness of that telescope as the quality of its optics. The main aim of site selection is to minimise the effects of the Earth's atmosphere, of which the most important are usually scattering, absorption and scintillation. It is a well-known phenomenon to astronomers that the best observing conditions at any site occur immediately prior to the final decision to place a telescope there! Thus not all sites have fulfilled their expectations.

Scattering by dust and molecules causes the sky background to have a certain intrinsic brightness, and it is this that imposes a limit upon the faintest detectable object through the telescope. The main source of the scattered light is artificial light, and most especially street lighting. Thus a first requirement of a site is that it be as far as possible from built-up areas. If an existing site is deteriorating due to encroaching suburbs etc. then some improvement may be possible for some types of observation by the use of a light pollution rejection (LPR) filter which absorbs in the regions of the most intense sodium and mercury emission lines. Scattering can be worsened by the presence of industrial areas upwind of the site or by proximity to deserts, both of which inject dust into the atmosphere.

Absorption is due mostly to the molecular absorption bands of the gases forming the atmosphere. The two well-known windows in the spectrum, wherein radiation passes through the atmosphere relatively unabsorbed, extend from about 360 nm to 100 μm and from 10 mm to 100 m. But even in these regions there is some absorption, so that visible light is decreased in its intensity by 10 to 20% for vertical incidence. The infrared region is badly affected by water vapour and other molecules to the extent that portions of it are completely obscured. Thus the second requirement of a site is that it be at as high an altitude as possible to reduce the air paths to a minimum, and that the water content of the air be as low as possible. We have already seen how balloons and spacecraft are used as a logical extension of this requirement, and a few high-flying aircraft are also used for this purpose.

Scintillation is the change in the wave front due to the varying densities of the atmospheric layers. It is sometimes divided into seeing, due to low level turbulence, and scintillation proper, arising at higher levels. It causes the image to shimmer when viewed through a telescope, changing its size, shape, position and brightness at frequencies from one to a thousand hertz. It is the primary cause of the low resolution of large telescopes, since the image of a point source is rarely less than half a second of arc across due to this blurring. Thus the third requirement for a good site is a steady atmosphere. The ground-layer effects of the structures and landscape in the telescope's vicinity may worsen scintillation. A rough texture to the ground around the dome, such as low-growing bushes, seems to reduce scintillation when compared with that found for smooth (e.g. paved) and very rough (e.g. tree-covered) surfaces. Great care also needs to be taken to match the dome, telescope and mirror temperatures to the ambient temperature or the resulting convection currents can worsen the scintillation by an order of magnitude or more. Cooling the telescope during the day to the predicted night-time temperature is helpful, provided that the weather forecast is correct! Forced air circulation by fans and thermally insulating and siting any ancillary equipment that generates heat as far from the telescope as possible, or away from the dome completely, can also reduce convection currents.

These requirements for an observing site restrict the choice very considerably and lead to a clustering of telescopes on the comparatively few optimum choices. Most are now found at high altitudes and on oceanic islands, or with the prevailing wind from the ocean. Built-up areas tend to be small in such places and the water vapour is usually trapped near sea level by an inversion layer. The long passage over the ocean by the winds tends to minimize dust and industrial pollution. The remaining problem is that there is quite a correlation of good observing sites and noted tourist spots so that it is somewhat difficult to persuade non-astronomical colleagues that one is going to, say, Tenerife or Hawaii, in order to work! This latter problem disappears for infrared and sub-millimetre wave observers, since their optimum site is the Antarctic plateau with its cold, dry and stable atmospheric conditions. The Antarctic plateau is also optimum for radio studies of the aurora etc. with, for example, the EISCAT (European Incoherent SCATter) fixed and movable dishes operating a wavelengths of a few hundreds of millimetres.

The recent trend in the reduction of the real cost of data transmission lines, whether these are via cables or satellites, is likely to lead to the greatly increased use of remote control of telescopes. In a decade or two most major observatories are likely to have only a few permanent staff physically present at the telescope wherever that might be in the world, and the astronomer would only need to travel to a relatively nearby control centre for his or her observing shifts. Since the travel costs for UK astronomers alone would suffice to buy two 1 m telescopes per year, such a development is likely to be greeted with relief by the funding agencies, if not by the more jet-setting astronomers. There are also a few observatories with completely robotic telescopes, in which all the operations are computer controlled, with no staff on site at all during their use. Many of these robotic instruments are currently used for long-term photometric monitoring programmes or for teaching purposes. However, their extension to more exacting observations such as spectroscopy is now starting to occur. The Liverpool telescope, for example, is due to start operating shortly and will be a 2 m fully robotic telescope sited in the Canary isles with CCD and infrared imaging and two spectroscopes eventually available. It will be available for schools to use for part of the time.

Exercises

1.1.1   Calculate the effective theoretical resolutions of the normal and the dark-adapted eye, taking their diameters to be 2 and 6 mm, respectively.

1.1.2   Calculate the focal lengths of the lenses of a cemented achromatic doublet, whose overall focal length is 4 m. Assume the contact surfaces are of equal radius, the second surface of the diverging lens is flat and that the components are made from crown and dense flint glasses. Correct it for wavelengths of 486 and 589 nm.

1.1.3   By ray tracing calculate the physical separation in the focal plane of two rays that are incident on to a parabolic mirror of focal length 4 m. The rays are parallel prior to reflection and at an angle of 1° to the optical axis. The first ray intersects the mirror at its centre, while the second intersects it 0.5 m from the optical axis, and in the plane containing the rays and the optical axis (figure 1.1.59).

Figure 1.1.59. Optical arrangement for exercise 1.1.3.

1.1.4 Calculate the maximum usable eyepiece focal length for the 5 m f3.3/f16 Mount Palomar telescope at its Cassegrain focus, and hence its lowest magnification.

1.1.5 If the original version Mount Hopkins multi-mirror telescope were to be used as a Michelson interferometer (section 2.5), what would be its resolution when used visually?

1.2 Radio and microwave detection

Introduction

Radio astronomy is the oldest of the 'new' astronomies since it is now three quarters of a century since Karl Jansky first observed radio waves from the galaxy. It has passed beyond the developmental stage wherein the other 'new' astronomies (infrared, ultraviolet, x-ray, neutrino astronomy etc.) still remain. It thus has quite well established instruments and techniques that are not likely to change overmuch.

The reason why radio astronomy developed relatively early is that radio radiation penetrates to ground level. For wavelengths from about 10 mm to 10 m, the atmosphere is almost completely transparent. The absorption becomes almost total at about 0.5 mm wavelength, and between 0.5 and 10 mm there are a number of absorption bands that are mainly due to oxygen and water vapour, with more or less transparent windows between the bands. The scale height for water vapour in the atmosphere is about 2000 m, so that observing from high altitudes reduces the short wave absorption considerably. Radiation with wavelengths longer than about 50 m again fails to penetrate to ground level, but this time the cause is reflection by the ionosphere. Thus this section is concerned with the detection of radiation with wavelengths longer than 0.1 mm, i.e. frequencies less than 3×10^{12} Hz, or photon energies less than 2×10^{-21} J (0.01 eV). The detection of radiation in the 0.1 to a few mm wavelength region using bolometers is covered under infrared detectors in section 1.1.

The unit of intensity that is commonly used at radio wavelengths is the jansky (Jy)

$$1 \, \text{Jy} = 10^{-26} \, \text{W} \, \text{m}^{-2} \, \text{Hz}^{-1} \tag{1.2.1}$$

and detectable radio sources vary from about 10^{-3} to 10^{6} Jy. Most radio sources of interest to astronomers generate their radio flux as thermal radiation, when the Rayleigh–Jeans law gives their spectrum

$$F_\nu = \frac{2\pi k}{c^2} T\nu^2 \tag{1.2.2}$$

or as synchrotron radiation from energetic electrons spiralling around magnetic fields, when the spectrum is of the form

$$F_\nu \propto \nu^{-\alpha} \tag{1.2.3}$$

where F_ν is the flux per unit frequency interval at frequency ν. α is called the spectral index of the source, and is related to the energy distribution of the electrons. For many sources $0.2 \leq \alpha \leq 1.2$.

Detectors and receivers

The detection of radio signals is a two-stage process in which the sensor produces an electrical signal that then has to be processed until it is in a directly usable form. Coherent detectors, which preserve the phase information of the signal, are available for use over the whole of the radio spectrum, in contrast to the optical and infrared detectors discussed in section 1.1 that respond only to the total power of the signal. In the MHz radio region, the sensor is normally a dipole placed directly at the focus of the telescope, such as the half-wave dipole shown in figure 1.2.1, although this can only be optimized for one wavelength and has very restricted bandwidths. The two halves of such a dipole are each a quarter of a wavelength long. Connection to the remainder of the system is by coaxial cable.

In the GHz and higher frequency regions a horn antenna is normally used to collect the radiation, usually with waveguides for the connection to the rest of the system, though plastic and quartz lenses may be used at very high frequencies. Recent developments in the design of horn antennas have enabled them to have much wider bandwidths than earlier designs. These developments include using a corrugated internal surface for the horn and reducing its diameter in a series of steps rather than smoothly, giving a bandwidth that covers a factor of two in terms of frequency. Bandwidths with factors of nearly eight in frequency are possible with dielectric loaded horns. These are smooth-walled horns filled with an appropriate

Conducting strips
$\frac{1}{4}$ of a wavelength long

Cables to the rest of
the system

Figure 1.2.1. A half-wave dipole.

dielectric except for a small gap between the filling and the walls of the horn through which the wave propagates. At high frequencies the feed horns may need cooling to cryogenic temperatures.

The sensor at the higher frequencies is nowadays normally a super-conductor–insulator–superconductor (SIS) device. In an SIS detector, an electron in one superconducting film absorbs a photon, giving the electron enough energy to tunnel through the insulating barrier into the other super-conducting film (cf. STJs, section 1.1). This process, known as photon-assisted tunnelling, produces one electron for every absorbed photon. Modern devices are based upon two niobium layers separated by an insulating region of aluminium oxide around 1 nm thick and the whole cooled to 4 K or less. Such SIS devices can operate up to about 700 GHz. SIS devices can be used up to 1.2 THz using niobium titanium nitride. Even higher frequencies require the use of Schottky diodes, or more recently TES devices (section 1.1) based upon niobium, and generally known among radio astronomers as hot electron bolometers (HEBs).

The signal from the sensor is carried to the receiver whose purpose is to convert the high frequency electrical currents into a convenient form. The behaviour of the receiver is governed by five parameters: sensitivity, amplification, bandwidth, receiver noise level and integration time.

The sensitivity and the other parameters are very closely linked, for the minimum detectable brightness, B_{min}, is given by

$$B_{min} = \frac{2k\nu^2 KT_s}{c^2\sqrt{t\Delta\nu}} \qquad (1.2.4)$$

where T_s is the noise temperature of the system, t is the integration time, $\Delta\nu$ is the frequency bandwidth and K is a constant close to unity that is a function of the type of receiver. The bandwidth is usually measured between output frequencies whose signal strength is half the maximum when the input signal power is constant with frequency. The amplification and integration time are self-explanatory, so that only the receiver noise level remains to be explained. This noise originates as thermal noise within the electrical components of the receiver, and may also be called Johnson or Nyquist noise (see also section 1.1). The noise is random in nature and is related to the temperature of the component. For a resistor, the r.m.s. voltage of the noise per unit frequency interval, $\overline{V}$, is given by

$$\overline{V} = 2\sqrt{kTR} \qquad (1.2.5)$$

where R is the resistance and T is the temperature. The noise of the system is then characterized by the temperature T_s that would produce the same noise level for the impedance of the system. It is given by

$$T_s = T_1 + \frac{T_2}{G_1} + \frac{T_3}{G_1 G_2} + \cdots + \frac{T_n}{G_1 G_2 \cdots G_{n-1}} \qquad (1.2.6)$$

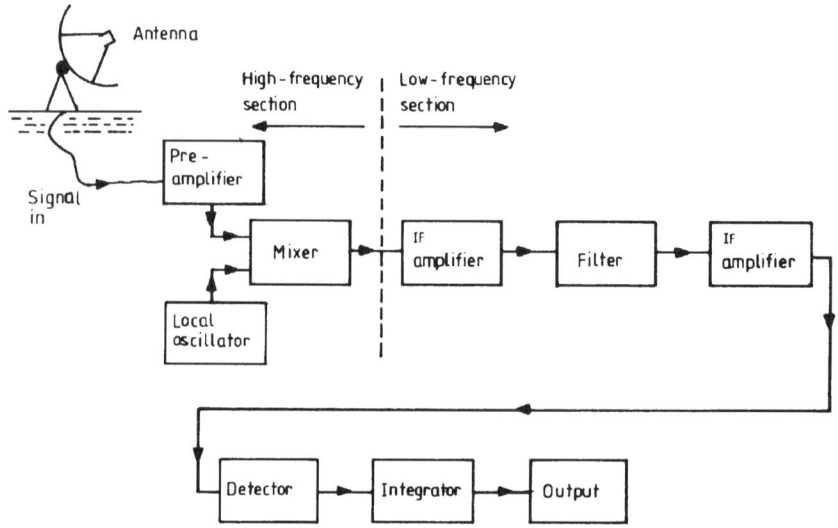

Figure 1.2.2. Block diagram of a basic heterodyne receiver.

where T_n is the noise temperature of the nth component of the system and G_n is the gain (or amplification) of the nth component of the system. It is usually necessary to cool the initial stages of the receiver with liquid helium in order to reduce T_s to an acceptable level. Other noise sources that may be significant include shot noise resulting from random electron emission, g–r noise due to a similar effect in semiconductors (section 1.1), noise from the parabolic reflector or other collector that may be used to concentrate the signal on to the antenna, radiation from the atmosphere, and last but by no means least, spill-over from radio taxis, microwave ovens and other artificial sources.

Many types of receiver exist; the simplest is a development of the heterodyne system employed in the ubiquitous transistor radio. The basic layout of a heterodyne receiver is shown in block form in figure 1.2.2. The pre-amplifier operates at the signal frequency and will typically have a gain of 10 to 1000. It is often mounted close to the feed and cooled to near absolute zero to minimize its contribution to the noise of the system (see above). The most widely used amplifiers today are based upon cooled gallium arsenide and indium phosphide heterostructure field-effect transistors (HFETs), also known as high electron mobility transistors (HEMTs), in which the current-carrying electrons are physically separated from the donor atoms. The current-carrying electrons are restricted to a thin (10 nm) layer of undoped material producing a fast, low noise device. Above 40 GHz the mixer must precede the pre-amplifier in order to decrease the frequency before it can be amplified; a second lower frequency local oscillator is then employed to reduce the frequency even further.

The local oscillator produces a signal that is close to but different from the main signal in its frequency. Thus when the mixer combines the main signal and the local oscillator signal, the beat frequency between them (intermediate frequency or IF) is at a much lower frequency than that of the original signal. The relationship is given by

$$\nu_{signal} = \nu_{LO} \pm \nu_{IF} \qquad (1.2.7)$$

where ν_{signal} is the frequency of the original signal (i.e. the operating frequency of the radio telescope), ν_{LO} is the local oscillator frequency, and ν_{IF} is the intermediate frequency. Normally, at lower frequencies, only one of the two possible signal frequencies given by equation (1.2.7) will be picked up by the feed antenna or passed by the pre-amplifier. At high frequencies, both components may contribute to the output.

The power of the intermediate frequency emerging from the mixer is directly proportional to the power of the original signal. The IF amplifiers and filter determine the pre-detector bandwidth of the signal and further amplify it by a factor of 10^6 to 10^9. The detector is normally a square-law device; that is to say, the output voltage from the detector is proportional to the square of the input voltage. Thus the output *voltage* from the detector is proportional to the input *power*. In the final stages of the receiver, the signal from the detector is integrated, usually for a few seconds, to reduce the noise level. Then it is fed to an output device, usually an analogue-to-digital input to a computer for further processing. Further advances in speed and noise reduction are expected soon from combining the sensor and much of the electronics on to MMICs (monolithic microwave integrated circuits).

The basic heterodyne receiver has a high system temperature, and its gain is unstable. The temperature may be lowered by applying an equal and opposite voltage in the later stages of the receiver, and the stability of the gain may be greatly improved by switching rapidly from the antenna to a calibration noise source and back again, with a phase-sensitive detector (section 3.1) to correlate the changes. Such a system is then sometimes called a Dicke radiometer. The radiometer works optimally if the calibration noise source level is the same as that of the signal, and so it may be further improved by continuously adjusting the noise source to maintain the balance, and it is then termed a null-balancing Dicke radiometer. Since the signal is only being detected half the time the system is less efficient than the basic receiver, but using two alternately switched receivers will restore its efficiency. The value of T_s for receivers varies from 10 K at metre wavelengths to 10 000 K at millimetre wavelengths. The noise sources must therefore have a comparable range, and at long wavelengths are usually diodes, while at the shorter wavelengths a gas discharge tube inside the waveguide and inclined to it by an angle of about $10°$ is used.

Receivers are generally sky background limited just like terrestrial optical telescopes. The Earth's atmosphere radiates at 100 K and higher temperatures below a wavelength of about 3 mm. Only between 30 and 100 mm does its temperature fall as low as 2 K. Then, at longer wavelengths, the galactic emission becomes important, rising to temperatures of 10^5 K at wavelengths of 30 m.

Spectrographs at radio frequencies can be obtained in several different ways. In the past, the local oscillator has been tuned, producing a frequency-sweeping receiver, or the receiver has been a multichannel device so that it registered several different discrete frequencies simultaneously. For pulsar observations such filter banks may have several hundred channels over a bandwidth of a few megahertz. Today, most radio spectroscopy is carried out by auto-correlation, even at the highest frequencies. Successive delays are fed into the signal that is then cross-correlated with the original signal in a computer. The spectrum is obtained from the Fourier transform of the result. The polarization of the original signal is determined by separately detecting the orthogonal components, and cross-correlating the electrical signals later within the receiver.

Alternatively the radio signal may be converted into a different type of wave, and the variations of this secondary wave studied instead. This is the basis of the acousto-optical radio spectrometer (AOS). The radio signal is converted into an ultrasonic wave whose intensity varies with that of the radio signal and whose frequency is also a function of that of the radio signal. In the first such instruments water was used for the medium in which the ultrasound propagated, and the wave was generated by a piezo-electric crystal driven either directly from the radio signal or by a frequency-reduced version of the signal. More recently materials such as fused silica, lithium niobate and lead molybdate have replaced water as the acoustic medium in order to improve the available spectral range. A laser illuminates the cell containing the acoustic medium and a part of the light beam is diffracted by the sound wave. The angle of diffraction depends upon the sound wave's frequency, while the intensity of the diffracted light depends on the sound wave's intensity. Thus the output from the device is a fan beam of light, position within which ultimately depends upon the observed radio frequency, and whose intensity at that position ultimately depends upon the radio intensity. The fan beam may then simply be detected by a linear array of detectors or by scanning, and the spectrum inferred from the result. The AOS initially found application to the observation of solar radio bursts (section 5.3) at metre wavelengths, but is now employed more in the sub-millimetre and infrared regions, since auto-correlation techniques have replaced it at the longer wavelengths, and are even starting to do so in the millimetre region.

A major problem at all frequencies in radio astronomy is interference from artificial noise sources. In theory, certain regions of the spectrum (see table 1.2.1) are reserved partially or exclusively for use by radio astronomers. But leakage from devices such as microwave ovens, incorrectly tuned

Table 1.2.1. Radio astronomy reserved frequencies. The wave bands listed in this table are compiled from the World Administrative Radio Conference 1992 (W.A.R.C. 92) and the World Radio Conference 2000 (W.R.C. 2000). Not all frequencies are kept free in all places, some allocations are shared with other users and the allocations are continually changing. The reader should consult an up-to-date local source for specific information relative to his or her requirements. These can often be found by searching the internet.

Lower frequency	Upper frequency	Lower frequency	Upper frequency
13.36 MHz	13.41 MHz	93.07 GHz	93.27 GHz
25.55 MHz	25.67 MHz	97.88 GHz	98.08 GHz
37.50 MHz	38.25 MHz	105.00 GHz	116.00 GHz
73.00 MHz	74.60 MHz	140.69 GHz	140.98 GHz
150.05 MHz	153.00 MHz	144.68 GHz	144.98 GHz
322.00 MHz	328.60 MHz	145.45 GHz	145.75 GHz
406.10 MHz	410.00 MHz	146.82 GHz	147.12 GHz
608.00 MHz	614.00 MHz	150.00 GHz	151.00 GHz
1.330 GHz	1.427 GHz	164.00 GHz	168.00 GHz
1.660 GHz	1.670 GHz	174.42 GHz	175.02 GHz
1.7188 GHz	1.7222 GHz	177.00 GHz	177.40 GHz
2.655 GHz	2.700 GHz	178.20 GHz	178.60 GHz
3.260 GHz	3.267 GHz	181.00 GHz	181.46 GHz
3.332 GHz	3.339 GHz	182.00 GHz	185.00 GHz
3.3458 GHz	3.3525 GHz	186.20 GHz	186.6 GHz
4.800 GHz	5.000 GHz	217.00 GHz	231.00 GHz
6.650 GHz	6.6752 GHz	250.00 GHz	251.00 GHz
10.60 GHz	10.70 GHz	257.50 GHz	258.00 GHz
14.47 GHz	14.50 GHz	261.00 GHz	265.00 GHz
15.35 GHz	15.40 GHz	262.24 GHz	262.76 GHz
22.01 GHz	22.50 GHz	265.00 GHz	275.00 GHz
22.81 GHz	22.86 GHz		
23.07 GHz	23.12 GHz		
23.60 GHz	24.00 GHz		
31.10 GHz	31.80 GHz		
36.43 GHz	36.50 GHz		
42.50 GHz	43.50 GHz		
48.94 GHz	49.04 GHz		
72.77 GHz	72.91 GHz		
86.00 GHz	92.00 GHz		

receivers, and illegal transmissions often overlap into these bands. The Russian GLONASS satellite navigation system, for example, overlapped into the band reserved for the interstellar OH lines at 1.61 GHz. The use of highly directional antennae (see the next subsection) reduces the problem to some extent. But it is likely that radio astronomers will have to follow their optical colleagues to remote parts of the globe or place their antennae in space if their work is to continue in the future. Even the latter means of escape may be threatened by solar-power satellites with their potentially enormous microwave transmission intensities.

Radio telescopes

The antenna and receiver, whilst they are the main active portions of a radio detecting system, are far less physically impressive than the large structures that serve to gather and concentrate the radiation and to shield the antenna from unwanted sources. Before going on, however, to the consideration of these large structures that form most people's ideas of what comprises a radio telescope, we must look in a little more detail at the physical background of antennae.

The theoretical optics of light and radio radiation are identical, but different traditions in the two disciplines have led to differences in the mathematical and physical formulations of their behaviours. Thus the image in an optical telescope is discussed in terms of its diffraction structure (figure 1.1.26 for example), while that of a radio telescope is discussed in terms of its polar diagram. However, these are just two different approaches to the presentation of the same information. The polar diagram is a plot, in polar coordinates, of the sensitivity or voltage output of the telescope, with the angle of the source from the optical axis. (Note that the polar diagrams discussed herein are all far-field patterns, i.e. the response for a distant source; near-field patterns for the antennae may differ from those shown here.) The polar diagram may be physically realized by sweeping the telescope past a point source, or by using the telescope as a transmitter and measuring the signal strength around it.

The simplest antenna, the half-wave dipole (figure 1.2.1), accepts radiation from most directions, and its polar diagram is shown in figure 1.2.3. This is only a cross section through the beam pattern; the full three-dimensional

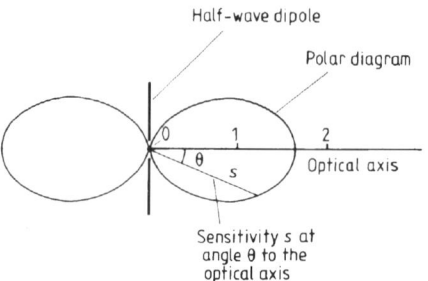

Figure 1.2.3. Polar diagram of a half-wave dipole.

polar diagram may be obtained by rotating the pattern shown in figure 1.2.3 about the dipole's long axis, which then has the appearance of a toroid that is filled-in to the centre. The polar diagram, and hence the performance of the antenna, may be described by four parameters: the beam width at half-power points (BWHP), the beam width at first nulls (BWFN), the gain, and the effective area. The first nulls are the positions either side of the optical axis where the sensitivity of the antenna first decreases to zero, and the BWFN is just the angle between them. Thus the value of the BWFN for the half-wave dipole is 180°. The first nulls are the direct equivalent of the first fringe minima in the diffraction pattern of an optical image, and for a dish antenna type of radio telescope, their position is given by equation (1.1.27) thus

$$\text{BWFN} = 2 \times \frac{1.22\lambda}{D}. \tag{1.2.8}$$

The Rayleigh criterion of optical resolution may thus be similarly applied to radio telescopes: two point sources are resolvable when one is on the optical axis and the other is in the direction of a first null. The half-power points may be best understood by regarding the radio telescope as a transmitter; they are then the directions in which the broadcast power has fallen to one half of its peak value. The BWHP is just the angular separation of these points. For a receiver they are the points at which the output voltage has fallen by a factor of the square root of two, and hence the output power has fallen by half. The maximum gain or directivity of the antenna is also best understood in terms of a transmitter. It is the ratio of the peak value of the output power to the average power. In a receiver it is a measure of the output from the system compared with that from a comparable (and hypothetical) isotropic receiver. The effective area of an antenna is the ratio of its output power to the strength of the incoming flux of the radiation that is correctly polarized to be detected by the antenna, i.e.

$$A_{\text{e}} = \frac{P_\nu}{F_\nu} \tag{1.2.9}$$

where A_{e} is the effective area, P_ν is the power output by the antenna at frequency ν and F_ν is the correctly polarized flux from the source at the antenna at frequency ν. The effective area and the maximum gain, g, are related by

$$g = \frac{4\pi}{c^2} \nu^2 A_{\text{e}}. \tag{1.2.10}$$

For the half-wave dipole, the maximum gain is about 1.6, and so there is very little advantage over an isotropic receiver.

The performance of a simple dipole may be improved by combining the outputs from several dipoles that are arranged in an array. In a collinear

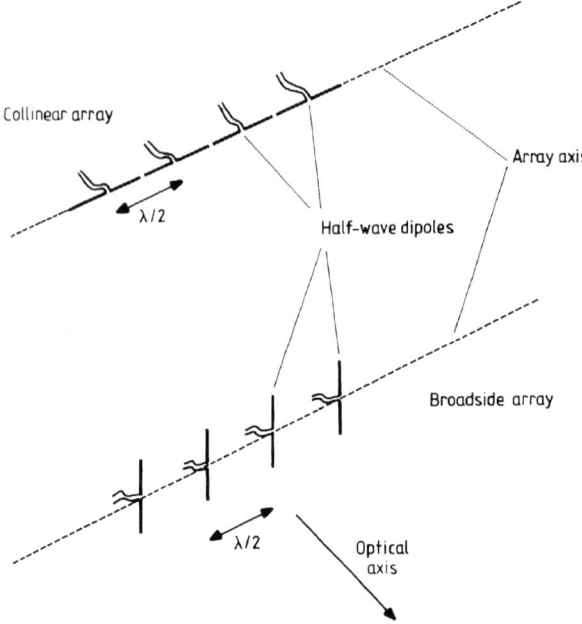

Figure 1.2.4. Dipole arrays.

array, the dipoles are lined up along their axes and spaced at intervals of half a wavelength (figure 1.2.4). The arrangement is equivalent to a diffraction grating and so the sensitivity at an angle θ to the long axis of the array, $s(\theta)$, is given by

$$s(\theta) = s_0 \left(\frac{\sin(n\pi \sin \theta)}{\sin(\pi \sin \theta)} \right) \qquad (1.2.11)$$

where n is the number of half-wave dipoles and s_0 is the maximum sensitivity (cf. equation (4.1.8)). Figure 1.2.5 shows the polar diagrams for one-, two- and four-dipole arrays; their three-dimensional structure can be obtained by rotating these diagrams around a vertical axis so that they become lenticular toroids. The resolution along the axis of the array, measured to the first null, is given by

$$\alpha = \sin^{-1} \left(\frac{1}{n} \right). \qquad (1.2.12)$$

The structure of the polar diagrams in figure 1.2.5 shows a new development. Apart from the main lobe whose gain and resolution increase with n as might be expected, a number of smaller side lobes have appeared. Thus the array has a sensitivity to sources that are at high angles of inclination to the optical axis. These side lobes correspond precisely to the fringes surrounding

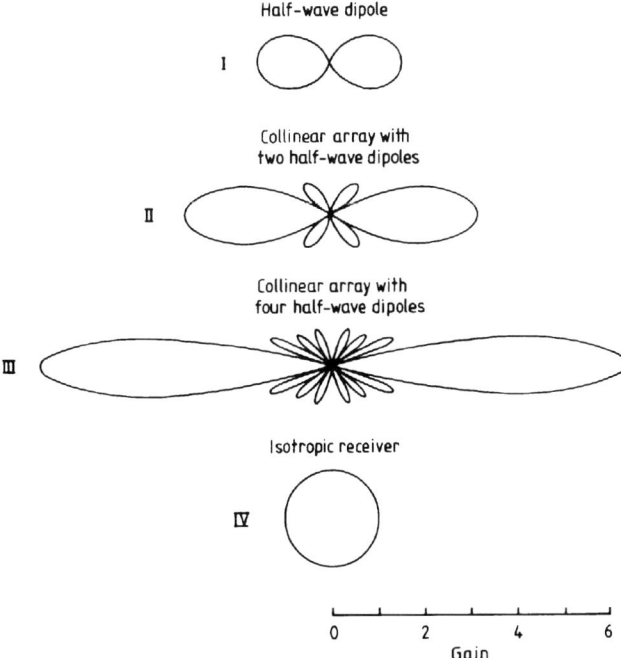

Figure 1.2.5. Polar diagrams for collinear arrays.

the Airy disc of an optical image (figure 1.1.26 etc.). Although the resolution of an array is improved over that of a simple dipole along its optical axis, it will still accept radiation from any point perpendicular to the array axis. The use of a broadside array in which the dipoles are perpendicular to the array axis and spaced at half wavelength intervals (figure 1.2.4) can limit this 360° acceptance angle somewhat. For a four-dipole broadside array, the polar diagram in the plane containing the optical and array axes is given by polar diagram number II in figure 1.2.5, while in the plane containing the optical axis and the long axis of an individual dipole, the shape of the polar diagram is that of a single dipole (number I in figure 1.2.5), but with a maximum gain to match that in the other plane. The three-dimensional shape of the polar diagram of a broadside array thus resembles a pair of squashed balloons placed end to end. The resolution of a broadside array is given by

$$\alpha = \sin^{-1}\left(\frac{2}{n}\right)$$

(1.2.13)

along the array axis, and is that of a single dipole, i.e. 90°, perpendicular to this. Combinations of broadside and collinear arrays can be used to limit the beam width further if necessary.

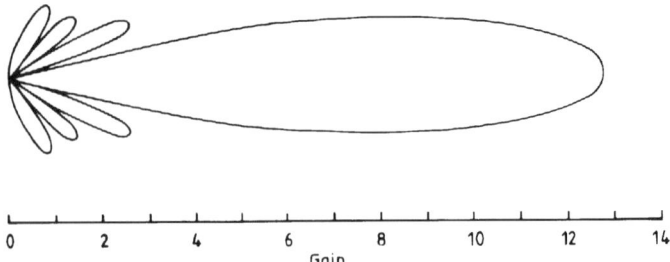

Gain

Figure 1.2.6. Polar diagram for a four-element collinear array with a mesh reflector.

With the arrays as shown, there is still a twofold ambiguity in the direction of a source that has been detected, however narrow the main lobe may have been made, due to the forward and backward components of the main lobe. The backward component may easily be eliminated, however, by placing a reflector behind the dipole. This is simply a conducting rod about 5% longer than the dipole and unconnected electrically with it. It is placed parallel to the dipole and about one eighth of a wavelength behind it. For an array, the reflector may be a similarly placed electrically conducting screen. The polar diagram of a four-element collinear array with a reflector is shown in figure 1.2.6 to the same scale as the diagrams in figure 1.2.5. It will be apparent that not only has the reflector screened out the backward lobe, but it has also doubled the gain of the main lobe. Such a reflector is termed a parasitic element since it is not a part of the electrical circuit of the antenna. Similar parasitic elements may be added in front of the dipole to act as directors. These are about 5% shorter than the dipole. The precise lengths and spacings for the parasitic elements can only be found empirically, since the theory of the whole system is not completely understood. With a reflector and several directors we obtain the parasitic or Yagi antenna, familiar from its appearance on so many rooftops as a television antenna. A gain of up to 12 is possible with such an arrangement and the polar diagram is shown in figure 1.2.7. The Rayleigh resolution is 45° and it has a bandwidth of about 3% of its operating frequency. The main use of parasitic antennae in radio astronomy is as the receiving element (sometimes called the feed) of a larger reflector such as a parabolic dish.

The use of a single dipole, or even several dipoles in an array, is the radio astronomy equivalent of a naked-eye observation. Just as in optical astronomy, the observations may be greatly facilitated by the use of some means of concentrating the signal from over a wide area on to the antenna. The most familiar of these devices are the large parabolic dishes that are the popular conception of a radio telescope. These are directly equivalent to an optical reflecting telescope. They are usually used at the prime focus or at the Cassegrain focus (section 1.1). The gain may be found roughly by substituting the dishes' area for the effective area in equation (1.2.10). The Rayleigh

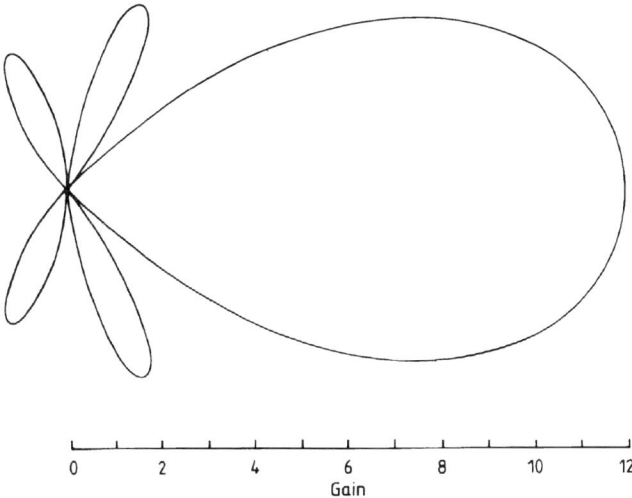

0   2   4   6   8   10   12
Gain

Figure 1.2.7. Polar diagram of a parasitic antenna.

resolution is given by equation (1.1.28). The size of the dishes is so large
because of the length of the wavelengths being observed; for example, to
obtain a resolution of 1° at a wavelength of 0.1 m requires a dish 7 m
across, which is larger than most optical reflectors for a resolution over
10^4 times poorer. The requirement on surface accuracy is the same as that
for an optical telescope: deviations from the paraboloid must be less than
$\lambda/8$ if the Rayleigh resolution is not to be degraded. Now, however, the
longer wavelength helps since it means that the surface of a telescope
working at 0.1 m, say, can deviate from perfection by over 10 mm without
seriously affecting the performance. In practice a limit of $\lambda/20$ is often
used, for as we have seen the Rayleigh limit does not represent the ultimate
limit of resolution. These less stringent physical constraints on the surface of
a radio telescope ease the construction problems greatly, but more importantly it also means that the surface need not be solid, as a wire mesh with
spacings less than $\lambda/20$ will function equally well as a reflector. The weight
and wind resistance of the reflector are thus reduced by very large factors.
At the shorter radio wavelengths a solid reflecting surface may be more convenient, however, and at very short wavelengths (<1 mm) active surface
control to retain the accuracy is used along somewhat similar lines to the
methods discussed in section 1.1 for optical telescopes. The shape of the
mirror is monitored holographically. The dishes are usually of very small
focal ratio (f0.5 is not uncommon) and the reason for this is so that the
dish acts as a screen against unwanted radiation as well as concentrating
the desired radiation. Fully steerable dishes up to 100 m across have been
built such as the Green Bank and Effelsberg telescopes, while the Arecibo

telescope in Puerto Rico is a fixed dish 300 m across. This latter instrument acts as a transit telescope and has some limited ability to track sources and look at a range of declinations by moving the feed antenna around the image plane. Such fixed telescopes may have spherical surfaces in order to extend this facility and use a secondary reflector to correct the resulting spherical aberration. The feed antenna for such dishes may be made of such a size that it intercepts only the centre lobe of the telescope response (i.e. the Airy disc in optical terms). The effects of the side lobes are then reduced or eliminated. This technique is known as tapering the antenna and is the same as the optical technique of apodization (section 4.1). The tapering function may take several forms and may be used to reduce background noise as well as eliminating the side lobes. Tapering reduces the efficiency and resolution of the dish, but this is usually more than compensated for by the improvement in the shape of the response function. In the microwave region the largest dishes are currently the 30 m IRAM instrument on Pico Veleta in Spain and the 45 m telescope at Nobeyama in Japan. A 50 m dish (the Large Millimetre Telescope) is currently under construction near Mexico City.

With a single feed, the radio telescope is a point-source detector only. Images have to be built up by scanning (section 2.4) or by interferometry (section 2.5). Scanning used to be accomplished by pointing the telescope at successive points along the raster pattern and observing each point for a set interval. This practice, however, suffers from fluctuations in the signal due to the atmosphere and the electronics. Current practice is therefore to scan continuously or 'on the fly'. The atmospheric or electronic variations are then on a longer timescale than the changes resulting from moving over a radio source and can be separated out.

True imaging can be achieved through the use of cluster or array feeds. These are simply multiple individual feeds arranged in a suitable array at the telescope's focus. Each feed is then the equivalent of a pixel in a CCD or other type of detector. The number of elements in such cluster feeds currently remains small compared with their optical equivalents. For example, the 64 m Parkes radio telescope uses a 13-beam receiver for 21 cm radiation, and the 14 m telescope at the Five College Radio Astronomy Observatory (FCRAO) in Massachusetts uses a 32-beam array operating at millimetre wavelengths (see also SCUBA, section 1.1, Bolometers).

Very many other systems have been designed to fulfil the same function as a steerable paraboloid but which are easier to construct. The best known of these are the multiple arrays of mixed collinear and broadside type, or similar constructions based upon other antenna types. They are mounted on to a flat plane which is oriented east–west and which is tiltable in altitude to form a transit telescope. Another system such as the 600 m RATAN telescope uses an off-axis paraboloid that is fixed and which is illuminated by a tiltable flat reflector. Alternatively the paraboloid may be cylindrical and tiltable itself around its long axis. For all such reflectors some form of feed antenna is required. A parasitic antenna is a common choice at longer wavelengths, while a horn antenna, which is essentially a flared end to a waveguide, may be used at higher frequencies.

A quite different approach is used in the Mills Cross type of telescope. This uses two collinear arrays oriented north–south and east–west. The first provides a narrow fan beam along the north–south meridian, while the second provides a similar beam in an east–west direction. Their intersection is a narrow vertical pencil beam, typically 1° across. The pencil beam may be isolated from the contributions of the remainders of the fan beams by comparing the outputs when the beams are added in phase with when they are added out of phase. The in-phase addition is simply accomplished by connecting the outputs of the two arrays directly together. The out-of-phase addition delays one of the outputs by half a wavelength before the addition, and this is most simply done by switching in an extra length of cable to one of the arrays. In the first case radiation from objects within the pencil beam will interfere constructively, while in the second case there will be destructive interference. The signals from objects not within the pencil beam will be mutually incoherent and so will simply add together in both cases. Thus, looking vertically down on to the Mills Cross, the beam pattern will alternate between the two cases shown in figure 1.2.8. Subtraction of the one from the other will then just leave the pencil beam. The pencil beam may be displaced by an angle θ from the vertical by introducing a phase shift between each dipole. Again the simplest method is to switch extra cable into the connections between each dipole, the lengths of the extra portions, L, being given by

$$L = d \sin \theta \qquad (1.2.14)$$

where d is the dipole separation. The pencil beam may thus be directed around the sky as wished. In practice, the beam is only moved along the

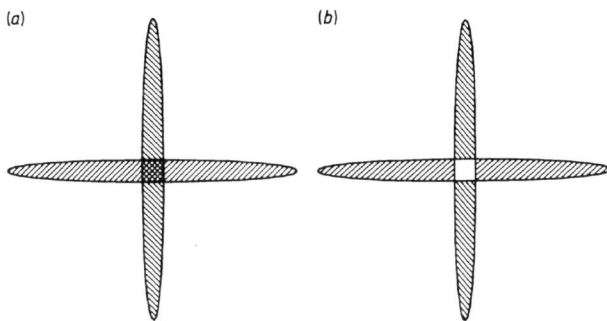

Figure 1.2.8. Beam patterns of a Mills Cross radio telescope, (a) with the beams added together (i.e. in phase) and (b) with the beams subtracted from each other (i.e. 180° out of phase).

north–south plane, and the telescope is used as a transit telescope, since the alteration of the cable lengths between the dipoles is a lengthy procedure. The resolution of a Mills Cross is the same as that of a parabolic dish whose diameter is equal to the array lengths. The sensitivity, however, is obviously much reduced from that of a dish, since only a tiny fraction of the aperture is filled by the dipoles. As well as being cumbersome to operate, the Mills Cross suffers from the disadvantage that it can only operate at a single wavelength unless its dipoles are all changed. Furthermore, confusion of sources may arise if a strong source is in one of the pencil beams formed by the side lobes, since it will have the same appearance as a weak source in the main beam. An alternative and related system uses a single fan beam and observes the object at many position angles. The structure of the source is then retrievable from the observations in a relatively unambiguous manner.

The phasing of two dipoles, as we have seen, can be used to alter the angle of the beam to their axis. By continuously varying the delay the lobe can be swept across the sky, or accurately positioned on an object and then moved to track that object. This is an important technique for use with interferometers (section 2.5), and for solar work. It also forms the basis of a phased array. This latter instrument is basically a Mills Cross in which the number of dipoles has been increased until the aperture is filled. It provides great sensitivity since a large area may be covered at a relatively low cost.

A radically different approach to concentrating the radiation is based upon refraction. The Luneburg lens has yet to find much application to radio astronomy, but may be used in the Square Kilometre Array (SKA, section 2.5). The Luneburg lens is a solid sphere within which the refractive index increases linearly inwards from unity at the surface. With a central refractive index of 2, the focus is on the surface of the lens. Since there is no axis of symmetry, the lens can be used to observe in many directions simultaneously, simply by having numerous feeds distributed around it. Many materials potentially can be used for the lens construction, but to date high density polystyrene is the one that has been used in practice.

A number of spacecraft carrying microwave detectors have been launched. These include COBE (Cosmic Background Explorer Satellite) that included Dicke radiometers operating at 31.5, 53 and 90 GHz, and the current MAP (Microwave Anisotropy Probe) mission observing from 22 to 90 GHz. Few longer-wave spacecraft-borne telescopes have been used, since the atmosphere is then transparent. However, the Japanese Halca spacecraft carries an 8 m dish operating at centimetre wavelengths as part of an aperture synthesis system (section 2.5).

Great improvements in resolution and sensitivity of radio telescopes may be obtained through the use of interferometers and aperture synthesis, and these devices and their associated receivers and detectors are discussed in detail in section 2.5.

Construction

The large dishes that we have been discussing pose major problems in their construction. Both the gravitational and wind loads on the structure can be very large, and shadowing of parts of the structure can lead to inhomogeneous heating and hence expansion- and contraction-induced stresses.

The worst problem is due to wind, since its effect is highly variable. The force can be very large: 1.5×10^6 N (150 tonnes) for a 50 m dish facing directly into a gale-force wind, for example. A rough rule of thumb to allow scaling the degree of wind distortion between dishes of different sizes is that the deflection, Δe, is given by

$$\Delta e \propto \frac{D^3}{A} \tag{1.2.15}$$

where D is the diameter of the dish and A is the cross-sectional area of its supporting struts. Thus, doubling the diameter of the design of a dish would require the supporting members' sizes to be tripled if it were to still work at the same wavelength. There are only two solutions to the wind problem: to enclose the dish, or to cease using it when the wind load becomes too great. Some smaller dishes, especially those working at short wavelengths where the physical constraints on the surface accuracy are most stringent, are enclosed in radomes, or space-enclosing structures built from non-conducting materials. But this is not usually practicable for the larger dishes. These must generally cease operating and be parked in their least wind-resistant mode once the wind speed rises above 10–$15 \, \mathrm{m \, s^{-1}}$.

The effects of gravity are easier to counteract. The problem arises from the varying directions and magnitudes of the loads placed upon the reflecting surface as the orientation of the telescope changes. Three approaches have been tried successfully for combating the effects of gravity. The first is the brute force approach, whereby the dish is made so rigid that its deformations remain within the surface accuracy limits. Except for very small dishes, this is an impossibly expensive option. The second approach is to compensate for the changing gravity loads as the telescope moves. The surface may be kept in adjustment by systems of weights, levers, guy ropes, springs etc. or more recently by computer-controlled hydraulic jacks. The final approach is much more subtle and is termed the homological transformation system. The dish is allowed to deform, but its supports are designed so that its new shape is still a paraboloid, although in general an altered one, when the telescope is in its new position. The only active intervention that may be required is to move the feed antenna to keep it at the foci of the changing paraboloids.

There is little that can be done about inhomogeneous heating, other than painting all the surfaces white so that the absorption of the heat is minimized. Fortunately it is not usually a serious problem in comparison with the first two.

The supporting framework of the dish is generally a complex, cross-braced skeletal structure, whose optimum design requires a computer for its calculation. This framework is then usually placed on to an alt–az mounting (section 1.1), since this is generally the cheapest and simplest structure to build, and also because it restricts the gravitational load variations to a single plane, and so makes their compensation much easier. A few, usually quite small, radio telescopes do have equatorial mountings, though.

Exercises

1.2.1   Show that the HPBW of a collinear array with n dipoles is given by

$$\text{HPBW} \approx 2\sin^{-1}\left(\frac{6n^2 - 3}{2n^4\pi^2 - \pi^2}\right)^{1/2}$$

when n is large.

1.2.2   Calculate the dimensions of a Mills Cross to observe at a wavelength of 0.3 m with a Rayleigh resolution of 0.25°.

1.2.3   Show that the maximum number of separable sources using the Rayleigh criterion, for a 60 m dish working at a wavelength of 0.1 m, is about 3.8×10^5 over a complete hemisphere.

1.3   X-ray and gamma-ray detection

Introduction

The electromagnetic spectrum comprises radiation of an infinite range of wavelengths, but the intrinsic nature of the radiation is unvarying. There is, however, a tendency to regard the differing wavelength regions from somewhat parochial points of view, and this tends to obscure the underlying unity of the processes that may be involved. The reasons for these attitudes are many: some are historical hang-overs from the ways in which the detectors for the various regions were developed, others are more fundamental in that different physical mechanisms predominate in the radiative interactions at different wavelengths. Thus high-energy gamma rays may interact directly with nuclei, at longer wavelengths we have resonant interactions with atoms and molecules producing electronic, vibrational and rotational transitions, while in the radio region currents are induced directly into conductors. But primarily the reason may be traced to the academic backgrounds of the workers involved in each of the regions. Because of the earlier reasons, workers involved in investigations in one spectral region will tend to have different bias in their backgrounds compared with workers in a different spectral region. Thus there will be different traditions, approaches, systems of notation etc. with a consequent tendency to isolationizm and unnecessary failures in communication. The discussion of the point source image in terms of the Airy disc and its fringes by optical astronomers and in terms of the polar diagram by radio astronomers as already mentioned is one good example of this process, and many more exist. It is impossible to break out from the straitjacket of tradition completely, but an attempt to move towards a more unified approach has been made in this work by dividing the spectrum much more broadly than is the normal case. We only consider three separate spectral regions, within each of which the detection techniques bear at least a familial resemblance to each other. The overlap regions are fairly diffuse with some of the techniques from each of the major regions being applicable. We have already discussed two of the major regions, radio and microwaves, and optical and infrared. The third region, x-rays and gamma rays, is the high-energy end of the spectrum and is the most recent area to be explored. This region also overlaps to a very considerable extent, in the nature of its detection techniques, with the cosmic rays that are discussed in the next section. None of the radiation discussed in this section penetrates down to ground level, so its study had to await the availability of observing platforms in space, or near the top of the Earth's atmosphere. Thus significant work on these photons has only been possible since the 1960s, and many of the detectors and observing techniques are still under development.

The high-energy spectrum is fairly arbitrarily divided into:

the extreme ultraviolet (EUV or XUV region): 10 to 100 nm wavelengths (12 to 120 eV photon energies)
soft x-rays: 1 to 10 nm (120 to 1200 eV)
x-rays: 0.01 to 1 nm (1.2 to 120 keV)
soft gamma rays: 0.001 to 0.01 nm (120 to 1200 keV)
gamma rays: less than 0.001 nm (greater than 1.2 MeV).

We shall be primarily concerned with the last four regions in this section, i.e. with wavelengths less than 10 nm and photon energies greater than 120 eV. (Note that the electron volt, eV, is 1.6×10^{-19} J and is a convenient unit for use in this spectral region and also when discussing cosmic rays in the next section.) The main production mechanisms for high-energy radiation include electron synchrotron radiation, the inverse Compton effect, free–free radiation, and pion decay, while the sources include the Sun, supernova remnants, pulsars, bursters, binary systems, cosmic rays, the intergalactic medium, galaxies, Seyfert galaxies, and quasars. Absorption of the radiation can be by ionization with a fluorescence photon or an Auger electron produced in addition to the ion and electron, by Compton scattering, or in the presence of matter, by pair production. This latter process is the production of a particle and its anti-particle, and not the pair production process discussed in section 1.1 which was simply the excitation of an electron from the valence band. The interstellar absorption in this spectral region varies roughly with the cube of the wavelength, so that the higher energy radiation can easily pass through the whole galaxy with little chance of being intercepted. At energies under about 2 keV, direct absorption by the heavier atoms and ions can be an important process. The flux of the radiation varies enormously with wavelength. The solar emission alone at the lower energies is sufficient to produce the ionosphere and thermosphere on the Earth. At 1 nm wavelength, for example, the solar flux is 5×10^9 photons m^{-2} s^{-1}, while the total flux from all sources for energies above 10^9 eV is only a few photons per square metre per day.

Detectors

Geiger counters

The earliest detection of high energy radiation from a source other than the Sun took place in 1962 when soft x-rays from a source that later became known as Sco X-1 were detected by large area Geiger counters flown on a sounding rocket. Geiger counters and the related proportional counters, which are collectively known as gas-filled ionization detectors, are still among the most frequently used detectors for high-energy radiation.

The principle of the Geiger counter is well known. Two electrodes inside an enclosure are held at such a potential difference that a discharge in the medium filling the enclosure is on the point of occurring. The entry of ionizing radiation triggers this discharge, resulting in a pulse of current between the electrodes that may then be amplified and detected. The electrodes are usually arranged as the outer wall of the enclosure containing the gas and as a central coaxial wire (figure 1.3.1). The medium inside the tube is typically

Figure 1.3.1. A typical arrangement for a Geiger counter.

argon at a low pressure with a small amount of an organic gas, such as alcohol vapour, added. The electrons produced in the initial ionization are accelerated towards the central electrode by the applied potential; as these electrons gain energy they cause further ionization, producing more electrons, which in turn are accelerated towards the central electrode, and so on. The amplification factor can be as high as 10^8 electrons arriving at the central electrode for every one in the initial ionization trail. The avalanche of electrons rapidly saturates, so that the detected pulse is independent of the original energy of the photon. This is a serious disadvantage if the device is used as the primary detector and not as a trigger for a more sophisticated detector such as a cloud chamber (section 1.4), and so Geiger counters have been replaced by proportional counters (see below) for this purpose. Another disadvantage of Geiger counters, that also applies to many of the other detectors discussed in this and the following section, is that a response to one event leaves the detector inoperative for a short interval, known as the dead time. In the Geiger counter the cause of the dead time is that a discharge lowers the potential between the electrodes, so that it is momentarily insufficient to cause a second avalanche of electrons should another x-ray enter the device. The length of the dead time is typically $200\,\mu s$.

Proportional counters

These devices are very closely related to Geiger counters and are in effect Geiger counters operated at less than the trigger voltage. By using a lower voltage, saturation of the pulse is avoided, and its strength is then proportional to the energy of the original interaction. The gain of the system operated in this way is reduced to about 10^4 or 10^5, but it is still sufficient for further detection by conventional amplifiers etc. Provided that all the energy of the ionizing radiation is absorbed within the detector, its original total energy may be obtained from the strength of the pulse, and we have a proportional counter. At low photon energies, a window must be provided for the radiation. These are typically made from thin sheets of plastic, mica or beryllium, and absorption in the windows limits the detectors to energies above a few hundred electron volts. When a window is used, the gas in the detector has to be continuously replenished because of losses by diffusion through the window. At high photon energies the detector is limited by the requirement that all the energy of the radiation be absorbed within the detector's enclosure. To this end, proportional counters for high-energy detection may have to be made very large. About 30 eV on average are required to produce one ion–electron pair, so that a 1 keV photon produces about 36 electrons, and a 10 keV photon about 360 electrons. The spectral energy resolution to two and a half standard deviations from the resulting statistical fluctuations of the electron numbers is thus about 40% at 1 keV and 12% at

10 keV. The quantum efficiencies of proportional counters approach 100% for energies up to 50 keV.

The position of the interaction of the x-ray along the axis of the counter may be obtained through the use of a resistive anode. The pulse is abstracted from both ends of the anode and a comparison of its strength and shape from the two ends then leads to a position of the discharge along the anode. The anode wires are very thin, typically 20 μm across, so that the electric field is most intense very near to the wire. The avalanche of electrons thus develops close to the wire, limiting its spread, and giving a precise position. The concept may easily be extended to a two-dimensional grid of anodes to allow genuine imaging. Spatial resolutions of about 0.1 mm are possible. In this form the detector is called a position-sensitive proportional counter. They are also known as multi-wire chambers, especially in the context of particle physics (see also drift chambers, section 1.4).

Many gases can be used to fill the detector: argon, methane, xenon, carbon dioxide, and mixtures thereof at pressures near that of the atmosphere, are among the commonest ones. The inert gases are to be preferred since there is then no possibility of the loss of energy into the rotation or vibration of the molecules.

Scintillation detectors

The ionizing photons do not necessarily knock out only the outermost electrons from the atom or molecule with which they interact. Electrons in lower energy levels may also be removed. When this happens, a 'hole' is left behind into which one of the higher electrons may drop, with a consequent emission of radiation. Should the medium be transparent to this radiation, the photons may be observed, and the medium may be used as an x-ray detector. Each interaction produces a flash or scintilla of light, from which the name of the device is obtained. There are many materials that are suitable for this application. Commonly used ones include sodium iodide doped with an impurity such as thallium, and caesium iodide doped with sodium or thallium. For these materials the light flashes are detected by a photomultiplier (figure 1.3.2). There is no dead time for a scintillation detector, and the strength of the flash depends somewhat upon the original photon energy, so that some spectral resolution is possible. The noise level, however, is quite high since only about 3% of the x-ray's energy is converted into detectable radiation, with a consequent increase in the statistical fluctuations in their numbers. The spectral resolution is thus about 6% at 1 MeV. Sodium iodide or caesium iodide are useful for x-ray energies up to several hundred keV, organic scintillators such as stilbene ($C_{14}H_{14}N_2$) can be used up to 10 MeV and bismuth germanate (BGO, $Bi_4Ge_3O_{12}$) for energies up to 30 MeV or more. Organically doped plastics are also used. Both sodium iodide and bismuth germanate are to be used for the burst monitor on

Figure 1.3.2. Schematic experimental arrangement of a scintillation counter.

board the Gamma-ray Large Area Space Telescope (GLAST), due for launch in 2006, to provide continuous detection from a few keV to 25 MeV.

Discrimination of the x-ray's arrival direction can be obtained by using sodium iodide and caesium iodide in two superimposed layers. The decay time of the pulses differs between the two compounds so that they may be separately identified, and the direction of travel of the photon inferred. This arrangement for a scintillation detector is frequently called a 'phoswich' detector. Several gases such as argon, xenon, nitrogen and their mixtures can also be used as scintillators, and combined with an optical system to produce another imaging device.

X-rays and cosmic rays (section 1.4) may be distinguished by the differences in their resulting pulse shapes. X-rays are rapidly absorbed, and their pulses are sharp and brief. Cosmic ray particles will generally have a much longer path and so their pulses will be comparatively broader and smoother than the x-ray pulse.

Gas scintillation proportional counters

A combination of the above two types of detector leads to a significant improvement in the low-energy spectral resolution. Resolutions as good as 8% at 6 keV have been achieved in practice with these devices. The x-radiation produces ion–electron pairs in an argon- or xenon-filled chamber. The electrons are then gently accelerated until they cause scintillations of their own in the gas. These scintillations can then be observed by a conventional scintillation counter system. These have been favoured detectors for launch on several of the more recent x-ray satellites because of their good spectral and positional discrimination, though in some cases their lifetimes in orbit have been rather short.

Charge coupled devices

The details of the operating principles for CCDs for optical detection are to be found in section 1.1. They are also, however, becoming increasingly widely used as primary detectors at EUV and x-ray wavelengths. The Chandra spacecraft, for example, uses CCDs with $24\,\mu m$ pixel size giving $0.5''$ resolution. CCDs become insensitive to radiation in the blue and ultraviolet parts of the spectrum because of absorption in the electrode structure on their surfaces. They regain sensitivity at shorter wavelengths as the radiation is again able to penetrate that structure ($\lambda < 10\,nm$ or so). As with optical CCDs, the efficiency of the devices may be improved by using a very thin electrode structure, an electrode structure that only partially covers the surface (virtual phase CCDs), or by illuminating the device from the back. The reader is referred to section 1.1 for further details of CCDs.

Superconducting tunnel junction detectors

The details of the operating principles of STJs may be found in section 1.1. They are, however, also sensitive to x-rays, and indeed their first applications were for x-ray detection. Since an x-ray breaks about a thousand times as many Cooper pairs as an optical photon, their spectral discrimination is about 0.1%. The need to cool them to below 1 K, and the small sizes of the arrays currently available, means that they have yet to be flown on a spacecraft.

Compton interaction detectors

Very-high-energy photons produce electrons in a scintillator through the Compton effect, and these electrons can have sufficiently high energies to produce scintillations of their own. Two such detectors separated by a metre or so can provide directional discrimination when used in conjunction with pulse analysers and time-of-flight measurements to eliminate other unwanted interactions.

Spark detectors

These are discussed in more detail in section 1.4 (figure 1.4.3). They are used for the detection of gamma rays of energies above 20 MeV, when the sparks follow the ionization trail left by the photon. Alternatively a layer of tungsten or some similar material may be placed in the spark chamber. The gamma ray then interacts in the tungsten to produce electron–positron pairs, and these in turn can be detected by their trails in the spark chamber. Angular resolutions of 2°, energy resolutions of 50%, and detection limits of 10^{-3} photons $m^{-2} s^{-1}$ are currently achievable.

Čerenkov detectors

Čerenkov radiation and the resulting detectors are referred to in detail in section 1.4. For x-ray and gamma radiation their interest lies in the detection of particles produced by the Compton interactions of the very-high-energy photons. If those particles have a velocity higher than the velocity of light in the local medium then Čerenkov radiation is produced. Čerenkov radiation produced in this manner high in the Earth's atmosphere is detectable. Thus the CANGAROO-II Čerenkov telescope[15] is a 10 m optical telescope in Australia formed from a hundred and fourteen 0.8 m mirrors. It has a field of view of 2.5° and observes the outer reaches of the Earth's atmosphere with 552 photomultipliers, detecting TeV gamma rays from active galaxies.

Very-high-energy primary gamma rays (10^{15} eV and more) may also produce cosmic ray showers and so be detectable by the methods discussed in section 1.4, especially through the fluorescence of atmospheric nitrogen.

Solid-state detectors

There are a number of different varieties of these detectors including CCDs and STJs (see above and section 1.1), so that suitable ones may be found for use throughout most of the x- and gamma-ray regions.

Germanium may be used as a sort of solid proportional counter. A cylinder of germanium cooled by liquid nitrogen is surrounded by a cylindrical cathode and has a central anode (figure 1.3.3). A gamma ray scatters off electrons in the atoms until its energy has been consumed in electron–hole pair production. The number of released electrons is proportional to the energy of the gamma ray, and these are attracted to the anode where they may be detected. The spectral resolution is very high (0.2% at 1 MeV) so that detectors of this type are especially suitable for gamma-ray line spectroscopy. Other materials that may replace the germanium include germanium doped with lithium, cadmium telluride and mercury–iodine. At lower energies (0.4 to 4 keV) silicon-based solid-state detectors may be used similarly. Their energy resolution ranges from 4 to 30%.

Closely related detectors consist of a thick layer (up to 5 mm) of a semiconductor, such as silicon (known as silicon strip detectors) or more recently cadmium (about 90%) and zinc (about 10%) doped with tellurium (known as CZT detectors), since this latter material has a higher stopping power for the radiation. The semiconductor has a high resistivity by virtue of being back biased to 100 V or so. The silicon surface is divided into 20 to 25 μm wide

[15] Collaboration of Australia and Nippon for a GAmma Ray Observatory in the Outback.

Incoming
gamma
ray

Germanium

Anode

Cathode

Liquid
nitrogen
jacket

Figure 1.3.3. Germanium gamma-ray detector.

strips by thin layers doped with boron. Aluminium electrodes are then deposited on the surface between these layers. When an x-ray, gamma ray or charged particle passes into the material, electron–hole pairs are again produced via the photo-electric effect, Compton scattering or collisional ionization. The pulse of current is detected as the electron–hole pairs are collected at the electrodes on the top and bottom of the semiconductor slice. Processing of the pulses can then proceed by any of the usual methods, and a positional accuracy for the interaction of $\pm 10\,\mu m$ can be achieved.

Solid-state detectors have several advantages that suit them particularly for use in satellite-borne instrumentation: a wide range of photon energies detected (from 1 keV to over 1 MeV), simplicity, reliability, low power consumption, high stopping power for the radiation, room temperature operation (some varieties), no entrance window needed, high counting rates possible etc. They also have an intrinsic spectral sensitivity since, provided the photon is absorbed completely, the number of electron–hole pairs produced is proportional to the photon's energy. About 5 eV are need to produce an electron–hole pair in CZT, so that the spectral sensitivity to two and a half standard deviations is potentially 18% at 1 keV and 0.5% at 1 MeV. The main disadvantages of these detectors are that their size is small compared with many other detectors, so that their collecting area is also small, and that unless the photon is stopped within the detector's volume the total energy cannot be determined. This latter disadvantage, however, also applies to most other detectors.

Solid-state detectors are increasing in popularity and recent or forthcoming examples include the INTErnational Gamma Ray Astrophysics

Laboratory (INTEGRAL), launched in 2002, that carried a spectrometer using germanium detectors. The Swift space craft (see also the section on collimation below) due for launch in 2003 will carry nearly 33 000 CZT detectors in its Burst Alert Telescope (BAT). Silicon strip detectors are to be used for the Large Area Telescope (LAT) on board the GLAST spacecraft. These provide positional and directional information on the gamma ray. The silicon strips are arranged orthogonally in two layers. A gamma ray, or the electron–positron pair produced within a layer of lead by the gamma ray, will be detected within one of the strips in each layer. The position of the ray is then localized to the crossover point of the two strips. By using a number of such pairs of layers of silicon strips piled on top of each other, the direction of the gamma ray can also be determined. The LAT will use 16 such 'towers', each comprising 16 layers of silicon strip pairs and lead plates, enabling it to determine gamma-ray source positions to within 0.5 to 5 minutes of arc.

Another type of detector altogether is a possibility for the future. This is a solid analogue of the cloud or bubble chamber (section 1.4). A superconducting solid would be held at a temperature fractionally higher than its critical temperature, or in a non-superconducting state and at a temperature slightly below the critical value. Passage of an ionizing ray or particle would then precipitate the changeover to the opposing state, and this in turn could be detected by the change in a magnetic field (see TES detectors, section 1.1).

Microchannel plates

For EUV and low-energy x-ray amplification and imaging, there is an ingenious variant of the photomultiplier (section 1.1). The devices are also known as Multi-Anode Micro-channel Arrays (MAMAs). A thin plate is pierced by numerous tiny holes, each perhaps only about 10 μm across or less. Its top surface is an electrode with a negative potential of some few thousand volts with respect to its base. The top is also coated with a photoelectron emitter for the x-ray energies of interest. An impinging photon releases one or more electrons that are then accelerated down the tubes. There are inevitably collisions with the walls of the tube during which further electrons are released, and these in turn are accelerated down the tube and so on (figure 1.3.4). As many as 10^4 electrons can be produced for a single photon, and this may be increased to 10^6 electrons in future devices when the problems caused by ion feedback are reduced. The quantum efficiency can be up to 20%. The electrons spray out of the bottom of each tube, where they may be detected by a variety of the more conventional imaging systems, or they may be fed into a second microchannel plate for further amplification. Early versions of these devices often used curved channels (figure 1.3.4). Modern versions now employ two plates with

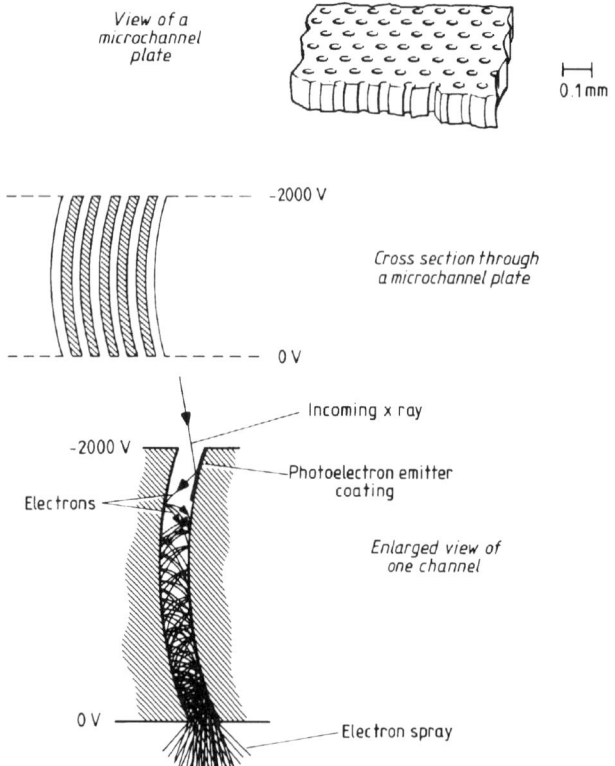

View of a
microchannel
plate

0.1 mm

Cross section through
a microchannel plate

-2000 V

0 V

Incoming x ray

-2000 V

Photoelectron emitter
coating

Electrons

Enlarged view of
one channel

0 V

Electron spray

Figure 1.3.4. Schematic view of the operation of a microchannel plate.

straight channels, the second set of channels being at an angle to the direction of the first set (known as a chevron microchannel plate). The plates are currently manufactured from a billet of glass that has an acid-resisting shell surrounding an acid-soluble core. The billet is heated and drawn out until it forms a glass 'wire' about 1 mm across. The wire is cut into short sections and the sections stacked to form a new billet. That billet in turn is heated and drawn out, and so on. The holes in the final plate are produced by etching away the acid-soluble portions of the glass. Holes down to 6 μm in diameter may be produced in this fashion and the fraction of the plate occupied by the holes can reach 65%. Plates can be up to 0.1 m square. For example, the high resolution x-ray camera on board the Chandra spacecraft uses a 93 mm square chevron microchannel plate detector, with sixty-nine million 10 μm holes, and can provide a resolution of 0.5″ of arc. A single x-ray results in about 30 million electrons that are collected by a grid of wires at the exit holes from the second stage plate. Microchannel plates can also be used in optical and near ultraviolet; however, they are then used simply to amplify the photoelectrons produced from a photo-cathode deposited on to the entrance window of the detector, and are thus similar to image intensifiers (section 2.3).

Nuclear emulsions

Although not normally used for high-energy photon detection, the photo-graphic emulsions used to detect the tracks of nuclear particles (sections 1.4 and 2.2) will respond to them. More correctly they pick up the tracks of the ejected electrons (or δ rays) produced by the passage of the photon. Direct interactions between the photon and a nucleus, or pair production, will also show up if they occur in the emulsion.

Shielding

Very few of the detectors that we have just reviewed are used in isolation. Usually several will be used in modes that allow the rejection of information on unwanted interactions. This is known as active shielding of the detector. The range of possible configurations is very wide, and a few examples will suffice to give an indication of those in current use.

1. The germanium solid-state detectors must intercept all the energy of an incoming photon if they are to provide a reliable estimate of its magnitude. If the photon escapes, then the measured energy will be too low. A thick layer of a scintillation crystal therefore surrounds the germanium. A detection in the germanium simultaneously with two in the scintillation crystal is rejected as an escapee.
2. The solid angle viewed by a germanium solid-state detector can be limited by placing the germanium at the bottom of a hole in a scintillation crystal. Only those detections not occurring simultaneously in the germanium and the crystal are used, and these are the photons that have entered down the hole. Any required degree of angular resolution can be obtained by sufficiently reducing the size of the entrance aperture.
3. Spark counters are surrounded by scintillation counters and Čerenkov detectors to discriminate between gamma-ray events and those due to cosmic rays.
4. A sodium iodide scintillation counter may be surrounded, except for an entrance aperture, by one formed from caesium iodide to eliminate photons from the unwanted directions.

Recently, nuclear power sources on board other spacecraft have caused interference with gamma-ray observations. These pulses have to be separated from the naturally occurring ones by their spectral signatures.

Passive shields are also used, and these are just layers of an absorbing material that screen out unwanted rays. They are especially necessary for the higher energy photon detectors in order to reduce the much greater flux of lower energy photons. The mass involved in an adequate passive shield, however, is often a major problem for satellite-based experiments with their very tight mass budgets arising from their launcher capabilities.

Imaging

Imaging of high-energy photons is a difficult task because of their extremely penetrating nature. Normal designs of telescope are impossible since in a reflector the photon would just pass straight through the mirror, while in a refractor it would be scattered or unaffected rather than refracted by the lenses. At energies under a few keV, forms of reflecting telescope can be made which work adequately, but at higher energies the only options are occultation, collimation and coincidence detection.

Collimation

A collimator is simply any device that physically restricts the field of view of the detector without contributing any further to the formation of an image. The image is obtained by scanning the system across the object. Almost all detectors are collimated to some degree, if only by their shielding. The simplest arrangement is a series of baffles (figure 1.3.5) that may be formed into a variety of configurations, but all of which follow the same basic principle. Their general appearance leads to their name, honeycomb collimator, even though the cells are usually square rather than hexagonal. They can restrict angles of view down to a few minutes of arc, but they are limited at low energies by reflection off their walls and at high energies by penetration of the radiation through the walls. At high energies, the baffles may be formed from a crystal scintillator and pulses from there used to reject detections of radiation from high inclinations (cf. active shielding). At the low energies the glancing reflection of the radiation can be used to advantage, and a more truly imaging collimator produced. This is called a 'lobster eye' focusing collimator, and is essentially a honeycomb collimator curved into a portion of a sphere (figure 1.3.6), with a position-sensitive detector at its focal surface. The imaging is not of very high quality, and there is a high background from the unreflected rays and the doubly reflected rays (the latter not shown in figure 1.3.6). But it is a cheap system to construct compared with the others that are discussed later in this section, and it has the potential for covering very wide fields of view (tens of degrees) at very high resolutions (seconds of arc). In practice the device is constructed by fusing together thousands of tiny glass rods in a matrix of a different glass. The composite is then heated and shaped and the glass rods etched away to leave the required hollow tubes in the matrix glass. This is the same process as that used to produce microchannel plates (see above), and the

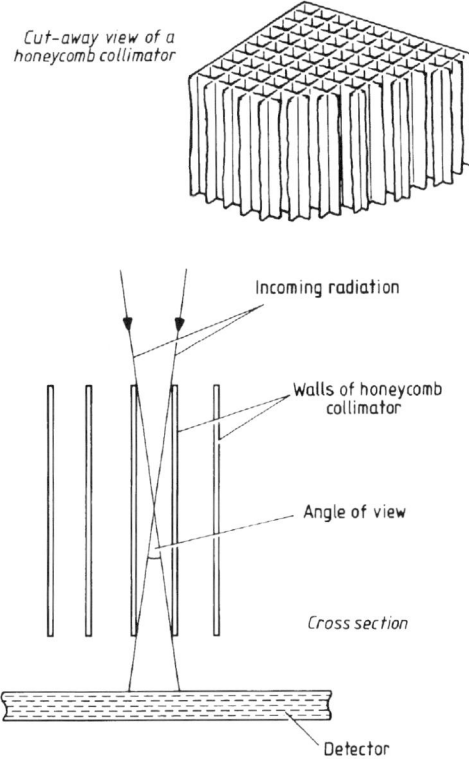

Figure 1.3.5. Honeycomb collimator.

only difference is that for a lobster eye collimator the channels are of square cross section, whereas they are circular in microchannel plates.

Another system, that is known as a modulation collimator or Fourier transform telescope, uses two or more parallel gratings that are separated by a short distance (figure 1.3.7). Since the bars of the gratings alternately obscure the radiation and allow it to pass through, the output as the system scans a point source is a sine wave (figure 1.3.7). The resolution is given by the angle α

$$\alpha = \frac{d}{s}. \tag{1.3.1}$$

To obtain unambiguous positions for the sources, or for the study of multiple or extended sources, several such gratings of different resolutions are combined. The image may then be retrieved from the Fourier components of the output (cf. aperture synthesis, section 2.5). Two such grating systems at right angles can give a two-dimensional image. With these systems resolutions of a few tens of seconds of arc can be realized even for the high-energy photons.

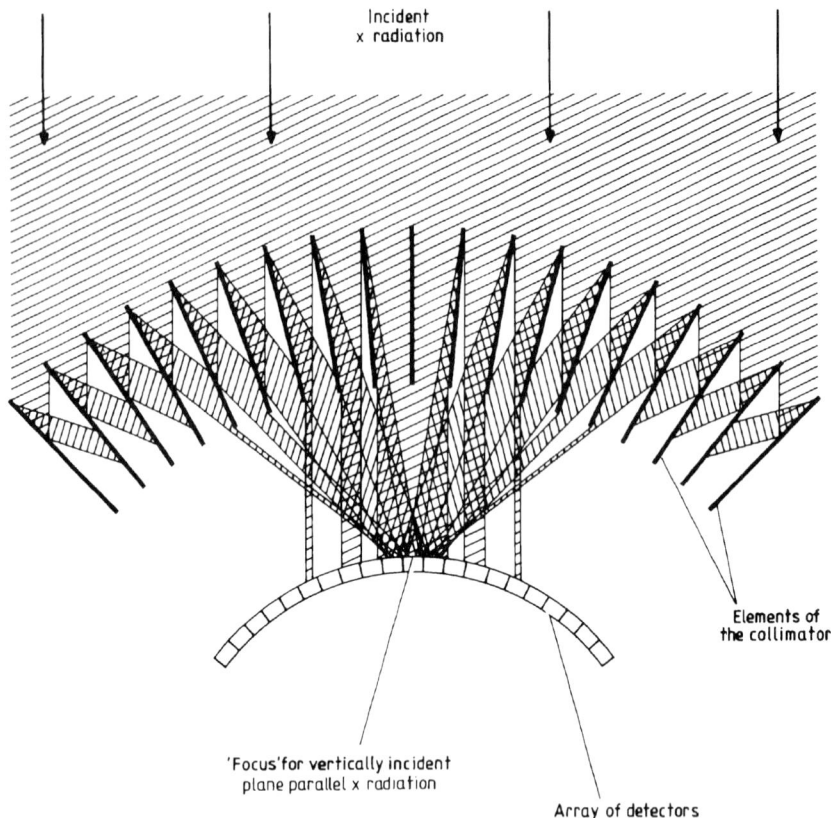

Incident
x radiation

Elements of
the collimator

'Focus'for vertically incident
plane parallel x radiation

Array of detectors

Figure 1.3.6. Cross section through a 'lobster eye' focusing wide angle x-ray collimator.

A third type of collimating imaging system is a simple pinhole camera. A position-sensitive detector, such as a resistive anode proportional counter, or a microchannel plate is placed behind a small aperture. The quality of the image is then just a function of the size of the hole and its distance in front of the detector. Unfortunately, the small size of the hole also gives a low flux of radiation so that such systems are limited to the brightest objects. A better system replaces the pinhole with a mask formed from clear and opaque regions. The pattern of the mask is known, so that when sources cast shadows of it on the detector, their position and structure can be reconstituted in a similar manner to that used for the modulation collimator. The technique is known as coded mask imaging, and resolutions of 10′ of arc or better can be achieved. Since only half the aperture is obscured, the technique obviously uses the incoming radiation much more efficiently than the pinhole camera. A somewhat related technique known as Hadamard mask imaging is discussed in section 2.4. The disadvantage of coded mask imaging is that the

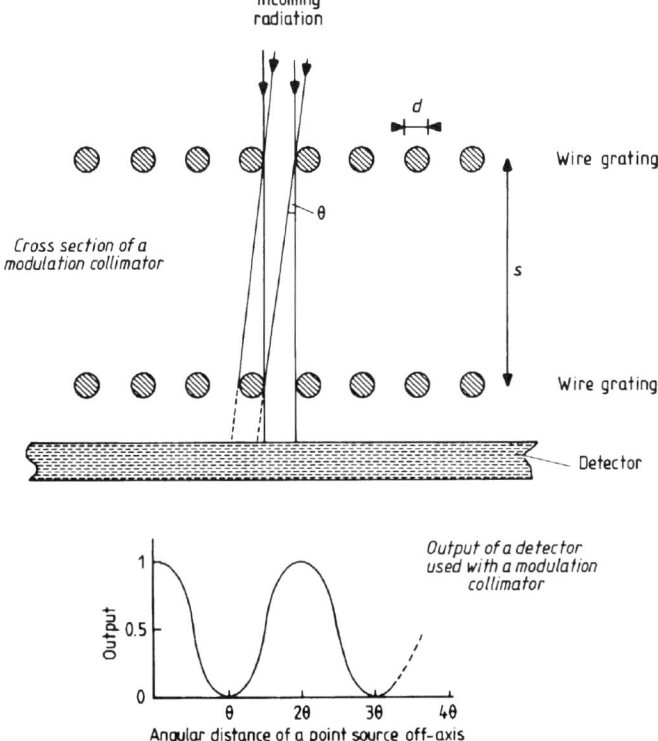

Figure 1.3.7. A modulation collimator.

detector must be nearly the same size as the mask. The Swift space craft, for example, due for launch in 2003, will use a coded mask with an area of $3.2\,m^2$, allied with nearly 33 000 CZT detectors (see solid-state detectors, above) covering an area of $0.52\,m^2$ in its Burst Alert Telescope (BAT). The BAT is for detecting gamma-ray bursters and it will have a field of view of 2 steradians, and be able to pinpoint the position of a gamma-ray burst to within $4'$ of arc within 15 s of its occurrence. The image from a coded aperture telescope is extracted from the data by a cross correlation between the detected pattern and the mask pattern.

Coincidence detectors

A telescope, in the sense of a device that has directional sensitivity, may be constructed for use at any energy, and with any resolution, by using two or more detectors in a line, and by rejecting all detections except those which occur in both detectors and separated by the correct flight time (see figure 1.4.2 for example). Two separated arrays of detectors can similarly provide a two-dimensional imaging system.

Occultation

Although it is not a technique that can be used at will on any source, the occultation of a source by the Moon or other object can be used to give very precise positional and structural information. The use of occultations in this manner is discussed in detail in section 2.7. The technique is less important now than in the past since other imaging systems are available, but it was used in 1964, for example, to provide the first indication that the x-ray source associated with the Crab nebula was an extended source.

Reflecting telescopes

At energies below about 100 keV, photons may be reflected with up to 50% efficiency off metal surfaces, when their angle of incidence approaches 90°. Mirrors in which the radiation just grazes the surface (grazing incidence optics) may therefore be built and configured to form reflecting telescopes. Several systems have been devised, but the one which has achieved most practical use is formed from a combination of annular sections of very deep paraboloidal and hyperboloidal surfaces (figure 1.3.8), and known as a Wolter type I telescope after its inventor. Other designs that are also due to Wolter are shown in figure 1.3.9, and have also been used in practice, although less often than the first design. A simple paraboloid can also be used, though the resulting telescope then tends to be excessively long. The aperture of such telescopes is a thin ring, since only the radiation incident on to the paraboloidal annulus is brought to the focus. To increase the effective aperture, and hence the sensitivity of the system, several such confocal systems of differing radii may be nested inside each other (figure 1.3.10). For the XMM-Newton spacecraft a total of 58 such nested telescope shells gave a total collecting area of about $0.5\,\text{m}^2$. A position-sensitive detector at the focal plane can then produce images with resolutions as high as $1''$ of arc. The limit of resolution is due to surface irregularities in the mirrors, rather than the diffraction limit of the system. The irregularities are about 0.3 nm in size for the very best of the current production techniques, i.e. comparable with the wavelength of photons of about 1 keV. The mirrors are produced by electro-deposition of nickel on to a mandrel of the inverse shape to that required for the mirror, the mirror shell then being separated from the mandrel by cooling. A coating of a dense metal such as gold, iridium or platinum may be applied to improve the x-ray reflectivity. At energies of a few keV, glancing incidence telescopes with low but usable resolutions can be made using foil mirrors. The incident angle is less than 1° so a hundred or more mirrors are needed. They may be formed from thin aluminium foil with a lacquer coating to provide the reflecting surface. The angular resolution of around $1'$ means that the mirror shapes can be simple cones, rather than the paraboloids and hyperboloids of a true Wolter telescope, and so fabrication costs are much reduced.

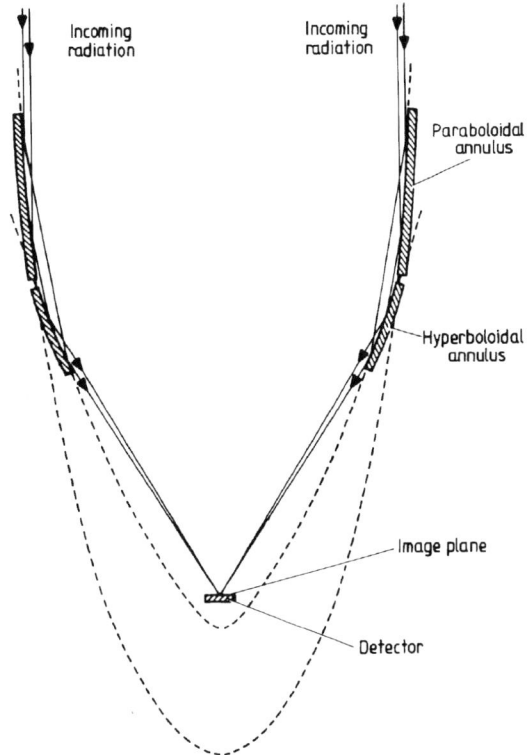

Figure 1.3.8. Cross section through a grazing incidence x-ray telescope.

An alternative glancing incidence system, known as the Kirkpatrick–Baez design, that has fewer production difficulties is based upon cylindrical mirrors. Its collecting efficiency is higher but its angular resolution poorer than the three-dimensional systems discussed above. Two mirrors are used which are orthogonal to each other (figure 1.3.11). The surfaces can have a variety of shapes, but two paraboloids are probably the commonest arrangement. The mirrors can again be stacked in multiples of the basic configuration to increase the collecting area.

At lower energies, in the EUV and soft x-ray regions, near normal incidence reflection with efficiencies up to 20% is possible using multilayer coatings. These are formed from tens, hundreds or even thousands of alternate layers of (for example) tungsten and carbon, aluminium and gold or magnesium and gold, each about 1 nm thick. The reflection is essentially monochromatic, with the wavelength depending upon the orientation of the crystalline structure of the layers, and upon their thickness. Reflection of several wavelengths may be accomplished by changing the thickness of the layers through the stack. The thickest layers are at the top and reflect

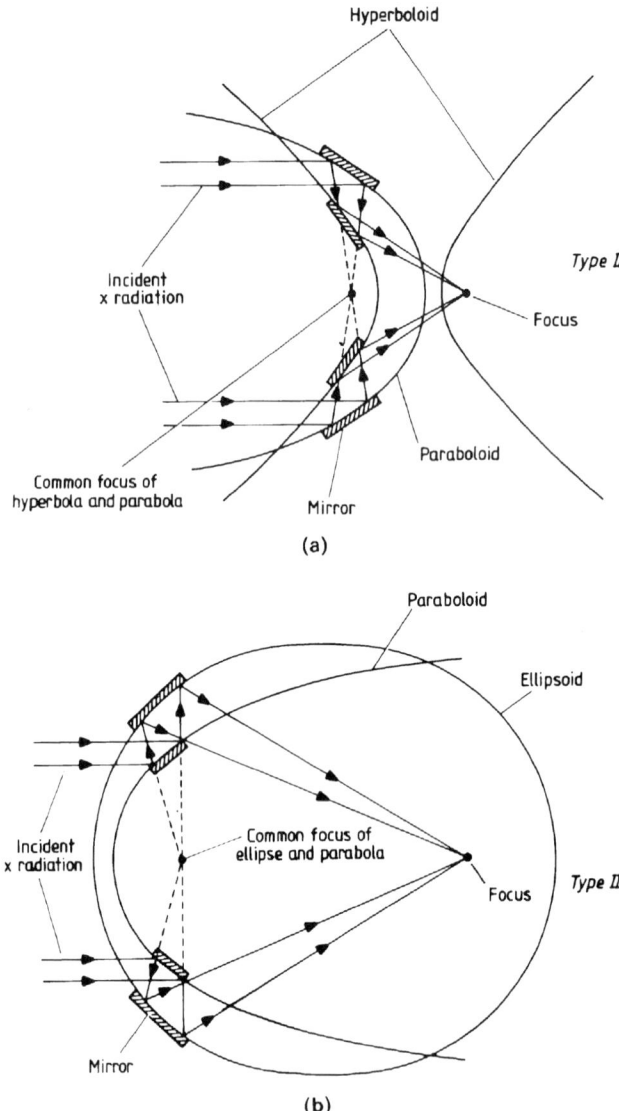

Figure 1.3.9. Cross sections through alternative designs for grazing incidence x-ray telescopes.

the longest, least penetrating wavelengths while, farther down, narrower layers reflect shorter wavelengths. Alternatively several small telescopes may be used each with a different wavelength response, or as in the TRACE (Transition Region And Corona Explorer) spacecraft, different multilayer coatings may be applied in each of four quadrants of the optics.

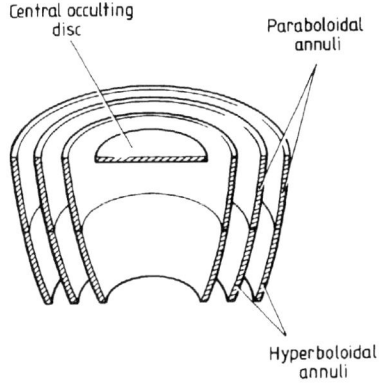

Figure 1.3.10. Section through a nested grazing incidence x-ray telescope.

Telescopes of relatively conventional design are used with these mirrors, and direct images of the Sun at wavelengths down to 4 nm can be obtained.

At energies of tens or hundreds of keV, Bragg reflection and Laue diffraction can be used to concentrate the radiation and to provide some limited imaging. For example, the Laue diffraction pattern of a crystal comprises a number of spots into which the incoming radiation has been concentrated (figure 1.3.12). With some crystals, such as germanium, only a few spots are produced. A number of crystals can be mutually aligned so that one of the spots from each crystal is directed towards the same point (figure 1.3.13), resulting in a crude type of lens. With careful design, efficiencies of 25% to 30% can be achieved at some wavelengths. As with the glancing incidence telescopes, several such ring lenses can be nested to

Figure 1.3.11. X-ray imaging by a pair of orthogonal cylindrical mirrors.

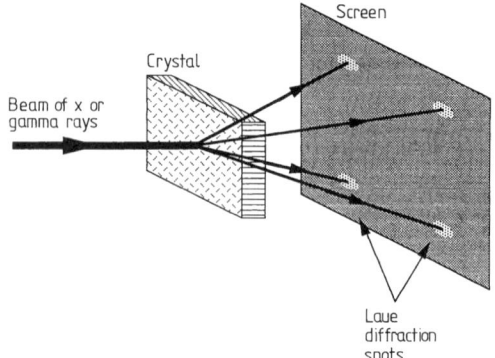

Figure 1.3.12. Laue diffraction of x- and gamma-rays.

increase the effective area of the telescope. A position-sensitive detector such as an array of germanium crystals (figure 1.3.3) can then provide direct imaging at wavelengths down to 0.001 nm (1 MeV).

A completely different approach to imaging gamma-ray sources is based upon cloud or bubble chambers (section 1.4). In these devices the tracks of the gamma-rays are seen directly and can be followed back to their point

Figure 1.3.13. A Laue diffraction lens for x- or gamma-rays (only three crystals shown illuminated for clarity).

of origin in the sky. Liquid xenon seems likely to be particularly useful at MeV energies since it has up to 50% detection efficiency and can provide information upon the spectral distribution and the polarization properties of the photons.

Resolution and image identification

At the low-energy end of the x-ray spectrum, resolutions and positional accuracies comparable with those of Earth-based optical systems are possible, as we have seen. There is therefore usually little difficulty in identifying the optical counterpart of an x-ray source, should it be bright enough to be visible. The resolution rapidly worsens, however, as higher energies are approached. The position of a source that has been detected can therefore only be given to within quite broad limits. It is customary to present this uncertainty in the position of a source by specifying an error box. This is an area of the sky, usually rectangular in shape, within which there is a certain specified probability of finding the source. For there to be a reasonable certainty of being able to find the optical counterpart of a source unambiguously, the error box must be smaller than about one square minute of arc. Since high-energy photon imaging systems have resolutions measured in tens of minutes of arc or larger, such identification is not usually possible. The exception to this occurs when an 'unusual' optical object lies within the error box. This object may then, rather riskily, be presumed also to be the x-ray source. Examples of this are the Crab nebula and Vela supernova remnant which lie in the same direction as two strong sources at 100 MeV and are therefore assumed to be those sources, even though the uncertainties are measured in many degrees.

Spectroscopy

Many of the detectors that were discussed earlier are intrinsically capable of separating photons of differing energies. The superconducting tunnel junction detector (section 1.1 and above) has an intrinsic spectral resolution in the x-ray region, that potentially could reach 10 000 (i.e. 0.0001 nm at 1 nm wavelength). However, the need to operate the devices at lower than 1 K seems likely to prevent their use on spacecraft for some time to come. At photon energies above about 10 keV, it is only this inherent spectral resolution which can provide information on the energy spectrum, although some additional resolution may be gained by using several similar detectors with varying degrees of passive shielding—the lower energy photons will not penetrate through to the most highly shielded detectors. At low and medium energies, the gaps between the absorption edges of the materials used for the windows of the detectors can form wideband filters. Devices akin to the more conventional idea of a spectroscope, however, can only be used at the low end of the energy spectrum and these are discussed below.

Grating spectrometers

Gratings may be either transmission or grazing incidence reflection. The former may be ruled gratings in a thin metal film which has been deposited on a substrate transparent to the x-rays, or they may be formed on a pre-ruled substrate by vacuum deposition of a metal from a low angle. The shadow regions where no metal is deposited then form the transmission slits of the grating. The theoretical background for x-ray gratings is identical with that for optical gratings and is discussed in detail in section 4.1. Typical transmission gratings have around 10^3 lines per millimetre. The theoretical resolution (section 4.1) is between 10^3 and 10^4, but is generally limited in practice to 50 to 100 by other aberrations.

Reflection gratings are also similar in design to their optical counterparts. Their dispersion differs, however, because of the grazing incidence of the radiation. If the separation of the rulings is d then, from figure 1.3.14, we may easily see that the path difference, ΔP, of two rays that are incident on to adjacent rulings is

$$\Delta P = d[\cos\theta - \cos(\theta + \phi)]. \tag{1.3.2}$$

We may expand this via the Taylor series, and neglect powers of θ and ϕ higher than two since they are small angles, to obtain

$$\Delta P = \tfrac{1}{2}d(\phi^2 - 2\theta\phi). \tag{1.3.3}$$

In the mth order spectrum, constructive interference occurs for radiation of wavelength λ if

$$m\lambda = \Delta P \tag{1.3.4}$$

so that

$$\phi = \left(\frac{2m\lambda}{d} + \theta^2\right)^{1/2} - \theta \tag{1.3.5}$$

and

$$\frac{d\phi}{d\lambda} = \left(\frac{m}{2d\lambda}\right)^{1/2} \tag{1.3.6}$$

Figure 1.3.14. Optical paths in a grazing incidence reflection grating.

where we have neglected θ^2, since θ is small. The dispersion for a glancing incidence reflection grating is therefore inversely proportional to the square root of the wavelength, unlike the case for near normal incidence, when the dispersion is independent of wavelength. The gratings may be plane or curved, and may be incorporated into many different designs of spectrometer (section 4.2). Resolutions of up to 10^3 are possible for soft x-rays, but again this tends to be reduced by the effects of other aberrations. Detection may be by scanning the spectrum over a detector (or vice versa), or by placing a position-sensitive detector in the image plane so that the whole spectrum is detected in one go.

Bragg spectrometers

Background.   Distances ranging from 0.1 to 10 nm separate the planes of atoms in a crystal. This is comparable with the wavelengths of x-rays, and so a beam of x-rays interacts with a crystal in a complex manner. The details of the interaction were first explained by the Braggs (father and son). Typical radiation paths are shown in figure 1.3.15. The path differences for rays such as a, b and c, or d and e, are given by multiples of the path difference, ΔP, for two adjacent layers

$$\Delta P = 2d \sin \theta. \tag{1.3.7}$$

There will be constructive interference for path differences that are whole numbers of wavelengths. So that the reflected beam will consist of just those wavelengths, λ, for which this is true:

$$m\lambda = 2d \sin \theta. \tag{1.3.8}$$

Figure 1.3.15. Bragg reflection.

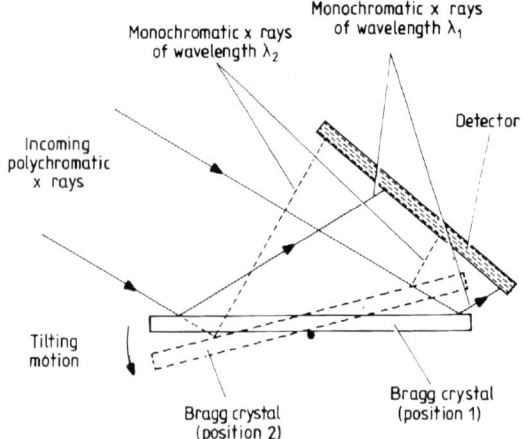

Figure 1.3.16. Scanning Bragg crystal x-ray spectrometer.

The spectrometer. The Bragg spectrometer uses a crystal to produce monochromatic radiation of known wavelength. If a crystal is illuminated by a beam of x-rays of mixed wavelengths, at an approach angle, θ, then only those x-rays whose wavelength is given by equation (1.3.8) will be reflected. The first-order reflection ($m = 1$) is by far the strongest and so the reflected beam will be effectively monochromatic with a wavelength of

$$\lambda_\theta = 2d \sin \theta. \tag{1.3.9}$$

The intensity of the radiation at that wavelength may then be detected with, say, a proportional counter, and the whole spectrum scanned by tilting the crystal to alter θ (figure 1.3.16). An improved version of the instrument uses a bent crystal and a collimated beam of x-rays so that the approach angle varies over the crystal. The reflected beam then consists of a spectrum of all the wavelengths (figure 1.3.17), and this may then be detected by a single observation using a position-sensitive detector. The latter system has the major advantage for satellite-borne instrumentation of having no moving parts, and so significantly higher reliability, and it also has good time resolution. High spectral resolutions are possible (up to 10^3 at 1 keV) but large crystal areas are necessary for good sensitivity, and this may present practical difficulties on a satellite. Many crystals may be used. Among the commonest are lithium fluoride, lithium hydride, tungsten disulphide, graphite, and potassium acid phthalate (KAP).

Many variants upon the basic spectrometer can be devised. It may be used as a monochromator and combined with a scanning telescope to

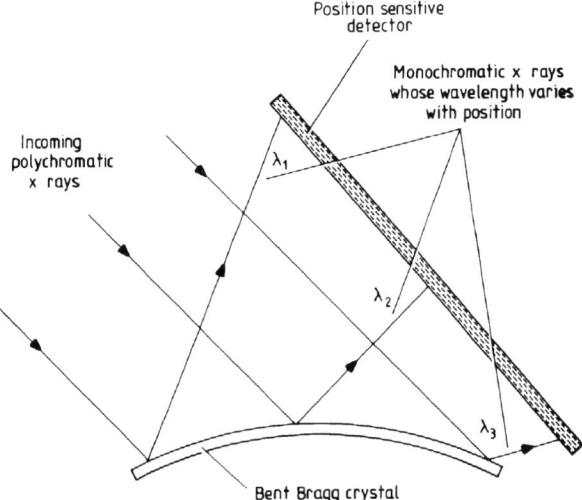

Figure 1.3.17. Bent Bragg crystal x-ray spectrometer.

produce spectroheliograms, for faint sources it may be adapted for use at the focus of a telescope and so on. Designs are rapidly changing and the reader desiring completely up-to-date information must consult the current literature.

Polarimetry

Bragg reflection of x-rays is polarization dependent. For an angle of incidence of 45°, photons that are polarized perpendicularly to the plane containing the incident and reflected rays will be reflected, while those polarized in this plane will not. Thus a crystal and detector at 45° to the incoming radiation and which may be rotated around the optical axis will function as a polarimeter. The efficiency of such a system would be very low, however, due to the narrow energy bandwidth of Bragg reflection. This is overcome by using many randomly oriented small crystals. The crystal size is too small to absorb radiation significantly. If it should be aligned at the Bragg angle for the particular wavelength concerned, however, that radiation will be reflected. The random orientation of the crystals ensures that overall, many wavelengths are reflected, and we have a polarimeter with a broad bandwidth.

A second type of polarimeter looks at the scattered radiation due to Thomson scattering in blocks of lithium or beryllium. If the beam is polarized, then the scattered radiation is asymmetrical, and this may be measured by surrounding the block with several pairs of detectors.

Observing platforms

The Earth's atmosphere completely absorbs x- and gamma-ray radiation, so that all the detectors and other apparatus discussed above, with the exception of the Čerenkov detectors, have to be lifted above at least 99% of the atmosphere. The three systems for so lifting equipment are balloons, rockets and satellites. Satellites give the best results in terms of their height, stability and the duration of the mission. Their cost is very high, however, and there are weight and space restrictions to be complied with; nonetheless there have been quite a number of satellites launched for exclusive observation of the x- and gamma-ray region, and appropriate detectors have been included on a great many other missions as secondary instrumentation. Balloons can carry much heavier equipment far more cheaply than satellites, but only to a height of about 40 km, and this is too low for many of the observational needs of this spectral region. Their mission duration is also comparatively short, being a few days to a week for even the most sophisticated of the self-balancing versions. Complex arrangements have to be made for communication and for retrieval of the payload because of the unpredictable drift of the balloon during its mission. The platform upon which the instruments are mounted has to be actively stabilized to provide sufficient pointing accuracy for the telescopes etc. Thus there are many drawbacks to set against the lower cost of a balloon. Sounding rockets are even cheaper still, and they can reach heights of several hundred kilometres without difficulty. But their flight duration is measured only in minutes and the weight and size restrictions are far tighter even than for most satellites. Rockets still, however, find uses as rapid response systems for events such as solar flares. A balloon- or satellite-borne detector may not be available when the event occurs, or may take some time to bear on to the target, whereas a rocket may be held on stand-by on Earth, during which time the cost commitment is minimal, and then launched when required at only a few minutes notice.

1.4   Cosmic ray detectors

Background

Cosmic rays comprise two quite separate populations of particles: primary cosmic rays, and secondary cosmic rays. The former are the true cosmic rays, and consist mainly of atomic nuclei in the proportions 84% hydrogen, 14% helium, 1% other nuclei, and 1% electrons and positrons, moving at velocities from a few per cent of the speed of light to 0.999 999 999 999 999 999 9999c. That is, the energies per particle range from 10^{-13} to 50 J. As previously mentioned (section 1.3), a convenient unit for measuring the energy of sub-nuclear particles is the electron volt. This is the energy gained by an electron in falling through a potential difference of one volt, and is about 1.602×10^{-19} J. Thus the primary cosmic rays have energies in the range 10^6 to 10^{20} eV, with the bulk of the particles having energies near 10^9 eV, or a velocity of $0.9c$. These are obviously relativistic velocities for most of the particles, for which the relationship between kinetic energy and velocity is

$$v = \frac{(E^2 + 2Emc^2)^{1/2}c}{E + mc^2} \qquad (1.4.1)$$

where E is the kinetic energy of the particle and m is the rest mass of the particle. The flux in space of the primary cosmic rays of all energies is about 10^4 particles m^{-2} s^{-1}. A limit, the GZK cut-off,[16] in the cosmic ray energy spectrum is expected above about 4×10^{19} eV (6 J). At this energy the microwave background photons are blue-shifted to gamma-rays of sufficient energy to interact with the cosmic ray particle to produce pions. This will slow the particle to energies lower than the GZK limit in around 10^8 years. However, surprisingly, higher energy cosmic rays are observed (see below), implying that there must be one or more sources of ultra-high energy cosmic rays within 10^8 light years (30 million pc) of the Earth.

Secondary cosmic rays are produced by the interaction of the primary cosmic rays with nuclei in the Earth's atmosphere. A primary cosmic ray has to travel through about 800 kg m^{-2} of matter on average before it collides with a nucleus. A column of the Earth's atmosphere 1 m^2 in cross section contains about 10^4 kg, so that the primary cosmic ray usually collides with an atmospheric nucleus at a height of 30 to 60 km. The interaction results in numerous fragments, nucleons, pions, muons, electrons, positrons, neutrinos, gamma-rays etc. and these in turn may still have sufficient energy to cause further interactions, producing more fragments, and so on. For high energy primaries ($>10^{11}$ eV), a few such secondary particles will survive down to sea level and be observed as secondary cosmic rays. At higher energies ($>10^{13}$ eV), large numbers of the secondary particles ($>10^9$) survive to sea level, and a cosmic ray shower or extensive air shower is produced. At altitudes higher than sea level, the secondaries from the lower energy primary particles may also be found. Interactions at all energies produce electron and muon neutrinos as a component of the shower. Almost all of these survive to the surface, even those passing through the whole of the Earth, though some convert to tau neutrinos during that passage. The detection of these particles is considered in section 1.5.

[16] Named for Ken Greisen, Timofeevich Zatsepin and Vadim Kuzmin.

Detectors

The methods of detecting cosmic rays may be divided into:

(a) *Real-time methods*. These observe the particles instantaneously and produce information on their direction as well as their energy and composition. They are the standard detectors of nuclear physicists, and there is considerable overlap between these detectors and those discussed in section 1.3. They are generally used in conjunction with passive and/or active shielding so that directional and spectral information can also be obtained.

(b) *Residual track methods*. The path of a particle through a material may be found some time (hours to millions of years) after its passage.

(c) *Indirect methods*. Information on the flux of cosmic rays at large distances from the Earth, or at considerable times in the past, may be obtained by studying the consequent effects of their presence.

Real-time methods

Cloud chamber

This is the standard cloud or bubble chamber of the nuclear physicist. It has numerous variants, but they are all based upon the same principle, i.e. initiation of droplet or bubble formation on to ions produced along the track of a rapidly moving charged particle (or an x- or gamma-ray). In a cloud chamber there is a supersaturated vapour that condenses out on to the ions because of their electrostatic attraction. Although the vapour is supersaturated before the particles pass through it, condensation does not occur until the ions or other nucleation centres are present because very small droplets are not in equilibrium with the vapour pressure if there is only a slight degree of supersaturation. Thus they evaporate or fail to form in the first place. As an example, a vapour pressure of twice the saturation pressure is required for the continued existence of water droplets of radius 10^{-9} m. In a bubble chamber a liquid, such as liquid hydrogen, is maintained a little above its boiling point, and in a similar manner to the cloud chamber, bubble formation does not start until nucleation centres are present. This time the reason why nucleation centres are required arises from the very high pressures inside small bubbles; the presence of the centres increases the size and reduces the pressure required for bubble initiation. If the chamber is illuminated, the tracks of the particles are visible as lines of tiny droplets or bubbles and are usually recorded by being photographed. The droplets or bubbles may be evaporated or recondensed by lowering or raising the pressure of the medium. If the pressure is then restored to its previous value, the material will again be supersaturated or superheated, and the chamber is ready for the next detection. The pressure may be conveniently changed using a piston, and cycle times as short as half a second are attainable.

Figure 1.4.1. Schematic experimental arrangement for the observation of cosmic rays using a cloud chamber.

In cosmic ray work the chamber is activated by the passage of the cosmic ray particle so that photographic film is not wasted on featureless images. The particle is detected by one or more pairs of Geiger counters (section 1.3) above and below the chamber (figure 1.4.1), and the piston is actuated when a near simultaneous detection occurs. Applying magnetic or electro-static fields and measuring the resulting curvature of the tracks may provide information on the nature of the particle.

Techniques such as this one and others in this section might seem to fit more appropriately into imaging in chapter 2. However, the imaging of the tracks in a bubble chamber is the imaging of the information carrier and not of the source, and so counts simply as detection. In any case the

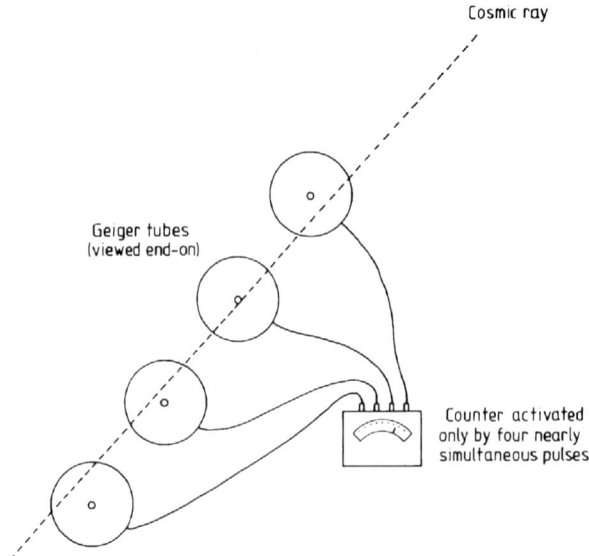

Figure 1.4.2. A cosmic ray telescope utilizing Geiger counters.

sources of cosmic rays cannot currently be imaged because of the extremely tortuous paths of the particles through the galactic and solar magnetic fields.

Geiger counters

We have already reviewed the principles and operational techniques of Geiger counters in section 1.3. Little change in them is required for their use in the detection of cosmic rays. They are now primarily used to trigger the operation of more informative detectors, but have in the past been used as primary detectors.

The azimuthal variation in the cosmic ray intensity was first found by using several counters in a line (figure 1.4.2) and by registering only those particles that triggered all the detectors, so that some directivity was conferred to the system. They have also found application, along with other detectors, within muon energy spectrometers. Arrays of the detectors are stacked above and below a powerful magnet to detect the angular deflection of the muons.

Proportional counters

Position-sensitive proportional counters (section 1.3) are also used for secondary cosmic ray detection, where they are generally known as multi-wire chambers. A variation, known as the drift chamber, is currently widely used by particle physicists, and may find application in detecting astronomical x- and gamma-rays and cosmic rays in the future. The electron

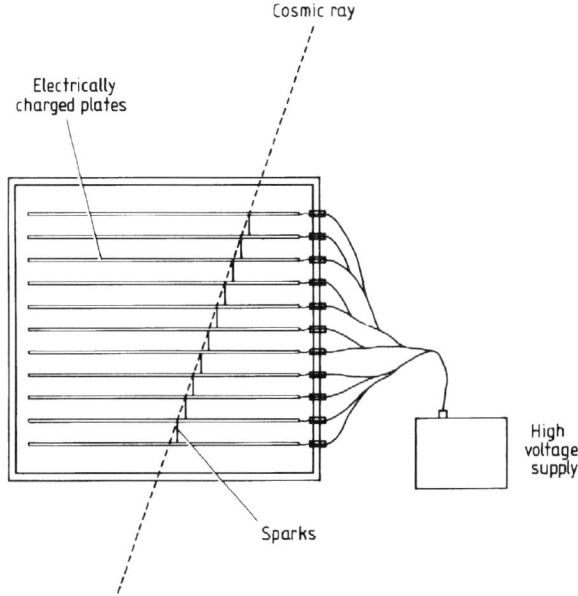

Figure 1.4.3. A spark chamber.

avalanche is confined to a narrow plane by two parallel negatively charged meshes. Sensing wires are spaced at distances of several centimetres. The accelerating voltage is designed so that the electrons gain energy from it at the same rate that they lose it via new ionizations, and they thus move between the meshes at a constant velocity. The time of the interaction is determined using scintillation counters around the drift chamber. The exact position of the original interaction between the sensing wires is then found very precisely by the drift time that it takes the electrons to reach the sensing wire.

Spark detectors

A closely related device to the Geiger counter in terms of its operating principle is the spark chamber. This comprises a series of plates separated by small gaps. Alternate plates are connected together and one set is charged to a high potential while the other set is earthed. The passage of an ionizing particle through the stack results in a series of sparks that follow the path of the particle (figure 1.4.3). The system is also used for the detection of gamma-rays (section 1.3). Then a plate of a material of a high atomic weight such as tantalum may be interposed, and the ray detected by the resulting electron–positron pairs rather than by its own ionizations. The device requires another detector to act as a trigger for the high voltage to be applied, since spontaneous discharges prevent the use of a permanent

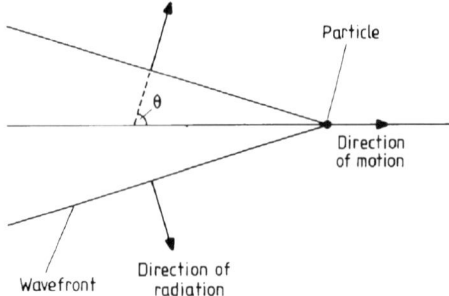

Figure 1.4.4. Čerenkov radiation.

high voltage. For the EGRET[17] detector on the Compton gamma-ray observatory, the trigger was two sheets of plastic scintillator. The trigger is also
needed to actuate cameras to image the spark tracks.

Flash tubes

If a high voltage pulse is applied to a tube filled with a low-pressure gas,
shortly after a particle has passed through it, then the electrons along the
charged particle's path will initiate a glow discharge in the tube. A photomultiplier or other optical detector can then easily detect this. Arrays of
such tubes can provide some limited pointing accuracy to the detection, as
in the case of Geiger tubes (figure 1.4.2).

Scintillation detectors

See section 1.3 for a discussion of these devices. There is little change required
in order for them to detect cosmic ray particles.

Čerenkov detectors

Background. When a charged particle is moving through a medium with a
speed greater than the local speed of light in that medium, it causes the atoms
of the medium to radiate. This radiation is known as Čerenkov radiation,
and it arises from the abrupt change in the electric field near the atom as
the particle passes by. At subphotic speeds, the change in the field is smoother
and little or no radiation results. The radiation is concentrated into a cone
spreading outward from the direction of motion of the particle (figure
1.4.4), whose half angle, θ, is

$$\theta = \tan^{-1}\left[\left(\mu_\nu^2 \frac{v^2}{c^2} - 1\right)^{1/2}\right] \qquad (1.4.2)$$

---

[17] Energetic Gamma Ray Experiment Telescope.

where μ_ν is the refractive index of the material at frequency ν and v is the particle's velocity ($v > c/\mu_\nu$). Its spectrum is given by

$$I_\nu = \frac{e^2\nu}{2\varepsilon_0 c^2}\left(1 - \frac{c^2}{\mu_\nu^2 v^2}\right) \tag{1.4.3}$$

where I_ν is the energy radiated at frequency ν per unit frequency interval, per unit distance travelled by the particle through the medium. The peak emission depends upon the form of the variation of refractive index with frequency. For a proton in air, peak emission occurs in the visible when its energy is about 2×10^{14} eV.

Detectors. Čerenkov detectors are very similar to scintillation detectors except in the manner in which their flashes of visible radiation are produced (see above). A commonly used system employs a tank of very pure water surrounded by photomultipliers for the detection of the heavier cosmic ray particles, while high-pressure carbon dioxide is used for the electrons and positrons. With adequate observation and the use of two or more different media, the direction, energy and possibly the type of the particle may be deduced.

As mentioned previously (section 1.3), the flashes produced in the atmosphere by the primary cosmic rays can be detected. A large light 'bucket' and a photomultiplier are needed, preferably with two or more similar systems observing the same part of the sky so that non-Čerenkov events can be eliminated by anti-coincidence discrimination. Thus the 10 m Whipple telescope on Mount Hopkins is formed from around a hundred small hexagonal mirrors feeding a common focus where there is a 150 pixel detector giving a 3.5° field of view. The stellar intensity interferometer (section 2.4) was also an example of such a system, although there the Čerenkov events were a component of its noise spectrum, and not usually detected for their own interest. The Fly's Eye detector in Utah used sixty-seven 1.5 m mirrors feeding photomultipliers on two sites 4 km apart to monitor the whole sky. It operated from 1982 to 1992 and in 1991 it recorded the highest energy cosmic ray yet found: 3×10^{20} eV. This is well above the GZK limit (see above). Fly's Eye has now been upgraded to HiRes which has a baseline of 12.5 km, 64 quadruple mirror units feeding 256 photo-multipliers.

An intriguing aside to Čerenkov detectors arises from the occasional flashes seen by astronauts when in space. These are thought to be Čerenkov radiation from primary cosmic rays passing through the material of the eyeball. Thus cosmic ray physicists could observe their subjects directly. It is also just possible that they could listen to them as well! A large secondary cosmic ray shower hitting a water surface will produce a 'click' sound that is probably detectable by the best of the current hydrophones. Unfortunately a great many other events produce similar sounds so that the cosmic ray clicks are likely to be well buried in the noise. Nonetheless the very highest energy cosmic ray showers might even be audible directly to a skin diver.

Solid-state detectors

These have been discussed in section 1.3. No change to their operation is needed for cosmic ray detection. Their main disadvantages are that their size is small compared with many other detectors, so that their collecting area is also small, and that unless the particle is stopped within the detector's volume, only the rate of energy loss may be found and not the total energy of the particle. This latter disadvantage, however, also applies to most other detectors.

Nucleon detectors

Generally only the products of the high-energy primaries survive to be observed at the surface of the Earth. A proton will lose about 2×10^9 eV by ionization of atoms in the atmosphere even if it travels all the way to the surface without interacting with a nucleus, and there is only about one chance in ten thousand of it doing this. Thus the lower energy primaries will not produce secondaries that can be detected at ground level, unless a neutron is one of its products. There is then a similar chance of non-interaction with atmospheric nuclei, but since the neutron will not ionize the atoms, even the low energy ones will still be observable at sea level if they do miss the nuclei. Thus neutrons and their reaction products from cosmic ray primaries of any energy can be detected using nucleon detectors.

   These instruments detect low energy neutrons by, for example, their interaction with boron in boron fluoride (BF_3)

$$^{10}_{5}B + n \longrightarrow {}^{7}_{3}Li + {}^{4}_{2}He + \gamma. \qquad (1.4.4)$$

The reaction products are detected by their ionization of the boron fluoride when this is used as a proportional counter (section 1.3). The slow neutrons are produced by the interaction of the cosmic ray neutrons and the occasional proton with lead nuclei, followed by deceleration in a hydrogen-rich moderator such as polyethylene (figure 1.4.5).

Residual track detectors

Photographic emulsions

In a sense the use of photographic emulsion for the detection of ionizing radiation is the second oldest technique available to the cosmic ray physicist after the electroscope, since Henri Becquerel discovered the existence of ionizing particles in 1896 by their effect upon a photographic plate. However, the early emulsions only hinted at the existence of the individual

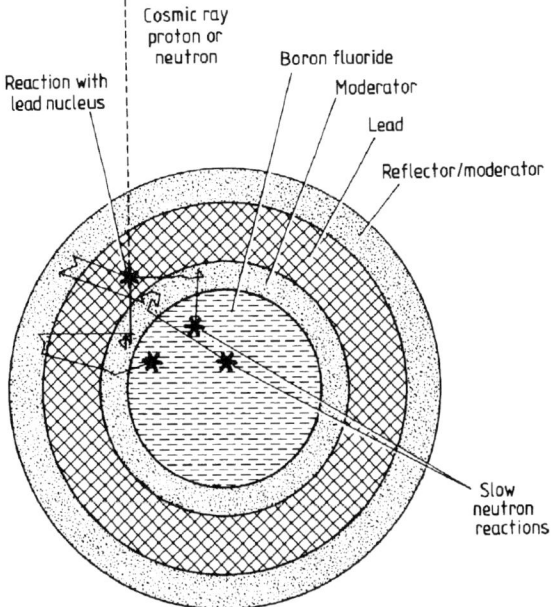

Figure 1.4.5. Cross section through a nucleon detector.

tracks of the particles, and were far too crude to be of any real use. The emulsions now used had thus to await the optimization of their properties for this purpose by C F Powell some 40 years later, before they became of possible use as detectors.

The grain size of the silver halide in nuclear emulsions is small, but not unusually so—ranging from 0.03 to 0.5 μm compared with the range for emulsions used to detect electromagnetic waves of 0.05 to 1.7 μm (an extensive discussion of the properties of the photographic emulsion will be found in section 2.2). Where the grains are unusual is in the very high degree of uniformity of their size that leads to a high degree of uniformity of response. Also very few of the grains are chemically fogged (section 2.2) so that long developing times can be used to show the fainter tracks. The emulsion has a very high silver halide content: up to 80% by weight, so that the emulsion has a very high spatial resolution. The high density of the emulsion also gives it a very high stopping power for the particles: their path lengths are only about 0.05% of the distances in air. Individual layers of the emulsion are quite thick, a millimetre or more not being uncommon. Many such layers are then stacked on top of each other to produce a block of emulsion whose volume may be several litres. Since 80% of such a block is silver, it is obviously a very costly item. However, since only a very tiny fraction of the silver is used to form the images of the particle tracks, the bulk of it may be recovered during processing. Emulsions may be produced to have

Figure 1.4.6. Response curves for nuclear emulsions (the tracks become opaque at a grain density of about 1.5 grains per micron).

differing sensitivities in the sense of their responses to differing rates of energy loss by the particles. A typical response curve is shown in figure 1.4.6. Using the least sensitive emulsions, energy loss values up to $1.5 \times 10^{10}\,\mathrm{eV\,m^{-1}}$ may be evaluated. For high-energy particles the rate of energy loss in a uniform medium, $\mathrm{d}E/\mathrm{d}x$, is given by a modification of the Bethe–Bloch formula

$$\frac{\mathrm{d}E}{\mathrm{d}x} = -\frac{e^2 Z^2 A N}{8\pi\varepsilon_0^2 m_e v^2}\left[\ln\left(\frac{2m_e v^2 W_m}{\chi^2[1-(v^2/c^2)]}\right) - \frac{2v^2}{c^2}\right] \qquad (1.4.5)$$

where Z is the charge of the incident particle, v the velocity of the incident particle, A the atomic number of the stopping atoms, N the number density of the stopping atoms, W_m the maximum amount of energy transferable from the particle to an electron and χ the average ionization potential of the stopping atoms. At low velocities the function is dominated by the v^{-2} term, while at high velocities v approaches asymptotically towards c, so is effectively constant and the $(1-v^2/c^2)^{-1}$ term then predominates. Thus there is a minimum in the rate of energy loss (figure 1.4.7) and hence also of the track density. This minimum can be used to identify particles since their velocities are then similar to each other, enabling the mass of the particle to be found. The energies at minimum track density are listed in table 1.4.1 for some of the commoner cosmic ray particles.

Equation (1.4.5) also shows that the rate of energy deposition is proportional to Z^2. Hence the tracks of heavy nuclei are enormously thicker than those of protons and so the nature of the nucleus can be determined from the track width. If particles with a wide range of energy and composition are to be adequately separated from each other, it may be necessary to interleave emulsions with different stopping powers. A practical measure of the track width that is frequently used is the number of high-energy electrons (often called δ rays) whose tracks extend more than $1.58\,\mu\mathrm{m}$ from the main

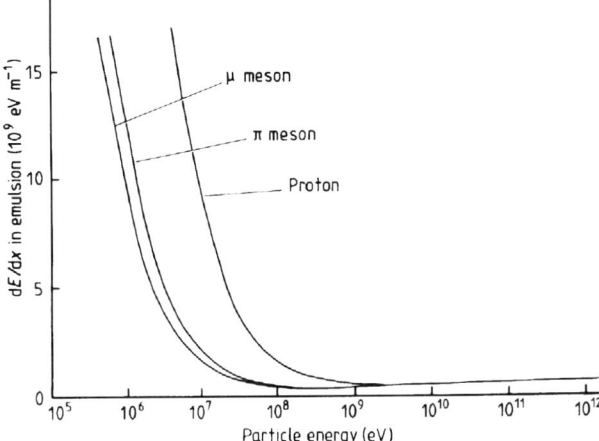

Figure 1.4.7. Rates of energy loss in nuclear emulsion.

track. Nuclei up to iron can then be unequivocally identified. The nuclei beyond iron become progressively harder to distinguish, but there is some evidence that elements up to and perhaps beyond uranium exist in the primary cosmic rays.

The precise mechanism of image formation in the emulsion is not fully understood (section 2.2). There seem to be two possibilities: uniform or discontinuous energy loss. A high-energy proton loses energy at a rate of about $5 \times 10^8 \text{ eV m}^{-1}$ in the emulsion. This corresponds to about 100 eV per 0.2 µm grain. The outermost electrons require some 6 eV for their ionization, so that at most 16 electrons will be available within each grain for the formation of the latent image. This is close to the threshold for stable latent image formation (section 2.2). But for most of the time fewer than the maximum number of electrons will be produced, so that continuous energy loss within the emulsion is a dubious mechanism for the track formation. An alternative hypothesis is that the particle loses its energy discontinuously in the production of occasional high-energy electrons. These electrons can then have an energy ranging up to many millions of electron

Table 1.4.1.

Particle	Energy (eV)
Electron	10^6
µ meson	2×10^8
π meson	3×10^8
Proton	2×10^9
Deuteron	4×10^9

volts depending upon the original interaction. Those with energies below a few keV will be halted within the grain by their own ionization of the atoms in the grain, so that several hundred or even a thousand electrons may be released and a stable latent image formed. Other grains along the particle's track, however, will not contribute to the image unless they also have high-energy electrons produced within them. Electrons with energies above a few keV will have path lengths longer than the size of a single grain, and so will leave the original grains and go on to produce their own short tracks. An electron with an energy of 10 keV, for example, has a path length of about 2 μm. The existence of such short electron tracks has already been mentioned, and they provide some support for this second hypothesis of the mode of image formation in nuclear emulsions.

The use of nuclear emulsions to detect cosmic rays is very simple. A block of the emulsion is exposed to cosmic rays, while shielded from light of course, for a period of time from several hours to several weeks. Exposure is achieved by the emulsion being left on top of a mountain or by being carried to higher altitudes by a balloon. The emulsion is then developed and the particle tracks studied through a microscope.

The processed emulsion must be examined layer-by-layer and measured, usually by hand while viewing through a microscope, since the process is not easily amenable to automation. It is obviously a long and tedious job to do this for the several square metres of emulsion in a typical stack but, nonetheless, the nuclear emulsion remains one of the most useful tools of the cosmic ray astronomer, and has provided most of the direct observations to date of primary cosmic rays.

Ionization damage detectors

These are a relatively recent development and they provide a selective detector for nuclei with masses above about 150 amu. Their principle of operation is allied to that of the nuclear emulsions in that it is based upon the ionization produced by the particle along its track through the material. The material is usually a plastic with relatively complex molecules. As an ionizing particle passes through it, the large complex molecules are disrupted, leaving behind short, chemically reactive segments, radicals etc. Etching the plastic reveals the higher chemical reactivity along the track of a particle and a conical pit develops along the line of the track. By stacking many thin layers of plastic, the track may be followed to its conclusion. The degree of damage to the molecules, and hence the characteristics of the pit which is produced, is a function of the particle's mass, charge and velocity. A particular plastic may be calibrated in the laboratory so that these quantities may be inferred from the pattern of the sizes and shapes of the pits along the particle's track. Cellulose nitrate and polycarbonate plastics are the currently favoured materials. The low weight of the plastic and the ease with which large-area detectors can be formed make this a particularly suitable method for use in space when there is an opportunity for returning the plastic to Earth for processing, as for example when the flight is a manned one.

Similar tracks may be etched into polished crystals of minerals such as feldspar and rendered visible by infilling with silver. Meteorites and lunar samples can thus be studied and provide data on cosmic rays which extend back into the past for many millions of years. The majority of such tracks appear to be attributable to iron group nuclei, but the calibration is very uncertain. Because of the uncertainties involved the evidence has so far been of more use to the meteoriticist in dating the meteorite, than to the cosmic ray astronomer.

Indirect detectors

Electroscope

The science department of any moderately well-equipped secondary school can repeat the earliest and simplest observation of cosmic ray effects. A simple gold leaf electroscope with an applied electric charge will show that charge slowly leaking away even if it is well isolated electrically. The leakage is due to the non-zero conductivity of air through the formation of ion–electron pairs by cosmic radiation. Near sea level this rate averages about $2 \times 10^6 \, \mathrm{m}^{-3} \, \mathrm{s}^{-1}$. The electroscope was used during the very early stages of the investigation of cosmic rays, and revealed, for example, the variation of their intensity with height, but now it is of historical interest only.

100 MeV gamma rays

Primary cosmic rays occasionally collide with nuclei in the interstellar medium. Even though the chance of this occurring is only about 0.1% if the particle were to cross the galaxy in a straight line, it happens sufficiently often to produce detectable results. In such collisions π^0 mesons will frequently be produced, and these will decay rapidly into two gamma rays, each with an energy of about 100 MeV. π^0 mesons may also be produced by the interaction of the cosmic ray particles and the 3 K microwave background radiation. This radiation when 'seen' by a 10^{20} eV proton is Doppler shifted to a gamma ray of 100 MeV energy, and neutral pions result from the reaction

$$p^+ + \gamma \longrightarrow p^+ + \pi^0. \qquad (1.4.6)$$

Inverse Compton scattering of starlight or the microwave background by cosmic ray particles can produce an underlying continuum around the line emission produced by the pion decay.

Gamma rays with energies as high as these are little affected by the interstellar medium, so that they may be observed, wherever they may originate within the galaxy, by artificial satellites (section 1.3). The 100 MeV gamma ray flux thus gives an indication of the cosmic ray flux throughout the galaxy and beyond.

Diffuse radio emission

Cosmic ray electrons are only a small proportion of the total flux, and the reason for this is that they lose significant amounts of energy by synchrotron emission as they interact with the galactic magnetic field. This emission lies principally between 1 MHz and 1 GHz and is observable as diffuse radio emission from the galaxy. However, the interpretation of the observations into electron energy spectra etc. is not straightforward, and is further complicated by the lack of a proper understanding of the galactic magnetic field.

Fluorescence

The very highest energy extensive air showers are detectable via weak fluorescent light from atmospheric nitrogen. This is produced through the excitation of the nitrogen by the electron–photon component of the cascade of secondary cosmic rays. The equipment required is a light bucket and a detector (cf. Čerenkov detectors), and detection rates of a few tens of events per year for particles in the 10^{19} to 10^{20} eV range are achieved by several automatic arrays.

Solar cosmic rays

Very high fluxes of low-energy cosmic rays can follow the eruption of a large solar flare. The fluxes can be high enough to lower the Earth's ionosphere and to increase its electron density. This in turn can be detected by direct radar observations, or through long-wave radio communication fade-outs, or through decreased cosmic radio background intensity as the absorption of the ionosphere increases.

Carbon-14

The radioactive isotope $^{14}_{6}C$ is produced from atmospheric $^{14}_{7}N$ by neutrons from cosmic ray showers:

$$^{14}_{7}N + n \longrightarrow {}^{14}_{6}C + p^{+}. \tag{1.4.7}$$

The isotope has a half-life of 5730 years and has been studied intensively as a means of dating archaeological remains. Its existence in ancient organic remains shows that cosmic rays have been present in the Earth's vicinity for at least 20 000 years. The flux seems, however, to have varied markedly at times from its present day value, particularly between about 4000 and 1000 years BC. But this is probably attributable to increased shielding of the Earth from the low-energy cosmic rays at times of high solar activity, rather than to a true variation in the number of primary cosmic rays.

Arrays

Primary cosmic rays may be studied by single examples of the detectors that we have considered above. The majority of the work on cosmic rays, however, is on the secondary cosmic rays, and for these a single detector is not very informative. The reason is that the secondary particles from a single high-energy primary particle have spread over an area of 10 km^2 or more by the time they have reached ground level from their point of production some 50 km up in the atmosphere. Thus, to deduce anything about the primary particle that is meaningful, the secondary shower must be sampled over a significant fraction of its area. Hence, arrays of detectors are used rather than single ones (see section 1.5 for the detection of cosmic-ray produced neutrinos). Plastic or liquid scintillators (section 1.3) and cloud chambers are frequently chosen as the detectors but any of the real-time instruments can be used and these are typically spread out over an area of several square kilometres. The resulting several hundred to a thousand individual detectors are all then linked to a central computer. For example, the Cygnus array in New Mexico has over 200 one-metre square plastic scintillators. The Akeno Giant Air Shower Array (AGASA) in Japan uses 111 detectors spread over 100 km^2, and has detected the second-highest-energy cosmic ray particle known at 2×10^{20} eV. The Pierre Auger observatory will have two 3000 km^2 arrays each with 1600 water-based Čerenkov detectors and 30 twelve square metre optical telescopes to detect nitrogen fluorescence at distances of up to 30 km.

The analysis of the data from such arrays is difficult. Only a very small sample, typically less than 0.01%, of the total number of particles in the shower is normally caught. The properties of the original particle have then to be inferred from this small sample. However, the nature of the shower varies with the height of the original interaction, and there are numerous corrections to be applied as discussed below. Thus normally the observations are fitted to a grid of computer-simulated showers. The original particle's energy can usually be obtained within fairly broad limits by this process, but its further development is limited by the availability of computer time, and by our lack of understanding of the precise nature of these extraordinarily high-energy interactions.

Correction factors

Atmospheric effects

The secondary cosmic rays are produced within the Earth's atmosphere, and so its changes may affect the observations. The two most important variations are caused by air mass and temperature.

The air mass depends upon two factors—the zenith angle of the axis of the shower, and the barometric pressure. The various components of the shower are affected in different ways by changes in the air mass. The muons are more penetrating than the nucleons and so the effect upon them is comparatively small. The electrons and positrons come largely from the decay of muons, and so their variation tends to follow that of the muons. The corrections to an observed intensity are given by

$$I(P_0) = I(P)\exp[K(P - P_0)/P_0]$$ (1.4.8)

$$I(0) = I(\theta)\exp[K(\sec\theta - 1)]$$ (1.4.9)

where P_0 is the standard pressure, P is the instantaneous barometric pressure at the time of the shower, $I(P_0)$ and $I(P)$ are the shower intensities at pressures P_0 and P respectively, $I(0)$ and $I(\theta)$ are the shower intensities at zenith angles of zero and θ respectively and K is the correction constant for each shower component. K has a value of 2.7 for the muons, electrons and positrons and 7.6 for the nucleons. In practice a given detector will also have differing sensitivities for different components of the shower, and so more precise correction factors must be determined empirically.

The temperature changes primarily affect the muon and electron components. The scale height of the atmosphere, which is the height over which the pressure changes by a factor e^{-1}, is given by

$$H = \frac{kR^2T}{GMm}$$ (1.4.10)

where R is the distance from the centre of the Earth, T the atmospheric temperature, M the mass of the Earth and m the mean particle mass for the atmosphere. Thus the scale height increases with temperature and so a given pressure will be reached at a greater altitude if the atmospheric temperature increases. The muons, however, are unstable particles and will have a longer time in which to decay if they are produced at greater heights. Thus the muon and hence the electron and positron intensity decreases as the atmospheric temperature increases. The relationship is given by

$$I(T_0) = I(T)\exp[0.8(T - T_0)/T]$$ (1.4.11)

where T is the atmospheric temperature, T_0 the standard temperature and $I(T_0)$ and $I(T)$ are the muon (or electron) intensities at temperatures T_0 and T respectively. Since the temperature of the atmosphere varies with height, equation (1.4.11) must be integrated up to the height of the muon formation in order to provide a reliable correction. Since the temperature profile will only be poorly known, it is not an easy task to produce accurate results.

Solar effects

The Sun affects cosmic rays in two main ways. First, it is itself a source of low-energy cosmic rays whose intensity varies widely. Second, the extended solar magnetic field tends to shield the Earth from the lower-energy primaries. Both these effects vary with the sunspot cycle and also on other timescales, and are not easily predictable.

Terrestrial magnetic field

The Earth's magnetic field is essentially dipolar in nature. Lower-energy charged particles may be reflected by it, and so never reach the Earth at all, or particles that are incident near the equator may be channelled towards the poles. There is thus a dependence of cosmic ray intensity on latitude (figure 1.4.8). Furthermore, the primary cosmic rays are almost all positively charged and they are deflected so that there is a slightly greater intensity from the west. The vertical concentration (equation (1.4.9) and figure 1.4.9) of cosmic rays near sea level makes this latter effect only of importance for high-altitude balloon or satellite observations.

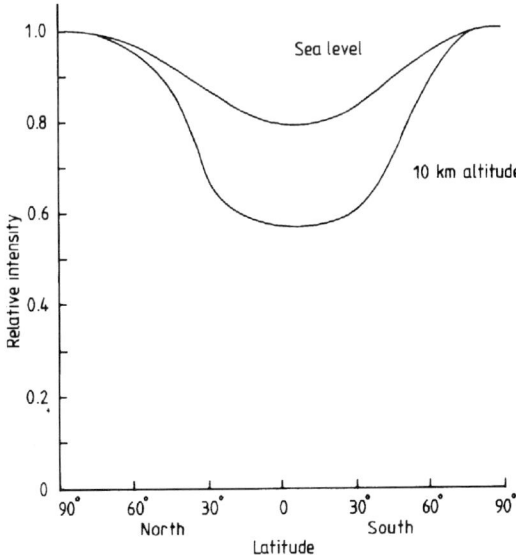

Figure 1.4.8. Latitude effect on cosmic rays.

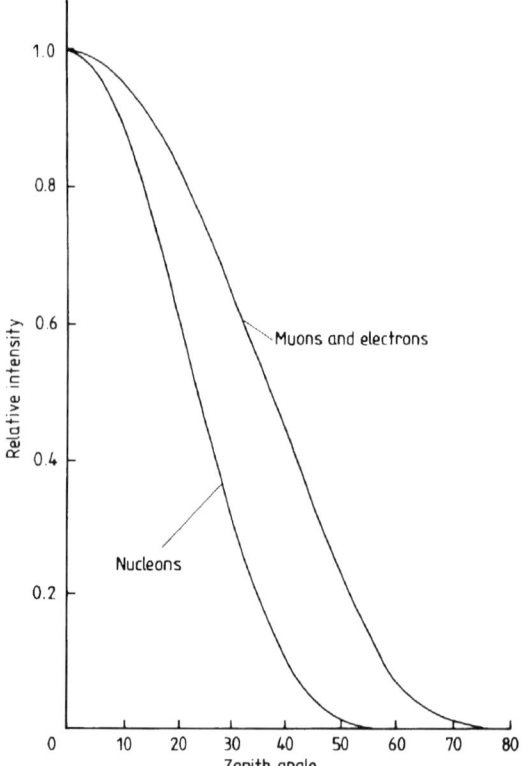

Figure 1.4.9. Zenithal concentration of secondary cosmic ray shower components as given by equation (1.4.9).

Exercises

1.4.1 Show that the true counting rate, C_t, of a Geiger counter whose dead time is of length Δt, is related to its observed counting rate, C_0, by

$$C_t = \frac{C_0}{1 - \Delta t C_0}$$

(section 1.3 is also relevant to this problem).

   If the effective range of a Geiger counter is limited to $C_t \leq 2C_0$, calculate the maximum useful volume of a Geiger counter used to detect secondary cosmic rays at sea level if its dead time is 250 μs.

1.4.2 The minimum particle energy required for a primary cosmic ray to produce a shower observable at ground level when it is incident vertically on to the atmosphere is about 10^{14} eV. Show that the minimum energy required to produce a shower, when the primary particle is incident at a zenith angle θ, is given by

$$E_{\min}(\theta) = 6.7 \times 10^{12} \exp(2.7 \sec \theta) \quad \text{eV}.$$

(Hint: use equation (1.4.9) for muons and assume that the number of particles in the shower at its maximum is proportional to the energy of the primary particle.)

The total number of primary particles, $N(E)$, whose energy is greater than or equal to E, is given at high energies by

$$N(E) \approx 10^{22} E^{-1.85} \quad \text{m}^{-2} \text{s}^{-1} \text{str}^{-1}$$

for E in eV. Hence show that the number of showers, $N(\theta)$, observable from the ground at a zenith angle of θ is given by

$$N(\theta) \approx 0.019 \exp(-5.0 \sec \theta) \quad \text{m}^{-2} \text{s}^{-1} \text{str}^{-1}.$$

1.4.3 By numerical integration, or otherwise, of the formula derived in problem 1.4.2, calculate the total flux of showers of all magnitudes on to a detector array covering one square kilometre. (Assume that the primary particle flux is isotropic.)

1.5 Neutrino detectors

Background

Wolfgang Pauli postulated the neutrino in 1930 in order to retain the principle of conservation of mass and energy in nuclear reactions. It was necessary in order to provide a mechanism for the removal of residual energy in some beta-decay reactions. From other conservation laws, its properties could be defined quite well: zero charge, zero or very small rest mass, zero electric moment, zero magnetic moment, and a spin of one half. Over a quarter of a century passed before the existence of this hypothetical particle was confirmed experimentally (in 1956). The reason for the long delay in its confirmation lay in the very low probability of the interaction of neutrinos with other matter (neutrinos interact via the weak nuclear force only). A neutrino originating at the centre of the Sun would have only one chance in 10^{10} of interacting with any other particle during the whole of its 700 000 km journey to the surface of the Sun. The interaction probability for particles is measured by their cross sectional area for absorption, σ, given by

$$\sigma = \frac{1}{\lambda N} \tag{1.5.1}$$

where N is the number density of target nuclei and λ is the mean free path of the particle. Even for high-energy neutrinos the cross section for the reaction

$$\nu + {}^{37}_{17}\text{Cl} \longrightarrow {}^{37}_{18}\text{Ar} + e^- \tag{1.5.2}$$

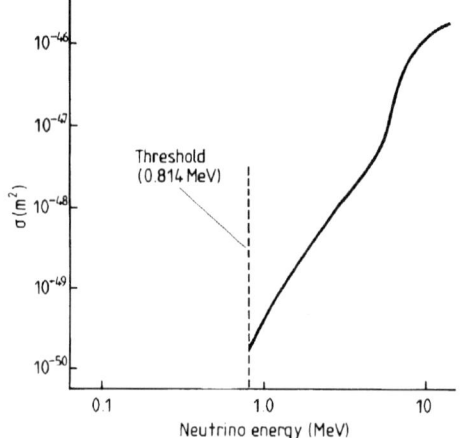

Figure 1.5.1. Neutrino absorption cross-sections for the $^{37}_{17}Cl \longrightarrow ^{37}_{18}Ar$ reaction.

that was originally used for their detection (see later in this section) is only $10^{-46} m^2$ (figure 1.5.1), so that such a neutrino would have a mean free path of over 1 pc even in pure liquid $^{37}_{17}Cl$!

Three varieties of neutrino are known, plus their anti-particles. The electron neutrino, ν_e, is the type originally postulated to save the β reactions, and which is therefore involved in the archetypal decay: that of a neutron

$$n \longrightarrow p^+ + e^- + \tilde{\nu}_e \qquad (1.5.3)$$

where $\tilde{\nu}_e$ is an anti-electron neutrino. The electron neutrino is also the type commonly produced in nuclear fusion reactions and hence to be expected from the Sun and stars. For example the first stage of the proton–proton cycle is

$$p^+ + p^+ \longrightarrow ^2_1H + e^+ + \nu_e \qquad (1.5.4)$$

and the second stage of the carbon cycle

$$^{13}_7N \longrightarrow ^{13}_6C + e^+ + \nu_e \qquad (1.5.5)$$

and so on.

The other two types of neutrino are the muon neutrino, ν_μ, and the tau neutrino, ν_τ. These are associated with reactions involving the heavy electrons called muons and tauons. For example, the decay of a muon

$$\mu^+ \longrightarrow e^+ + \nu_e + \tilde{\nu}_\mu \qquad (1.5.6)$$

involves an anti-muon neutrino among other particles. The muon neutrino was detected experimentally in 1962, but the tau neutrino was not found until 2000.

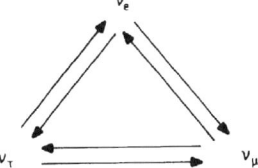

Figure 1.5.2. Neutrino oscillation.

The first neutrino detector started operating in 1968 (see chlorine-37 detectors below), and it quickly determined that the observed flux of solar neutrinos was too low compared with the theoretical predictions by a factor of 3.3. This deficit was confirmed later by other types of detectors such as SAGE, GALLEX and Kamiokande and became known as the solar neutrino problem. However, these detectors were only sensitive to electron neutrinos. Then in 1998, the Super Kamiokande detector found a deficit in the number of muon neutrinos produced by cosmic rays within the Earth's atmosphere. Some of the muon neutrinos were converting to tau neutrinos during their flight time to the detector. The Super Kamiokande and SNO detectors have since confirmed that all three neutrinos can exchange their identities (figure 1.5.2). The solar neutrino problem has thus disappeared, since two-thirds of the electron neutrinos produced by the Sun have changed to muon and tau neutrinos by the time that they reach the Earth, and so were not detected by the early experiments. The flux of solar neutrinos (of all types) at the Earth is now measured to be about $6 \times 10^{14} \, \mathrm{m}^{-2} \, \mathrm{s}^{-1}$, in good agreement with the theoretical predictions. The experiments also show that neutrinos have mass, although only upper limits have been determined so far. These are $<3 \, \mathrm{eV}$ (less than 0.0006% of the mass of the electron) for the electron neutrino and $<0.17 \, \mathrm{MeV}$ and $<18 \, \mathrm{MeV}$ for the muon and tau neutrinos respectively. It has been suggested that there might be a fourth type of neutrino, either extremely lightweight or super-massive. The lightweight neutrino would interact with ordinary matter only through gravity, and so be even less easy to detect than 'normal' neutrinos; they have thus been termed 'sterile' neutrinos. The super-massive ($>10^{21} \, \mathrm{eV}$) neutrino would oscillate into being from one of the other types for an imperceptibly brief instant of time only. However, recent experiments using beams of anti-electron neutrinos from nuclear reactors observed by the KamLAND detector have almost certainly ruled out the existence of a fourth neutrino.

Chlorine-37 detectors

The problems involved in any attempt to detect neutrinos arise mainly from the extreme rarity of the interactions that are involved. This not only dictates

Figure 1.5.3. Postulated solar neutrino spectrum.

the use of very large detectors and very long integration times, but also means that other commoner radioactive processes may swamp the neutrino-induced reactions. The first neutrino 'telescope' to be built is designed to detect electron neutrinos through the reaction given in equation (1.5.2). The threshold energy of the neutrino for this reaction is 0.814 MeV (see figure 1.5.1), so that 80% of the neutrinos from the Sun that are detectable by this experiment are expected to arise from the decay of boron in a low-probability side chain of the proton–proton reaction

$$^{8}_{5}\text{B} \longrightarrow {}^{8}_{4}\text{Be} + e^{+} + \nu_{e}. \tag{1.5.7}$$

The full predicted neutrino spectrum of the Sun, based upon the conventional astrophysical models, is shown in figure 1.5.3.

The detector began operating in 1968 under the auspices of Ray Davis Jr and his group. It consists of a large tank of over 600 tonnes of tetra-chloroethene (C_2Cl_4) located 1.5 km down in the Homestake gold mine in South Dakota. About one chlorine atom in four is the $^{37}_{17}\text{Cl}$ isotope; so on average each of the molecules contains one of the required atoms, giving a total of about 2×10^{30} $^{37}_{17}\text{Cl}$ atoms in the tank. Tetrachloroethene is chosen rather than liquid chlorine for its comparative ease of handling, and because of its cheapness: it is a common industrial solvent. The interaction of a neutrino (equation (1.5.2)) produces a radioactive isotope of argon. The half-life of the argon is 35 days, and it decays back to $^{37}_{17}Cl$ by capturing one of its own inner orbital electrons, at the same time ejecting a 2.8 keV electron. The experimental procedure is to allow the argon atoms to accumulate for some time, then to bubble helium through the tank. The argon is swept up by the helium and carried out of the tank. It can be separated from the helium by passing the gas stream through a charcoal cold trap. The number of $^{37}_{18}Ar$ atoms can then be counted by counting the 2.8 keV electrons as they are emitted.

The experimental procedure must be carried out with extraordinary care if the aim of finding a few tens of atoms in 10^{31} is to succeed (similar precautions have to be taken for all neutrino detectors). Pre-eminent among the precautions are:

(a) Siting the tank 1.5 km below the ground to shield it against cosmic rays and natural and artificial radiation sources. The neutrinos of course pass almost without effect through this depth of material; indeed they can pass right through the Earth without any measurable drop in their intensity.
(b) Surrounding the tank with a thick water jacket to shield against neutrons.
(c) Monitoring the efficiency of the argon extraction by introducing a known quantity of $^{36}_{18}Ar$ into the tank before it is swept with helium, and determining the percentage of this that is recovered.
(d) Shielding the sample of $^{37}_{18}Ar$ during counting of its radioactive decays and using anti-coincidence techniques to eliminate extraneous pulses.
(e) Use of long integration times (50 to 100 days) to accumulate the argon in the tank.
(f) Measurement of and correction for the remaining noise sources.

Currently about one neutrino is detected every other day. This corresponds to a detection rate of 2.23 ± 0.22 SNU. (A solar neutrino unit, SNU, is 10^{-36} captures per second per target atom.) The expected rate from the Sun is 7.3 ± 1.5 SNU, and the apparent disparity between the observed and predicted rates prior to the confirmation of the oscillation of neutrino types gave rise to the previously mentioned solar neutrino problem. In 2002 Ray Davis shared the Nobel prize for physics for his work on neutrinos.

Water-based detectors

The next generation of working neutrino telescopes did not appear until nearly two decades after the chlorine-based detector. Neither of these

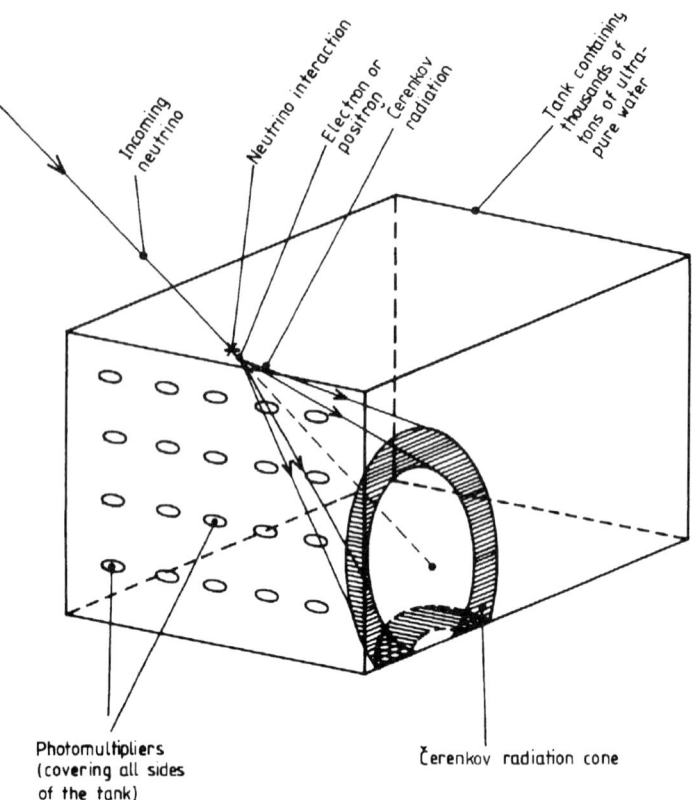

Photomultipliers
(covering all sides
of the tank)

Čerenkov radiation cone

Figure 1.5.4. Principle of neutrino detection by water-based detectors.

instruments, the Kamiokande[18] detector buried 1 km below Mount Ikenoyama in Japan nor the IMB[19] detector, 600 m down a salt mine in Ohio,
were built to be neutrino detectors. Both were originally intended to look
for proton decay, and were only later converted for use with neutrinos.
The design of both detectors is similar (figure 1.5.4); they differ primarily
only in size (Kamiokande, 3000 tonnes; IMB, 8000 tonnes).

The principle of their operation is the detection of Čerenkov radiation
from the products of neutrino interactions. These may take two forms,
electron scattering and inverse beta-decay. In the former process, a collision
between a high-energy neutrino and an electron sends the latter off at a speed
in excess of the speed of light in water ($225\,000\,\mathrm{km\,s^{-1}}$) and in roughly the
same direction as the neutrino. All three types of neutrino scatter electrons,
but the electron neutrinos are 6.5 times more efficient at the process. In

[18] Kamioka Neutrino DEtector.
[19] Irvine–Michigan–Brookhaven.

inverse beta-decay, an energetic positron is produced via the reaction

$$\tilde{\nu}_e + p \longrightarrow n + e^+. \tag{1.5.8}$$

Inverse beta decays are about 100 times more probable than the scattering events, but the positron can be emitted in any direction and thus gives no clue to the original direction of the incoming neutrino. Both positron and electron, travelling at superphotic speeds in the water, emit Čerenkov radiation (section 1.4) in a cone around their direction of motion. That radiation is then picked up by arrays of photomultipliers (section 1.1) that surround the water tank on all sides. The pattern of the detected radiation can be used to infer the energy of the original particle, and in the case of a scattering event, also to give some indication of the arrival direction. The minimum detectable energy is around 5 MeV due to background noise.

Both detectors were fortunately in operation in February 1987 and detected the burst of neutrinos from the supernova in the Large Magellanic Cloud (SN1987A). This was the first (and so far only) detection of neutrinos from an astronomical source other than the Sun, and it went far towards confirming the theoretical models of supernova explosions.

Both IMB and Kamiokande have now ceased operations,[20] but a number of detectors operating on the same principle are functioning or are under construction. Foremost among these is Super Kamiokande that contains 50 000 tonnes of pure water and uses 13 000 photomultiplier tubes in a tank buried 1 km below Mount Ikenoyama in Japan. Super Kamiokande is able to detect muon neutrinos as well as electron neutrinos. The muon neutrino interacts with a proton to produce a relativistic muon. The Čerenkov radiation from the muon produces a well-defined ring of light. The electron resulting from an electron neutrino scattering event by contrast generates a much fuzzier ring of light. This is because the primary electron produces gamma rays that in turn produce electron–positron pairs. The electron–positron pairs then create additional Čerenkov radiation cones, thus blurring the ring resulting from the primary electron.

This ability to distinguish between electron and muon neutrinos enabled Super Kamiokande to provide the first evidence of neutrino oscillations (see background section above). Muon neutrinos produced in the Earth's atmosphere by high-energy cosmic rays should be twice as abundant as electron neutrinos.[21] But in 1998 Super Kamiokande found that although this was the case for neutrinos coming from above (a distance of about 60 km),

---

[20] The IMB photomultipliers were re-used as an active shield for Super Kamiokande.

[21] The cosmic ray interaction first produces (amongst other particles) pions. A positively charged pion usually then decays in 2.6×10^{-8} s to a positive muon and a muon neutrino, and the muon decays in 2.2×10^{-6} s to a positron, a muon anti-neutrino and an electron neutrino. A negative pion likewise decays to an electron, a muon neutrino, a muon anti-neutrino and an electron anti-neutrino. There should thus be two muon neutrinos to each electron neutrino in a secondary cosmic ray shower (section 1.4).

there were roughly equal numbers of muon neutrinos and electron neutrinos coming from below (i.e. having travelled 12 000 km across the Earth). Since there were no extra electron neutrinos, some of the muon neutrinos must be oscillating to tau neutrinos during the flight time across the Earth to the detector. This result was confirmed in 2000 by a 30% shortfall in the number of muon neutrinos observed from an artificial source at the KEK laboratory at Tsukuba some 250 km away from the detector. Further confirmation of the oscillation of all three types of neutrino has recently come from comparison of the Super Kamiokande and SNO results and the KamLAND experiment (see below). The designer of the Kamiokande and Super Kamiokande detectors, Matatoshi Koshiba, shared the Nobel prize for physics in 2002 for his work.

Man-made detectors larger than Super Kamiokande are probably impractical. However, much greater quantities of water (or ice, which is equally good) may be monitored using parts of lakes, the sea or the Antarctic ice cap. Of the water detectors, NT-200 is currently operating while ANTARES[22] and NESTOR[23] are under construction, AMANDA2[24] utilizes the Antarctic ice cap and is currently operating with an upgrade being planned to be called ICECUBE.

The NT-200 detector comprises 192 sets of photomultipliers suspended on eight strings forming an array some 45 m across and 100 m long. It is immersed at a depth of 1.1 km in Lake Baikal and monitors 200 000 m^3 of water for multi-TeV muons coming from neutrino interactions within the rock below it. The neutrinos themselves originate from high-energy cosmic ray impacts near the top of the Earth's atmosphere. The higher-energy particles can be detected coming from interactions above the detector, since the background noise at that energy is low, but at lower energies, the detector observes upwardly moving particles originating in the atmosphere on the far side of the Earth. ANTARES and NESTOR are being constructed in the Mediterranean at depths of 2500 m and 4000 m off Marseilles and the Peloponnese respectively. The detectors are similar to NT200, using photomultipliers on long strings to monitor a volume of seawater for Čerenkov radiation, again arising from high-energy muons. ANTARES will be 150 m across by 300 m high and will use 900 photomultipliers to observe a volume of 10^6 m^3, while NESTOR will be 32 m across by 240 m high and will monitor 200 000 m^3 of water using 168 photomultipliers.

AMANDA2 has been in operation since February 2000, and succeeds a smaller trial version of the same type of detector called AMANDA. The ice at a depth of one or more kilometres in the Antarctic ice cap is highly transparent, since there are few impurities and the pressure of the overlying

[22] Astronomy with a Neutrino Telescope and Abyssal Environmental RESearch.
[23] Neutrino Extended Submarine Telescope with Oceanographic Research.
[24] Antarctic Muon And Neutrino Detector Array.

layers has squeezed out any air bubbles. So the Čerenkov radiation can easily be seen. Like the seawater detectors, AMANDA2 uses 680 photomultipliers suspended on 26 strings. There is an outer array 200 m across extending from a depth of 1150 m to 2350 m in the ice, surrounding a smaller array 120 m across that extends from 1500 m down to 2000 m, resulting in a total volume of ice of $40 \times 10^6 \, \mathrm{m}^3$ being monitored. Unlike the seawater detectors though, the strings of photomultipliers are frozen in place, having been lowered down holes drilled through the ice by a hot water drill. Plans are currently under discussion to upgrade AMANDA2 to 4800 photomultipliers on 80 strings and to monitor 1000 million cubic metres (1 km^3) of ice. The upgraded detector is to be called ICECUBE.

Although the peak Čerenkov emission produced by multi-TeV particles is in the optical region, radio waves are produced as well (equation (1.4.3)). At the temperature of the ice cap (about $-50\,^\circ\mathrm{C}$), the ice is almost completely transparent to metre-wavelength radio waves, and so these can be detected by receivers at the surface. There was a trial experiment, RAND,[25] a decade ago, and there is now a proposal to fly 36 antennas using a high-altitude balloon at the South Pole. This would enable the entire Antarctic ice cap to be watched: a volume of around $10^6 \, \mathrm{km}^3$.

A neutrino detector based upon 1000 tonnes of heavy water (D_2O)[26] has been operating at Sudbury in Ontario since late 1999. Known as SNO,[27] the heavy water is contained in a 12 m diameter acrylic sphere that is immersed in an excavated cavity filled with 7000 tonnes of highly purified normal water. The normal water shields the heavy water from radioactivity in the surrounding rock. It is located 2 km down within the Creighton copper and nickel mine and uses 9600 photomultipliers to detect Čerenkov emissions within the heavy water. Neutrinos may be detected through electron scattering as with other water-based detectors, but the deuterium in the heavy water provides two other mechanisms whereby neutrinos may be found. The first of these senses just electron neutrinos. An electron neutrino interacts with a deuterium nucleus to produce two protons and a relativistic electron

$$\nu_e + {}_1^2\mathrm{H} \longrightarrow \mathrm{p}^+ + \mathrm{p}^+ + \mathrm{e}^- \qquad (1.5.9)$$

and the electron is observed via its Čerenkov radiation. This mechanism is called the charged current reaction since the charged W boson mediates it. The second mechanism is mediated by the neutral Z boson, and is thus called the neutral current reaction. In it, a neutrino of any type simply splits the deuterium nucleus into its constituent proton and neutron. The neutron is thermalized in the heavy water, and eventually combines with another nucleus with the emission of gamma rays. The gamma rays in turn

---

[25] Radio Array Neutrino Detector.
[26] On loan from the Canadian government. It has a value of around $300 million.
[27] Sudbury Neutrino Observatory.

produce relativistic electrons via Compton scattering and these are finally detected from their Čerenkov radiation. Although the neutron can combine with a deuterium nucleus, the capture efficiency is low so that 75% of the neutrons will escape from the detector. The SNO detector can therefore add 2 tonnes of salt to the heavy water to enable the neutron to be captured more easily by a $^{35}_{17}$Cl nucleus (converting it to $^{36}_{17}$Cl) and reducing to 17% the number of escaping neutrons. The thermalized neutrons can also be detected using $^{3}_{2}$He proportional counters (section 1.3). The neutron combines with the $^{3}_{2}$He to produce a proton and tritium, and these then ionize some of the remaining gas to give an output pulse.

Since SNO can detect the number of electron neutrinos separately from the total for all types of neutrino, it can provide the definitive data for the solar neutrino problem (see above). The standard solar model predicts that SNO should detect about 30 charged current or neutral current reactions per day and about three electron scattering events. However, the reality of the oscillation of neutrinos between their different types, and hence the solution to the solar neutrino problem, was demonstrated by the very first results from SNO. It was initially operated in the charged current mode (i.e. without salt being added to the heavy water) to detect just the solar electron neutrinos. This gave a flux that was lower than that measured by Super Kamiokande, since the latter also detects a proportion of the muon and tau neutrinos as well as the electron neutrinos. Some solar electron neutrinos must therefore have oscillated to the other types during their journey from the centre of the Sun. SNO then made measurements of the total solar neutrino flux using the neutral current mode (i.e. with salt added to the heavy water), and in April 2002 finally demonstrated that the total neutrino flux is as predicted by the standard solar model. SNO is continuing to operate in order to refine its measurements.

Gallium-based detectors

There are two neutrino detectors based upon gallium. The SAGE[28] detector uses 60 tonnes of liquid metallic gallium buried beneath the Caucasus Mountains, and detects neutrinos via the reaction

$$^{71}_{31}\text{Ga} + \nu_e \longrightarrow {}^{71}_{32}\text{Ge} + e^-. \qquad (1.5.10)$$

The germanium product of the reaction is radioactive, with a half-life of 11.4 days. It is separated out by chemical processes on lines analogous to those of the chlorine-based detector. The detection threshold is only 0.236 MeV so that the p–p neutrinos (figure 1.5.3) are detected directly. Since gallium has a market price of some $500 000 per tonne the existence of the SAGE detector is under a continual threat arising from the Russian budgetary

---

[28] Soviet–American Gallium Experiment.

problems. The other gallium-based detector, GNO,[29] is based upon the same detection mechanism as SAGE, but uses 30 tonnes of gallium in the form of gallium trichloride (GaCl$_3$). It replaces the Gallex experiment and is located within the Gran Sasso Tunnel under the Apennines, 150 km east of Rome. Both detectors have measured a solar neutrino flux around 50% of that predicted by the standard solar model and are continuing to operate in order to cover a complete solar sunspot cycle.

Scintillator-based detectors

High-energy photons and particles resulting from neutrino interactions may be observed using scintillation detectors (section 1.3). The Soudan II detector, located 700 m below the surface in an old iron mine at Soudan in Minnesota, uses 1000 tonnes of alternating sheets of steel and a plastic scintillator. It was originally built to try and detect the decay of protons, but can also detect the muons resulting from muon neutrino interactions within the steel plates. It is currently used to detect neutrinos arising from cosmic ray interactions. Also in the Soudan mine, a new detector, MINOS,[30] operating on the same lines as Soudan II is currently being built. It will contain 5400 tonnes of steel plates and scintillator sheets. Together with an almost identical detector at Fermilab, it is intended to refine the measurements of the parameters of neutrino oscillations by observing a beam of muon neutrinos produced at Fermilab in Illinois, some 730 km away.

Also measuring the parameters of neutrino oscillations is the KamLAND experiment in Japan. This uses two nested plastic spheres, 13 and 18 m in diameter, surrounded by 2000 photomultipliers. The spheres are filled with isoparaffin mineral oil and the inner one is also doped with a liquid scintillator. KamLAND detects anti-electron neutrinos from nuclear reactors some 175 km away, and has recently shown that a fourth type of neutrino (see background section above) almost certainly does not exist.

Borexino is an experiment currently under construction aimed at detecting the solar neutrino emission at 0.863 MeV arising from the decay of $^7_4$Be to $^7_3$Li. It will detect all types of neutrino via their scattering products within 300 tonnes of liquid scintillator observed by 2200 photomultipliers and will have a detection threshold of 250 keV. The active scintillator will be surrounded by 1000 tonnes of a liquid buffer, and that in turn by 2400 tonnes of water. 400 photomultipliers pointing outwards into the water buffer will provide an active shield. It is located in the Grand Sasso tunnel in Italy.

---

[29] Gallium Neutrino Observatory.
[30] Main Injector Neutrino Oscillation Search.

Other types of detector

Detectors based upon $^{37}_{17}Cl$ etc. can only detect the high-energy neutrinos, while the bulk of the solar neutrinos are of comparatively low energies (figure 1.5.3). Furthermore, no indication of the direction of the neutrinos is possible. Water- and gallium-based detectors also look at the high-energy neutrinos. Numerous alternative detectors that are designed to overcome such restrictions are therefore proposed or imagined at the time of writing. A brief survey of the range of possibilities is given below. We may roughly divide the proposals into two types, direct interaction and geological, based upon the nature of their interactions.

Direct-interaction-type detectors

These are mostly variants of the Davis and the gallium experiments with another element used in place of the chlorine or gallium. LENS[31] is a proposal to undertake real-time neutrino spectroscopy using $^{176}_{70}Yb$. Interaction with a neutrino converts it to $^{176}_{71}Lu$ with the production of a high-energy electron and, after a 50 ns delay, two gamma rays at 72 and 144 keV respectively. So there is a clear signature to a neutrino detection enabling it to be distinguished from background reactions. The electron's energy depends upon that of the neutrino enabling the latter to be determined. The ytterbium would be dissolved in a liquid scintillator, and about 10 tonnes would be needed to give a detection every other day. It would be located in the Grand Sasso tunnel in Italy. Indium may also be used in a similar manner. The reaction in this case is

$$^{115}_{49}In + \nu_e \longrightarrow {}^{115}_{50}Sn + e^-. \tag{1.5.11}$$

The tin nucleus is in an excited state after the interaction and emits two gamma rays with energies of 116 and 498 keV, 3 µs, on average, after its formation. Like ytterbium, neutrino detection therefore has a characteristic signature of an electron emission followed after a brief delay by two distinctive gamma rays. The electron's energy and direction are related to those of the neutrino so allowing these to be found. The threshold is again low enough to capture some of the proton–proton neutrinos. About 10 tonnes of indium might be needed to detect one neutrino per day.

Another possibility uses lithium and is based upon the reaction

$$^{7}_{3}Li + \nu_e \longrightarrow {}^{7}_{4}Be + e^-. \tag{1.5.12}$$

It has the disadvantage of a high threshold (0.862 MeV), but the advantage of comparative cheapness, only about 15 tonnes being needed for one

---

[31] Low Energy Neutrino Spectroscope.

detection per day. There is also the possibility of the detection of neutrinos via the reaction

$$^{37}_{19}K + \nu_e \longrightarrow {}^{37}_{18}Ar + e^+ \tag{1.5.13}$$

and a detector using 6 tonnes of potassium hydroxide is currently operating alongside the chlorine detector in the Homestake mine.

Other direct interaction detectors which are based upon different principles have recently been proposed: for example, the detection of the change in the nuclear spin upon neutrino absorption by $^{115}_{49}In$ and its conversion to $^{115}_{50}Sn$. Up to 20 solar neutrinos per day might be found by a detector based upon 10 tonnes of superfluid helium held at a temperature below 0.1 K. Neutrinos would deposit their energy into the helium leading to the evaporation of helium atoms from the surface of the liquid. Those evaporated atoms would then be detected from the energy (heat) that they deposit into a thin layer of silicon placed above the helium.

There is also a suggestion to detect neutrinos from the Čerenkov radio radiation produced by their interaction products within natural salt domes in a similar manner to observing the Čerenkov radio emissions within ice (see above). Alternatively electrons may be stored within a superconducting ring, and coherent scattering events detected by the change in the current. With this latter system detection rates for solar neutrinos of one per day may be achievable with volumes for the detector as small as a few millilitres.

Geological detectors

This class of detectors offers the intriguing possibility of looking at the neutrino flux over the past few thousand years. They rely upon determining the isotope ratios for possible neutrino interactions in elements in natural deposits. This is a function of the half-life of the product and the neutrino flux in the recent past, assuming that a state of equilibrium has been reached. Possible candidate elements are listed in table 1.5.1.

Table 1.5.1.

Element	Reaction	Product half-life (years)	Threshold energy (MeV)
Thallium	$^{205}_{81}Tl + \nu_e \longrightarrow {}^{205}_{82}Pb + e^-$	3×10^7	0.046
Molybdenum	$^{98}_{42}Mo + \nu_e \longrightarrow {}^{97}_{43}Tc + n + e^-$	2.6×10^6	8.96
Bromine	$^{81}_{35}Br + \nu_e \longrightarrow {}^{81}_{36}Kr + e^-$	2.1×10^5	0.490
Potassium	$^{41}_{19}K + \nu_e \longrightarrow {}^{41}_{20}Ca + e^-$	8×10^4	2.36

Currently operating and planned neutrino detectors

Table 1.5.2. The status in 2002 of neutrino detectors that are intended wholly or partly for the detection of astronomically produced neutrinos.

Detector	Site	Operating principle	Status (2002)
AMANDA2	Antarctic	Antarctic ice/Čerenkov radiation	In operation
ANTARES	Mediterranean	Sea water/Čerenkov radiation	Under construction
Borexino	Apennines, Italy	Liquid scintillator	Nearing completion
DUMAND	Hawaii	Sea water/Čerenkov radiation	Abandoned 1996
GALLEX	Apennines, Italy	$^{71}_{31}$Ga	Replaced by GNO
GNO	Apennines, Italy	$^{71}_{31}$Ga	In operation
Homestake	South Dakota, USA	$^{37}_{17}$Cl	In operation
ICECUBE	Antarctic	Antarctic ice/Čerenkov radiation	Planned upgrade of AMANDA2
IMB	Ohio, USA	Water/Čerenkov radiation	Closed 1992
Kamiokande	Japan	Water/Čerenkov radiation	Replaced by Super Kamiokande
KamLAND	Japan	Oil/liquid scintillator	In operation
LENS		$^{176}_{70}$Yb/liquid scintillator	Proposal
MINOS	North Minnesota, USA	Steel/plastic scintillator	Under construction
NESTOR	Mediterranean	Sea water/Čerenkov radiation	Under construction
No name		Subterranean salt/radio Čerenkov radiation	Suggestion
NT-200	Lake Baikal	Sea water/Čerenkov radiation	In operation
RAND	Antarctic	Antarctic ice/radio Čerenkov radiation	Trial conducted/ proposal
SAGE	Caucasus	$^{71}_{31}$Ga	In operation
SNO	Ontario, Canada	Heavy water/Čerenkov radiation	In operation
Soudan II	North Minnesota, USA	Steel/plastic scintillator	In operation
Super Kamiokande	Japan	Water/Čerenkov radiation	In operation

Exercises

1.5.1   Show that if an element and its radioactive reaction product are in an equilibrium state with a steady flux of neutrinos, then the number of decays per second of the reaction product is given by

$$N_p T_{1/2}^{-1} \sum_{n=1}^{\infty} \frac{1}{n} \left(\frac{1}{2} \right)^n$$

when $T_{1/2} \gg 1$ s. N_p is the number density of the reaction product and $T_{1/2}$ is its half-life. Hence show that the equilibrium ratio of product to original element is given by

$$\frac{N_p}{N_e} = \sigma_\nu F_\nu T_{1/2} \left\{ \left[\sum_{n=1}^{\infty} \frac{1}{n} \left(\frac{1}{2} \right)^n \right]^{-1} \right\}$$

where N_e is the number density of the original element, σ_ν is the neutrino capture cross section for the reaction and F_ν is the neutrino flux.

1.5.2   Calculate the equilibrium ratio of $^{41}_{20}\mathrm{Ca}$ to $^{41}_{19}\mathrm{K}$ for $^{8}_{5}\mathrm{B}$ solar neutrinos ($T_{1/2} = 80\,000$ years, $\sigma_\nu = 1.45 \times 10^{-46}\,\mathrm{m}^{-2}$, $F_\nu = 3 \times 10^{10}\,\mathrm{m}^{-2}\,\mathrm{s}^{-1}$).

1.5.3   If Davis' chlorine-37 neutrino detector were allowed to reach equilibrium, what would be the total number of $^{37}_{18}\mathrm{Ar}$ atoms to be expected? (Hint: see problem 1.5.1, and note that $\sum_{n=1}^{\infty} (1/n)(\frac{1}{2})^n = 0.693$).

1.6   Gravitational radiation

Introduction

When the first edition of this book was written 20 years ago, this section began 'This section differs from the previous ones because none of the techniques that are discussed have yet indisputably detected gravitational radiation'. That is still the case, though it seems possible that if there is a fifth edition, the statement may then need to be changed. There has been a recent claim of detection of gravity waves coming from somewhere in the Milky Way by the bar-type (see below) detectors EXPLORER and NAUTILUS based in Geneva and Rome. However, the claim is disputed, and it will be at least another year before sufficient data have been accumulated to confirm or refute the result.

The basic concept of a gravity wave[32] is simple; if an object with mass changes its position then, in general, its gravitational effect upon another object will change, and the information on that changing gravitational

---

[32] The term 'gravity wave' is also used to describe oscillations in the Earth's atmosphere arising from quite different processes. There is not usually much risk of confusion.

field propagates outwards through the space–time continuum at the speed of light. The propagation of the changing field obeys equations analogous to those for electromagnetic radiation provided that a suitable adaptation is made for the lack of anything in gravitation that is equivalent to positive and negative electric charges. Hence we speak of gravitational radiation and gravitational waves. Their frequencies for astronomical sources are anticipated to run from a few kilohertz for collapsing or exploding objects to a few microhertz for binary star systems. A pair of binary stars coalescing into a single object will emit a burst of waves whose frequency rises rapidly with time: a characteristic 'chirp' if we could hear it.

Theoretical difficulties with gravitational radiation arise from the multitudinous metric, curved space–time theories of gravity that are currently extant. The best known of these are due to Einstein (general relativity), Brans and Dicke (scalar–tensor theory) and Hoyle and Narlikar (C-field theory), but there are many others. Einstein's theory forbids dipole radiation, but this is allowed by most of the other theories. Quadrupole radiation is thus the first allowed mode of gravitational radiation in general relativity, and will be about two orders of magnitude weaker than the dipole radiation from a binary system that is predicted by the other theories. Furthermore, there are only two polarization states for the radiation as predicted by general relativity, compared with six for most of the other theories. The possibility of decisive tests for general relativity is thus a strong motive for aspiring gravity wave observers, in addition to the information which may be contained in the waves on such matters as collapsing and colliding stars, supernovae, close binaries, pulsars, early stages of the big bang etc.

The detection problem for gravity waves is best appreciated with the aid of a few order-of-magnitude calculations on their expected intensities. The quadrupole gravitational radiation from a binary system of moderate eccentricity ($e \lesssim 0.5$) is given in general relativity by

$$L_G \approx \frac{(2 \times 10^{-63}) M_1^2 M_2^2 (1 + 30e^3)}{(M_1 + M_2)^{2/3} P^{10/3}} \quad \text{W} \tag{1.6.1}$$

where M_1 and M_2 are the masses of the components of the binary system, e is the orbital eccentricity and P is the orbital period. Thus for a typical dwarf nova system with

$$M_1 = M_2 = 1.5 \times 10^{30} \, \text{kg} \tag{1.6.2}$$

$$e = 0 \tag{1.6.3}$$

$$P = 10^4 \, \text{s} \tag{1.6.4}$$

we have

$$L_G = 2 \times 10^{24} \, \text{W}. \tag{1.6.5}$$

But for an equally typical distance of 250 pc, the flux at the Earth is only

$$F_G = 3 \times 10^{-15} \, \text{W m}^{-2}. \tag{1.6.6}$$

This energy will be radiated predominantly at twice the fundamental frequency of the binary with higher harmonics becoming important as the orbital eccentricity increases. Even for a nearby close binary such as ι Boo (distance 23 pc, $M_1 = 2.7 \times 10^{30}$ kg, $M_2 = 1.4 \times 10^{30}$ kg, $e = 0$, $P = 2.3 \times 10^4$ s), the flux only rises to

$$F_G = 5 \times 10^{-14} \, \text{W m}^{-2}. \tag{1.6.7}$$

Rotating elliptical objects radiate quadrupole radiation with an intensity given approximately by

$$L_G \approx \frac{GM^2 \omega^6 r^4 (A+1)^6 (A-1)^2}{64c^5} \quad \text{W} \tag{1.6.8}$$

where M is the mass, ω the angular velocity, r the minor axis radius and A the ratio of the major and minor axis radii, so for a pulsar with

$$\omega = 100 \, \text{rad s}^{-1} \tag{1.6.9}$$

$$r = 15 \, \text{km} \tag{1.6.10}$$

$$A = 0.99998 \tag{1.6.11}$$

we obtain

$$L_G = 1.5 \times 10^{26} \, \text{W} \tag{1.6.12}$$

which for a distance of 1000 pc leads to an estimate of the flux at the Earth of

$$F_G = 10^{-14} \, \text{W m}^{-2}. \tag{1.6.13}$$

Objects collapsing to form black holes within the galaxy or nearby globular clusters or coalescing binary systems may perhaps produce transient fluxes up to three orders of magnitude higher than these continuous fluxes.

Now these fluxes are relatively large compared with, say, those of interest to radio astronomers, whose faintest sources may have an intensity of $10^{-29} \, \text{W m}^{-2} \, \text{Hz}^{-1}$. But the gravitational detectors which have been built to date and those planned for the future have all relied upon detecting the strain ($\delta x / x$) produced in a test object by the tides of the gravitational wave rather than the absolute gravitational wave flux, and this in practice means that measurements of changes in the length of the test object of only 5×10^{-21} m, or about 10^{-12} times the diameter of the hydrogen atom, must be obtained even for the detection of the radiation from ι Boo.

In spite of the difficulties that the detection of such small changes must obviously pose, a detector was built by Joseph Weber that appeared to have detected gravity waves from the centre of the galaxy early in 1969. Unfortunately other workers did not confirm this, and the source of Weber's pulses remains a mystery, although they are not now generally attributed to gravity waves. The pressure to confirm Weber's results, however, has led to a wide variety of gravity wave detectors being proposed and built, and their rate of development is such that an unequivocal detection should occur within the next decade. The detectors so far built, under construction or proposed are of two types: direct detectors in which there is an attempt to detect the radiation experimentally, and indirect detectors in which the existence of the radiation is inferred from the changes that it incidentally produces in other observable properties of the object. The former group may be further subdivided into resonant, or narrow bandwidth detectors, and non-resonant or wideband detectors.

Detectors

Direct resonant detectors

The vast majority of the working gravity wave telescopes fall into this category. They are similar to, or improvements on, Weber's original system. This used a massive (>1 tonne) aluminium cylinder which was isolated by all possible means from any external disturbance, and whose shape was monitored by piezo-electric crystals attached around its 'equator'. Two such cylinders separated by up to 1000 km were utilized, and only coincident events were regarded as significant in order to eliminate any remaining external interference. With this system, Weber could detect a strain of 10^{-16} (cf. 5×10^{-21} for the detection of ι Boo). Since his cylinders had a natural vibration frequency of 1.6 kHz, this was also the frequency of the gravitational radiation that they would detect most efficiently. Weber detected about three pulses per day at this frequency with an apparent sidereal correlation and a direction corresponding to the galactic centre or a point 180° away from the centre. If originating in the galactic centre, some 500 solar masses would have to be totally converted into gravitational radiation each year in order to provide the energy for the pulses. Unfortunately (or perhaps fortunately from the point of view of the continuing existence of the galaxy) the results were not confirmed even by workers using identical equipment to Weber's, and it is now generally agreed that there have been no definite detections of gravitational waves to date.

   Adaptations of Weber's system have been made by Aplin, who used two cylinders with the piezo-electric crystals sandwiched in between them, and by Papini, who used large crystals with very low damping constants so that the vibrations would continue for days following a disturbance. The EXPLORER detector at CERN near Geneva and the NAUTILUS detector near Rome use 2300 kg bars of aluminium cooled to 2 K and 0.1 K, respectively. They have resonant frequencies around 900 Hz. Other systems based upon niobium bars have also been used. Ultimately bar detectors

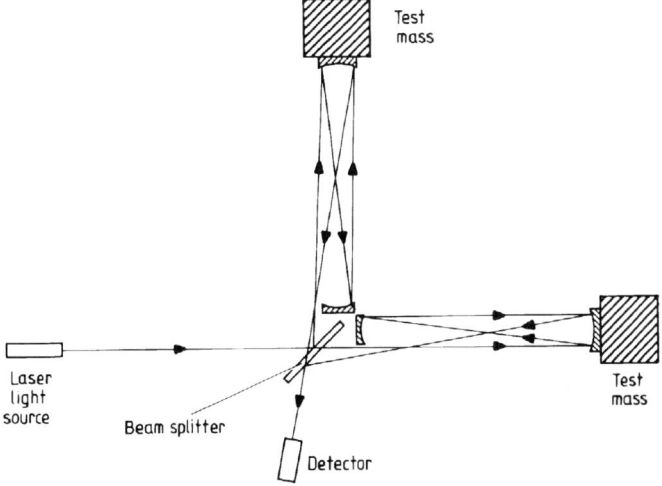

Figure 1.6.1. A possible layout for an interferometric gravity wave detector.

might achieve strain sensitivities of 10^{-21} or 10^{-22}. At Leiden, a detector based upon a spherical mass of copper–aluminium alloy, known as MiniGRAIL, is expected to reach a strain sensitivity of 4×10^{-21} when operated at 0.02 K, and will have the advantage of being equally sensitive to gravity waves from any direction.

Direct, non-resonant detectors

These are of two types and have the potential advantage of being capable of detecting gravity waves with a wide range of frequencies. The first uses a Michelson type interferometer (section 2.5) to detect the relative changes in the positions of two or more test masses. A possible layout for the system is shown in figure 1.6.1. Usually, however, the path length will be amplified by up to a factor of 300 by multiple reflections from the mirrors though these are not shown in the figure in order to preserve simplicity. The mirror surfaces have to be flat to 0.5% of the operating wavelength. The light sources are high-stability neodymium–yttrium–garnet lasers with outputs of 100 W or more. The mirrors are mounted on test masses on pendulum suspensions, and the whole system, including the light paths, operates in a vacuum. Path length changes are detected by looking at the interference pattern between the two orthogonal beams. Ultimate strain sensitivities of 10^{-22} over a bandwidth of 1 kHz are predicted for these systems, enabling detections of collapsing neutron stars, supernovae, and coalescing binaries to be made out to distances of 10^7 to 10^8 pc. Terrestrial detectors are limited to frequencies above about 10 Hz because of noise. A proposed space-based system, LISA, that will be able to detect much lower frequencies is discussed below.

Two Michelson-interferometer-type gravity wave detectors, known as LIGO (Laser Interferometer Gravitational-wave Observatory), have recently been completed in the USA. Their interferometer arms are 4 km long and they are sited 3000 km apart in Washington State, and in Louisiana, so that gravity waves may be distinguished from other disturbances through coincidence techniques. The Washington machine has a second inter-ferometer with 2 km arms that will also help to separate out gravity wave disturbances from other effects. This is because the gravity wave should induce changes in the 2 km machine that have half the amplitude of those in the 4 km machine, whereas other disturbances are likely to have compar-able effects on the two machines. It is marginal whether LIGO will initially be sensitive enough to detect gravity waves. An upgrade planned for 2006, in particular replacing the existing glass mirrors with 40 kg sapphire mirrors, however, should improve its sensitivity by an order of magnitude.

Also currently starting to operate or nearing completion are the Italian–French VIRGO with 3 km arms and sited near Pisa, the German–British GEO600 at Hanover with 600 m arms and the Japanese TAMA near Tokyo that has arms 300 m long. The GEO600 machine uses a technique called signal recycling to improve its sensitivity. This is based upon the result that the gravity wave side-bands do not lead to destructive interference of the laser beam. The emerging light can therefore be reflected back into the system, which then forms a resonant cavity and enhances the signal size.

The distribution of these detectors over the Earth will not only provide confirmation of detections, but also enable the arrival directions of the waves to be pinpointed to a few minutes of arc.

There is also a proposal for a space-based interferometer system called LISA (Laser Interferometer Space Antenna). This would employ three drag-free spacecraft (figure 1.6.3) at the corners of an equilateral triangle with 5 000 000 km sides. Each spacecraft would be able to operate

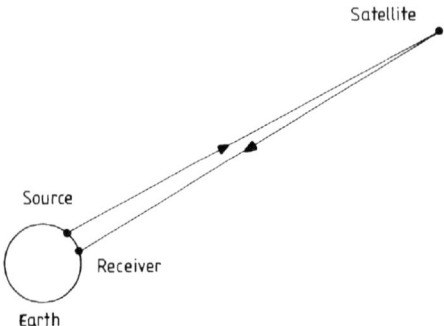

Figure 1.6.2. Arrangement for a satellite-based Doppler tracking gravity wave detector.

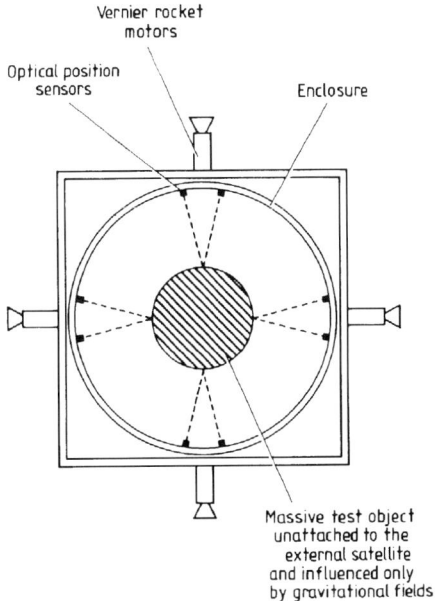

Figure 1.6.3. Schematic cross section through a drag-free satellite. The position of the massive test object is sensed optically, and the external satellite driven so that it keeps the test object centred.

as the vertex of the interferometer, and as the proof mass for the other spacecraft. LISA would thus have three separate interferometer systems. It would orbit at 1 AU from the Sun, but some 20° behind the Earth. Its sensitivity would probably be comparable with the best of the terrestrial gravity wave detectors, but its low noise would enable it to detect the important low frequency waves from binary stars etc. LISA is currently planned as a joint NAS-ESA mission for 2011.

An ingenious idea underlies the second method in this class. An artificial satellite and the Earth will be independently influenced by gravity waves whose wavelength is less than the physical separation of the two objects. If an accurate frequency source on the Earth is broadcast to the satellite, and then returned to the Earth (figure 1.6.2), its frequency will be changed by the Doppler shift as the Earth or satellite is moved by a gravity wave. Each gravity wave pulse will be observed three times enabling the reliability of the detection to be improved. The three detections will correspond to the effect of the wave on the transmitter, the satellite and the receiver, although not necessarily in that order. A drag-free satellite (figure 1.6.3) would be required in order to reduce external perturbations arising from the solar wind, radiation pressure etc. A lengthy series of X-band observations using the Ulysses spacecraft aimed at detecting gravity waves by this method have been made recently, but so far without any successful detections.

Indirect detectors

Proposals for these so far only involve binary star systems. The principle of the method is to attempt to observe the effect on the period of the binary of the loss of angular momentum due to the gravitational radiation. A loss of angular momentum by some process is required to explain the present separations of dwarf novae and other close binary systems with evolved components. These are so small that the white dwarf would have completely absorbed its companion during its earlier giant stage, thus the separation at that time must have been greater than it is now. Gravitational radiation can provide an adequate orbital angular momentum loss to explain the observations, but it is not the only possible mechanism. Stellar winds, turbulent mass transfer and tides may also operate to reduce the separation, so that these systems do not provide unequivocal evidence for gravitational radiation.

The prime candidate for study for evidence of gravitational radiation is the binary pulsar (PSR 1913 + 16). Its orbital period is 2.8×10^4 s, and it was found to be decreasing at a rate of 10^{-4} s per year soon after the system's discovery. This was initially attributed to gravitational radiation, and the 1993 Nobel prize was awarded to Russell Hulse and Joseph Taylor, for their discovery of the pulsar and their interpretation of its orbital decay. However, it now appears that a helium star may be a part of the system, and so tides or other interactions could again explain the observations. On the other hand, the effect of a rapid pulsar on the local interstellar medium might mimic the appearance of a helium star when viewed from a distance of 5000 pc. Thus the detection of gravity waves from the binary pulsar, and indeed by all other methods, remains 'not proven' at the time of writing.

Exercises

1.6.1 Show that the gravitational radiation of a planet orbiting the Sun is approximately given by

$$L_G \approx (7 \times 10^{21}) M^2 P^{-10/3} \text{ W}$$

where M is the planet's mass in units of the solar mass and P the planet's orbital period in days.

1.6.2 Calculate the gravitational radiation for each planet in the solar system and show that every planet radiates more energy than the combined total radiated by those planets whose individual gravitational luminosities are lower than its own.

1.6.3 Find the distance at which a typical dwarf nova system might be detectable by a Michelson interferometer type of direct detector, assuming that the hoped-for sensitivity of such detectors can be achieved.

Chapter 2

Imaging

2.1 The inverse problem

A problem that occurs throughout much of astronomy and other remote sensing applications is how best to interpret noisy data so that the resulting deduced quantities are real and not artefacts of the noise. This problem is termed the inverse problem.

For example, stellar magnetic fields may be deduced from the polarization of the wings of spectrum lines (section 5.2). The noise (errors, uncertainties) in the observations, however, will mean that a range of field strengths and orientations will fit the data equally well. The three-dimensional distribution of stars in a globular cluster must be found from observations of its two-dimensional projection on to the plane of the sky. Errors in the measurements will lead to a variety of distributions being equally good fits to the data. Similarly the images from radio and other inter-ferometers (section 2.5) contain spurious features due to the side lobes of the beams. These spurious features may be removed from the image if the effects of the side lobes are known. But since there will be uncertainty in both the data and the measurements of the side lobes, there can remain the possibility that features in the final image are artefacts of the noise, or are incompletely removed side lobe effects etc.

The latter illustration is an instance of the general problem of instru-mental degradation of data. Such degradation occurs from all measurements, since no instrument is perfect: even a faultlessly constructed telescope will spread the image of a true point source into the Airy diffraction pattern (figure 1.1.29). If the effect of the instrument and other sources of blurring on a point source or its equivalent is known (the point spread function, PSF, or instrumental profile) then an attempt may be made to remove its effect from the data. The process of removing instrumental effects from data can be necessary for any type of measurement, but is perhaps best studied in relation to imaging, when the process is generally known as deconvolution.

Deconvolution

This form of the inverse problem is known as deconvolution, because the true image *convolves* with the PSF to give the observed (or dirty) image. Inversion of its effect is thus deconvolution. Convolution is most easily illustrated in the one-dimensional case (such as the image of a spectrum or the output from a Mills Cross radio array), but applies equally well to two-dimensional images.

A one-dimensional image, such as a spectrum, may be completely represented by a graph of its intensity against the distance along the image (figure 2.1.1). The PSF may similarly be plotted and may be found, in the case of a spectrum, by observing the effect of the spectroscope on a monochromatic source.

If we regard the true spectrum as a collection of adjoining monochromatic intensities, then the effect of the spectroscope will be to broaden each monochromatic intensity into the PSF. At a given point (wavelength) in the observed spectrum, therefore, some of the original energy will have been displaced out to nearby wavelengths, while energy will have been added from the spreading out of nearby wavelengths (figure 2.1.2). The process may be expressed mathematically by the convolution integral

$$O(\lambda_1) = \int_0^\infty T(\lambda_2)I(\lambda_1 - \lambda_2)\,d\lambda_2 \qquad (2.1.1)$$

where $O(\lambda_1)$ is the intensity in the observed spectrum at wavelength λ_1, $T(\lambda_2)$ is the intensity in the true spectrum at wavelength λ_2 and $I(\lambda_1 - \lambda_2)$ is the response of the instrument (spectroscope) at a distance $(\lambda_1 - \lambda_2)$ from its centre.

Equation (2.1.1) is normally abbreviated to

$$O = I * I \qquad (2.1.2)$$

where $*$ is the convolution symbol.

The inversion of equation (2.1.1) to give the true spectrum cannot be accomplished directly but involves the use of Fourier transforms.

Figure 2.1.1. Representation of a one-dimensional image and the instrumental profile (PSF) plotted as intensity versus distance along image.

Figure 2.1.2. Convolution of the true spectrum with the PSF to produce the observed spectrum.

With the Fourier transform and its inverse in the form

$$F(s) = \mathscr{F}(f(x)) = \int_{-\infty}^{\infty} f(x)\,e^{-2\pi i x s}\,dx \qquad (2.1.3)$$

$$f(x) = \mathscr{F}^{-1}(F(s)) = \int_{-\infty}^{\infty} F(s)\,e^{2\pi i x s}\,ds \qquad (2.1.4)$$

then the convolution theorem states that:

Convolution of two functions corresponds to the *multiplication* of their Fourier transforms.

Thus taking Fourier transforms of equation (2.1.2) we have

$$\mathscr{F}(O) = \mathscr{F}(T * I) \qquad (2.1.5)$$

$$= \mathscr{F}(T) \times \mathscr{F}(I) \qquad (2.1.6)$$

and so the true spectrum (etc.) may be found from inverting equation (2.1.6) and taking its inverse Fourier transform:

$$T = \mathscr{F}^{-1}\left[\frac{\mathscr{F}(O)}{\mathscr{F}(I)}\right]. \qquad (2.1.7)$$

In practice, obtaining the true data (or source function) via equation (2.1.7) is complicated by two factors. First, data is sampled at discrete intervals and so is not the continuous function required by equations (2.1.3) and (2.1.4), and also it is not available over the complete range from $-\infty$ to $+\infty$. Second, the presence of noise will produce ambiguities in the calculated values of T.

The first problem may be overcome by using the discrete versions of the Fourier transform and inverse transform:

$$F_D(s_n) = \mathscr{F}_D(f(x))_n = \sum_{k=0}^{N-1} f(x_k) e^{-2\pi i k n/N} \Delta \qquad (2.1.8)$$

$$f(x_k) = \mathscr{F}_D^{-1}(F_D(s_n)) = \sum_{n=0}^{N-1} F_D(s_n) e^{2\pi i k n/N} \frac{1}{N} \qquad (2.1.9)$$

where $F_D(s_n)$ is the nth value of the discrete Fourier transform of $f(x)$, N is the total number of measurements, $f(x_k)$ is the kth measurement and Δ is the step length between measurements, and setting the functions to zero outside the measured range. Now a function that has a maximum frequency of f is completely determined by sampling at $2f$ (the sampling theorem). Thus, the use of the discrete Fourier transform involves no loss of information providing the sampling frequency $(1/\Delta)$ is twice the highest frequency in the source function. The highest frequency that can be determined for a given sampling interval $(1/2\Delta)$ is known as the Nyquist or critical frequency. If the source function contains frequencies higher than the Nyquist frequency, then these will not be determined by the measurements and the finer detail in the source function will be lost when it is reconstituted via equation (2.1.7). Rather more seriously, however, the higher frequency components may beat with the measuring frequency to produce spurious components at frequencies lower than the Nyquist frequency. This phenomenon is known as aliasing and can give rise to major problems in finding the true source function.

The actual evaluation of the transforms and inverse transforms may nowadays be relatively easily accomplished using the fast Fourier transform algorithm on even quite small computers. The details of this algorithm are outside the scope of this book, but may be found from books on numerical computing.

The one-dimensional case just considered may be directly extended to two or more dimensions, though the number of calculations involved then increases dramatically. Thus, for example, the two-dimensional Fourier transform equations are

$$F(s_1, s_2) = \mathscr{F}(f(x_1, x_2)) = \int_{-\infty}^{\infty} \int_{-\infty}^{\infty} f(x_1, x_2) e^{-2\pi i x_1 s_1} e^{-2\pi i x_2 s_2} \, dx_1 \, dx_2 \quad (2.1.10)$$

$$f(x_1, x_2) = \mathscr{F}^{-1}(F(s_1, s_2)) = \int_{-\infty}^{\infty} \int_{-\infty}^{\infty} F(s_1, s_2) e^{2\pi i x_1 s_1} e^{2\pi i x_2 s_2} \, ds_1 \, ds_2. \quad (2.1.11)$$

Some reduction in the noise in the data may be achieved by operating on its Fourier transform. In particular, cutting back on or removing the corresponding frequencies in the transform domain may reduce cyclic noise such as 50 or 60 Hz mains hum or the stripes on scanned images. Random noise may be reduced by using the optimal (or Wiener) filter defined by

$$W = \frac{[\mathscr{F}(O)]^2}{[\mathscr{F}(O)]^2 + [\mathscr{F}(N)]^2} \qquad (2.1.12)$$

where $\mathscr{F}(O)$ is the Fourier transform of the observations, without the effect of the random noise, and $\mathscr{F}(N)$ is the Fourier transform of the random noise. (The noise and the noise-free signal are separated by assuming the high-frequency tail of the power spectrum to be just due to noise, and then extrapolating linearly back to the lower frequencies.) Equation (2.1.7) then becomes

$$T = \mathscr{F}^{-1}\left[\frac{\mathscr{F}(O)W}{\mathscr{F}(I)} \right]. \qquad (2.1.13)$$

While processes such as those outlined above may reduce noise in the data, it can never be totally eliminated. The effect of the residual noise, as previously mentioned, is to cause uncertainties in the deduced quantities. The problem is often ill-conditioned, that is to say, the uncertainties in the deduced quantities may, proportionally, be very much greater than those in the original data.

A widely used technique, especially for reducing the instrumental broadening for images obtained via radio telescope, is due to W H Richardson and Leon Lucy. The RL algorithm is an iterative procedure in which the $(n + 1)$th approximation to the true image is related to the PSF and nth approximation by

$$T_{n+1} = T_n \int \frac{O}{T_n * I} I. \qquad (2.1.14)$$

The first approximation to start the iterative process is usually taken as the observed data. The RL algorithm has the advantages compared with some other iterative techniques of producing a normalized approximation to the true data without any negative values, and it also usually converges quite rapidly.

Recently, several methods have been developed to aid choosing the 'best' version of the deduced quantities from the range of possibilities. Termed 'non-classical' methods, they aim to stabilize the problem by introducing additional information not inherently present in the data as constraints, and thus to arrive at a unique solution. The best known of these methods is the maximum entropy method (MEM). The MEM introduces the external constraint that the intensity cannot be negative, and finds the solution that has the least structure in it that is consistent with the data. The name derives from the concept of entropy as the inverse of the structure (or information content) of a system. The maximum entropy solution is thus the one with the least structure (the least information content or the smoothest solution) that is consistent with the data. The commonest measure of the entropy is

$$s = -\sum p_i \ln p_i \qquad (2.1.15)$$

where

$$p_i = \frac{d_i}{\sum_j d_j} \qquad (2.1.16)$$

and d_i is the ith datum value, but other measures can be used. A solution obtained by an MEM has the advantage that any features in it must be real and not artefacts of the noise. However, it also has the disadvantages of perhaps throwing away information, and that the resolution in the solution is variable, being highest at the strongest values.

Other approaches can be used either separately or in conjunction with the MEM to try to improve the solution. The CLEAN method, much used on interferometer images, is discussed in section 2.5. The Method of Regularization stabilizes the solution by minimizing the norm of the second derivative of the solution as a constraint on the smoothness of the solution. The non-negative least-squares (NNLS) approach solves for the data algebraically, but subject to the added constraint that there are no negative elements. Myopic deconvolution attempts to correct the image when the PSF is poorly known by determining the PSF as well as the corrected image from the observed image, while Blind Deconvolution is used when the PSF is unknown and requires several images of the same object, preferably with dissimilar PSFs, such as occurs with adaptive optics images (section 1.1). Burger–Van Cittert deconvolution first convolves the PSF with the observed image and then adds the difference between the convolved and observed images from the observed image. The process is repeated with the new image and iterations continue until the result of convolving the image with the PSF matches the observed image. The $(n + 1)$th approximation is thus

$$T_{n+1} = T_n + (O - T_n * I). \qquad (2.1.17)$$

Compared with the RL approach, the Burger–van Cittert algorithm has the disadvantage of producing spurious negative values.

2.2 Photography

Introduction

When the first edition of this book appeared in 1984, photography was still the main means of imaging in astronomy throughout the visual region and into the very-near infrared (to about 1 μm). CCDs were starting to come on the scene, but were still small and expensive. The situation is now completely reversed. Photography is hardly used at all in professional astronomy—the last photograph on the AAT was taken in 1999 for example. CCDs now dominate imaging from the ultraviolet to the near infrared, though they are still expensive in the largest sizes and are themselves perhaps on the verge of replacement by wavelength-sensitive imaging detectors such as STJs. Among amateur astronomers, small CCDs are now quite common and can be purchased along with sophisticated software for image processing for much less than the cost of a 0.2 m Schmidt–Cassegrain telescope. Digital still and video cameras produced for the popular market can also be used on telescopes, although without the long exposures possible with cooled CCDs specifically manufactured for astronomical use.

However, photography is not quite dead yet! Many, perhaps most, archives are still in the form of photographs and even recent major surveys such as the MAST's[33] Digitized Sky Survey that produced the Hubble Guide Star Catalogue (GST) and the AAO/UKST Hα survey are based upon photographic images. Any web search for 'astronomy images' will also show that photography is still widely used among amateur astronomers (and many professional astronomers are amateur astronomers in their spare time) despite those who use CCDs. Thus some coverage of photography still needs to be included, if only so that a student working on archive photographs is aware of how they were produced and what their limitations and problems may be.

Photography's pre-eminence as a recording device throughout most of the 20th century arose primarily from its ability to provide a permanent record of an observation that is (largely) independent of the observer. But it also has many other major advantages including its cheapness, its ability to accumulate light over a long period of time and so to detect sources fainter than those visible to the eye, its very high information density and its ease of storage for future reference. These undoubted advantages coupled with our familiarity with the technique have, however, tended to blind us to its many disadvantages. The most important of these are its slow speed and very low quantum efficiency. The human eye has an effective exposure time of about a tenth of a second, but a photograph would need an exposure of ten to a hundred times longer in order to show the same detail. The impression that photography is far more sensitive than the eye only arises because most photographs of astronomical objects have exposures ranging from minutes to hours and so show far more detail than is seen directly. The processing of a photograph is a complex procedure and there are many possibilities for the introduction of errors. Furthermore, and by no means finally in this list of problems of the photographic method, the

---

[33] Multi-mission Archive for the Space Telescope.

final image is recorded as a density, and the conversion of this back to intensity is a tedious process. The whole business of photography in fact bears more resemblance to alchemy than to science, especially when it has to be used near the limit of its capabilities.

Colour photography has no place in astronomy, except for taking pretty pictures of the Orion nebula etc. The reason for this is that the colour balance of both reversal (slide) and print colour films is optimized for exposures of a few seconds or less. Long exposures, such as those needed in astronomy, therefore result in false colours for the object. When a colour image is needed, it is normally obtained by three monochromatic (black and white) images obtained through magenta, yellow and cyan filters that are then printed on to a single final image using the same filters.

Structure of the photographic emulsion

Many materials exhibit sensitivity to light that might potentially be usable to produce images. In practice, however, a compound of silver and one or more of the halogens is used almost exclusively. This arises from the relatively high sensitivity of such compounds allied to their ability to retain the latent image for long periods of time. For most purposes the highest sensitivity is realized using silver bromide with a small percentage of iodide ions in solid solution. The silver halogen is in the form of tiny crystals, and these are supported in a solid, transparent medium, the whole structure being called the photographic emulsion. The supporting medium is usually gelatine, which is a complex organic material manufactured from animal collagen, whose molecules may have masses up to half a million amu. Gelatine has the advantage of adding to the stability of the latent image by absorbing the halogens that are released during its formation. It also allows easy penetration of the processing chemicals, and forms the mixing medium during the manufacture of the emulsion

The size of the silver halide crystals, or grains as they are commonly known, is governed by two conflicting requirements. The first of these is resolution, or the ability of the film to reproduce fine detail. Resolution is affected by both the grain size and by scattering of the light within the emulsion, and is measured by the maximum number of lines per millimetre that can be distinguished in images of gratings. It ranges from about 20 to 2000 lines per millimetre, with generally the smaller the grain size the higher the resolution. The modulation transfer function (MTF) specifies more precisely the resolution of an emulsion. The definition of the MTF is in terms of the recording of an image that varies sinusoidally in intensity with distance. In terms of the quantities shown in figure 2.2.1, the modulation transfer, T, is given by

$$T = \frac{(A_R/I_R)}{(A_O/I_O)}. \tag{2.2.1}$$

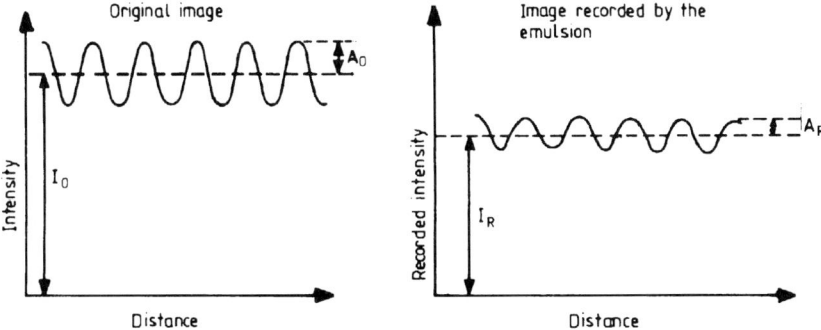

Figure 2.2.1. Schematic representation of the reproduction of an image varying spatially in a sinusoidal manner by an emulsion.

The MTF is then the manner in which T varies with spatial frequency. Schematic examples of the MTF for some emulsions are shown in figure 2.2.2. Using an MTF curve, the resolution may be defined as the spatial frequency at which the recorded spatial variations become imperceptible. For visual observation, this usually occurs for values of T near 0.05.

The second requirement is that of the speed of the emulsion, and this is essentially the reciprocal of a measure of the time taken to reach a given image density under some standard illumination. The two systems in common use for measuring the speed of an emulsion are ISO, based upon the exposure times for image densities of 0.1 and 0.9, and DIN, based only on the exposure time to reach an image density of 0.1. ISO is an arithmetic scale, while DIN is logarithmic. Their relationship is shown in figure 2.2.3, together with an indication of the speeds of normal films. The two numbers are frequently combined into the exposure index number (EI) that is simply the ISO number followed by the DIN number, as in EI $100/21°$. The speed

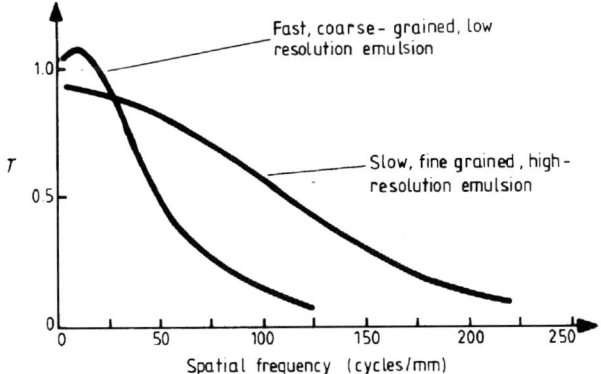

Figure 2.2.2. Schematic MTF curves for typical emulsions.

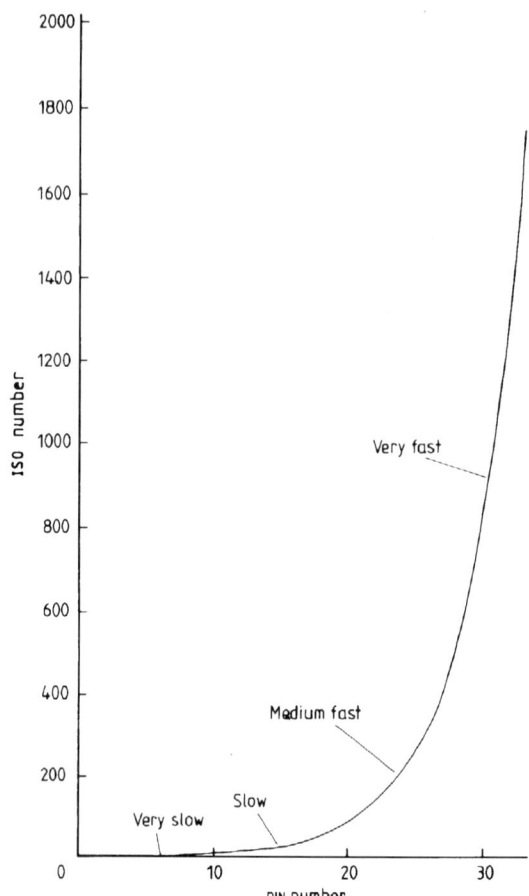

Figure 2.2.3. Emulsion speeds (approximate relationship).

of an emulsion is proportional to the volume of a grain for unsensitized emulsions, and to the surface area of a grain for dye-sensitized (see below) emulsions. Thus in either case, higher speed requires larger grains. The conflict between these two requirements means that high-resolution films are slow and that fast films have poor resolution. Grain sizes range from 50 nm for a very high resolution film, through 800 nm for a normal slow film to 1100 nm for a fast film. Not all the grains are the same size, except in nuclear emulsions (section 1.4), and the standard deviation of the size distribution curve ranges from about 1% of the particle diameter for very high resolution films to 50% or more for normal commercial films.

A normal silver halide emulsion is sensitive only to short wavelengths, i.e. to the blue, violet and near ultraviolet parts of the spectrum. Hermann Vogel made the fundamental discovery that was to render lifelike photography

Figure 2.2.4. Effects of dyes upon the spectral sensitivity of photographic emulsion.

possible and to extend the sensitivity into the red and infrared in 1873. That discovery was that certain dyes when adsorbed on to the surfaces of the silver halide grains would absorb light at their own characteristic frequencies, and then transfer the absorbed energy to the silver halide. The latent image would then be produced in the same manner as normal. The effect of the dye can be very great; a few spectral response curves are shown in figure 2.2.4 for unsensitized and sensitized emulsions. Panchromatic or isochromatic emulsions have an almost uniform sensitivity throughout the visible, while the orthochromatic emulsions mimic the response of the human eye. Colour film would of course be quite impossible without the use of dyes since it uses three emulsions with differing spectral responses. The advantages of dye sensitization are very slightly counteracted by the reduction in the sensitivity of the emulsion in its original spectral region, due to the absorption by the dye, and more seriously by the introduction of some chemical fogging.

In many applications the response of the film can be used to highlight an item of interest, for example comparison of photographs of the same area of sky with blue- and red-sensitive emulsions is a simple way of finding very hot and very cool stars.

The photographic image

When an emulsion is exposed to light, the effect is to form the latent image, i.e. an image that requires the further step of development in order to be visible. Its production on silver halide grains is thought to arise in the following manner.

(a) Absorption of a photon by an electron occurs in the valence band of the silver halide. The electron thereby acquires sufficient energy to move into the conduction band (see section 1.1 for a discussion of solid-state energy levels).

(b) The removal of the electron leaves a positive hole in the valence band.

(c) Both the electron and the hole are mobile. If they meet again, they will recombine and emit a photon. There will then have been no permanent effect. To avoid this recombination, the electron and the hole must be separated by their involvement in alternative reactions. The electrons are immobilized at chemical or physical traps formed from impurities or crystal defects. The positive hole's motion eventually carries it to the surface of the grain. There it can be consumed directly by reaction with the gelatine, or two holes can combine, releasing a halogen atom that will then in turn be absorbed by the gelatine.

(d) The electron, now immobilized, neutralizes a mobile silver ion leaving a silver atom within the crystal structure.

(e) The effect of the original electron trap is now enhanced by the presence of the silver atom, and it may more easily capture further electrons. In this way, a speck of pure silver containing from a few atoms to a few hundred atoms is formed within the silver halide crystal.

(f) Developers are reducing solutions that convert silver halide into silver. They act only very slowly, however, on pure silver halide, but the specks of silver act as catalysts, so that those grains containing silver specks are reduced rapidly to pure silver. Since a 1 μm grain will contain 10^{10} to 10^{11} silver atoms, this represents an amplification of the latent image by a factor of 10^9. Adjacent grains that do not contain an initial silver speck will be unaffected and will continue to react at the normal slow rate.

(g) Thus the latent image consists of those grains that have developable specks of silver on them. For the most sensitive emulsions, three to six silver atoms per grain will suffice, but higher numbers are required for less sensitive emulsions, and specks comprising ten to twelve silver atoms normally represent the minimum necessary for a stable latent image in normal emulsions. The latent image is turned into a visible one, therefore, by developing for a time which is long enough to reduce the grains containing the specks of silver, but which is not long enough to affect those grains without such specks. The occasional reduction of an unexposed grain occurs through other processes and is a component of chemical fog. After the development has stopped, the remaining silver halide grains are removed by a second chemical solution known as the fixer.

The final visible image thus consists of a number of silver grains dispersed in the emulsion. These absorb light, and so the image is darkest at those points where it was most brightly illuminated, i.e. the image is the negative of the original. Astronomers customarily work directly with negatives since the production of a positive image normally requires the formation of a photograph of the negative with all its attendant opportunities for the introduction of distortions and errors. This practice may seem a little strange at first, but the

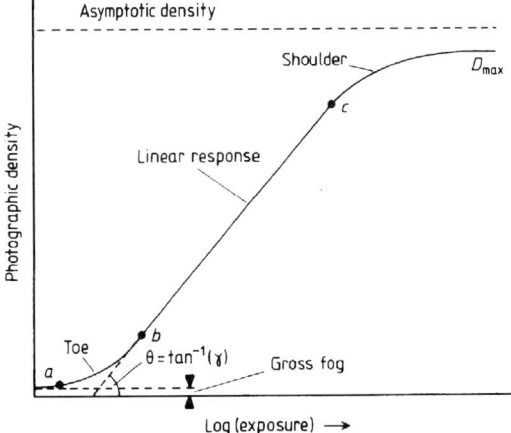

Figure 2.2.5. Schematic characteristic curve of an emulsion.

student will rapidly become so familiar with negatives that the occasional positive, when encountered, requires considerable thought for its interpretation.

One of the problems encountered in using photographic images to determine the original intensities of sources is that the response of the emulsion is nonlinear. The main features of the response curve, or characteristic curve as it is more commonly known, are shown in figure 2.2.5. Photographic density is plotted along the vertical axis. This is more properly called the optical density since it is definable for any partially transparent medium, and it is given by

$$D = \log_{10}\left(\frac{F_i}{F_t}\right) \tag{2.2.2}$$

where F_i is the incident flux and F_t is the transmitted flux (figure 2.2.6). The transmittance, T, of a medium is defined similarly by

$$T = \frac{F_t}{F_i} = 10^{-D} \tag{2.2.3}$$

and the opacity, A, by

$$A = \frac{F_i}{F_t} = T^{-1} = 10^{D}. \tag{2.2.4}$$

The intensity of the exposure is plotted on the horizontal axis as a logarithm. Since density is also a logarithmic function (equation (2.2.2)), the linear portion of the characteristic curve (b to c in figure 2.2.5) does represent a genuine linear response of the emulsion to illumination. The response is nonlinear above and below the linear portion. The gross fog level is the background fog of the unexposed but developed emulsion and of its supporting material. The point marked 'a' on figure 2.2.5 is called the threshold and is

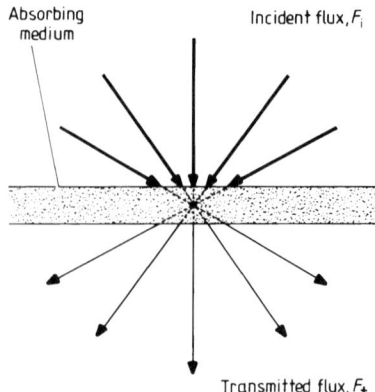

Figure 2.2.6. Quantities used to define photographic density.

the minimum exposure required for a detectable image. The asymptotic density is the ultimate density of which the emulsion is capable. The maximum density achieved for a particular developer, D_{max}, will generally be below this value. In practice the image is ideally formed in the linear portion of the response curve, although this may not always be possible for objects with a very wide range of intensities, or for very faint objects.

The contrast of a film is a measure of its ability to separate regions of differing intensity. It is determined by the steepness of the slope of the linear portion of the characteristic curve. A film with a high contrast has a steep linear section, and a low contrast film a shallow one. Two actual measures of the contrast are used: gamma, and the contrast index. Gamma is simply the slope of the linear portion of the curve:

$$\gamma = \tan\theta. \tag{2.2.5}$$

The contrast index is the slope of that portion of the characteristic curve that is used to form most images. It usually includes some of the toe, but does not extend above a density of 2.0. Contrast and response of emulsions both vary markedly with the type of developer and with the conditions during development. Thus for most astronomical applications a calibration exposure must be obtained that is developed along with the required photograph and which enables the photographic density to be converted back into intensity.

The resolution of an emulsion is affected by the emulsion structure and by the processing stages. In addition to the grain size and light scattering in the emulsion, there may also be scattering or reflection from the supporting material (figure 2.2.7). This is known as halation and it is usually reduced by the addition of an absorbing coating on the back of the film. Processing affects resolution through the change in the concentration of the processing solutions within the emulsion since these are more rapidly consumed in well-exposed regions and almost unused elsewhere. The basic effect of this on the

Figure 2.2.7. Halation.

image may best be demonstrated at a sharp boundary between a uniformly exposed region and an unexposed region. The availability of the developer from the unexposed region at the edge of the exposed region leads to its greater development and so greater density (figure 2.2.8). In astronomy this has important consequences for photometry of small images. If two small regions are given exposures of equal length and equal intensity per unit area, then the smaller image will be darker than the larger one by up to 0.3 in density units. This corresponds to an error in the finally estimated intensities by a factor of two, unless account is taken of the effect. A similar phenomenon occurs in the region between two small close images. The developer becomes exhausted by its work within the two images,

Figure 2.2.8. Edge effect in photographic emulsions.

leaving the region between them underdeveloped. Images of double stars, close spectrum emission lines etc. may therefore appear more widely separated because of this than is the true case.

The quantum efficiency of an emulsion and its speed are two related concepts. We may measure the quantum efficiency by the number of photons required to produce a detectable image compared with an ideal detector (see also section 1.1). Now all detectors have a certain level of background noise, and we may regard a signal as ideally detectable if it exceeds the noise level by one standard deviation of the noise. In a photographic image the noise is due to the granularity of the image and the background fog. For a Poisson variation in granularity, the standard deviation of the noise is given by

$$\sigma = (N\omega)^{1/2} \tag{2.2.6}$$

where σ is the noise equivalent power (section 1.1), N is the number of grains in the exposed region and ω is the equivalent background exposure in terms of photons per grain. (All sources of fogging may be regarded as due to a uniform illumination of a fog-free emulsion. The equivalent background exposure is therefore the exposure required to produce the gross fog level in such an emulsion.) Now for our practical, and non-ideal, detector let the actual number of photons per grain required for a detectable image be Ω. Then the detective quantum efficiency (DQE) (section 1.1) is given by

$$\text{DQE} = \left(\frac{\sigma}{N\Omega}\right)^{2} \tag{2.2.7}$$

or

$$\text{DQE} = \left(\frac{\omega}{N\Omega^2}\right). \tag{2.2.8}$$

A given emulsion may further reduce its quantum efficiency by failing to intercept all the photons that fall on it. Also, between two and twenty photons are required to be absorbed before any given grain reaches developability, as we have already seen. So the quantum efficiency rarely rises above 1% in practice (figure 2.2.9).

The speed of an emulsion and its quantum efficiency are related since the speed is a measure of the exposure required to reach a given density in the image. Aspects of the speed have already been mentioned (figure 2.2.3 etc), and it is inherent in the position of the characteristic curve along the horizontal axis (figure 2.2.5). It can also be affected by processing; the use of more active developers and/or longer developing times usually increases the speed. Hypersensitization can also improve the speed (see below). Other factors that affect the speed include temperature, humidity, the spectral region being observed, the age of the emulsion etc.

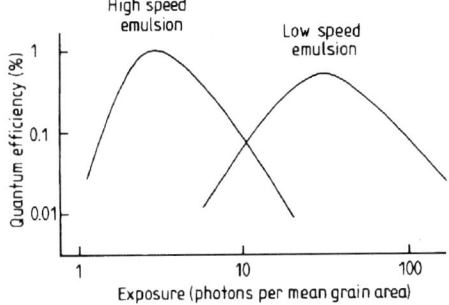

Figure 2.2.9. Quantum efficiency curves for photographic emulsions.

One important factor for astronomical photography that influences the speed very markedly is the length of the exposure. The change in the speed for a given level of total illumination as the exposure length changes is called reciprocity failure. If an emulsion always gave the same density of image for the same number of photons, then the exposure time would be proportional to the reciprocal of the intensity of the illumination. However, for very long and very short exposures this reciprocal relationship breaks down: hence the name reciprocity failure. A typical example of the variation caused by exposure time is shown in figure 2.2.10. Here the total exposure, i.e. intensity × time, IT, that is required to produce a given density in the final image is plotted against the illumination. Exposure times are also indicated. If the reciprocity law held, then the graph would be a horizontal straight line. The effect of reciprocity failure is drastic; some commonly available roll films, for example, may have a speed of a few per cent of their normal ratings for exposures of an hour or more.

The photographic emulsion is a weak and easily damaged material and it must therefore have a support. This is usually either a glass plate or a plastic film. Both are used in astronomy, but plates are preferred for applications requiring high dimensional stability and/or retention for comparative

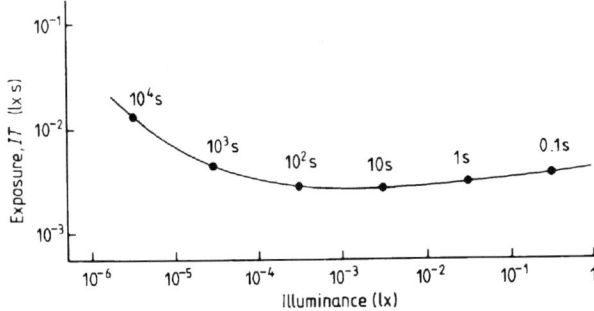

Figure 2.2.10. A typical reciprocity curve for a photographic emulsion.

purposes for long periods of time. In both cases great care must be taken in the storage of the material to avoid scratches, high humidity, mould, breakages, distortion etc. that can all too easily destroy the usefulness of the information on the photograph.

Processing

Processing is the means whereby the latent image is converted into a visible and permanent image. For monochrome work there are two main stages: development and fixing. Although black and white film can still be processed commercially, it is becoming increasingly hard to find laboratories that will undertake the work. Also astronomical emulsions are often hypersensitized and/or non-standard processing is used. It is also very irritating to be told that your magnificent star pictures have not come out because all the processing lab can see is the clear background and they have not noticed the scatter of black dots that are the stars. Thus most amateur astronomers using monochromatic film, and all professional astronomers still using it, undertake their own processing. Brief details of this are given below. For more information earlier editions of this book may be consulted, or specialist books on photography and the film supplier's leaflets used. Colour processing is much more complex and is rarely required for astronomical work. The interested reader is therefore referred to specialist books on the subject and to the manufacturer's literature for further information.

The action of a developer is to reduce to a grain of silver those silver halide grains that contain a latent image. Four factors principally affect the ability of a developer to carry out this function: the chemical activity of the developer, the length of the developing time, the temperature of the solution, and the degree of agitation (mixing) during development. Normally the manufacturer's guidelines should be followed, since these are usually optimized to minimize the background fog. However, the characteristic curve may be altered to some extent by varying the developing time. In particular the speed and contrast may be changed if so desired (figure 2.2.11), though both the fog and the granularity of the image are increased by increasing the developing time, so the technique is not without its drawbacks.

The second main stage of processing is fixing. The purpose of this is to remove the silver halide left in the emulsion after development. Commonly, sodium thiosulphate (known as 'hypo') or ammonium thiosulphate are used for this purpose. Their reaction products are water soluble and they do not attack the photographic materials. If fixation is over-extended, however, the fixer will start to attack the silver image, and so the manufacturer's recommended times should again be used.

In addition to these two active stages of processing, the chemicals must be thoroughly washed from the emulsion at each stage. It is also good practice to halt the development sharply by immersing the plate in a bath

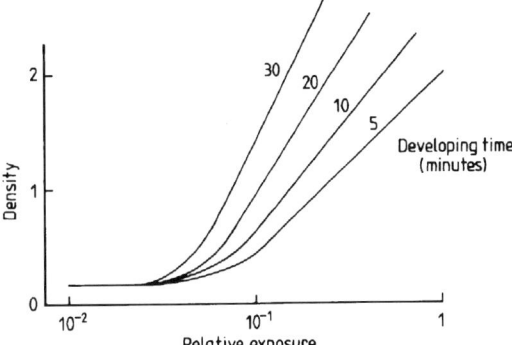

Figure 2.2.11. Effect of changing the developing time upon the characteristic curve of an emulsion.

of acetic acid, or one of the commercially available stop baths. The main stages of processing, in summary, are therefore: development; stop bath; washing; fixation; washing; drying in a dust-free atmosphere.

Most of the processing stages must be carried out in total darkness. Some emulsions do have safelights; that is, they may be illuminated in a wavelength region to which they are insensitive, but this is not possible with all emulsions, and in any case it is recommended to dispense with safelights if possible, since the emulsion usually has some residual sensitivity at all visible wavelengths. Small quantities of developing may be undertaken in commercially available developing tanks, and with makeshift solution preparation and water supplies. For larger quantities of emulsion and regular usage, however, a darkroom is almost essential. This should be a light-tight, dust-free room that is temperature controlled to 20 °C. It also needs adequate water and electricity supplies, drainage and timing facilities. The processing materials and the processing baths may then be arranged conveniently and permanently.

If a positive image is needed, then these are obtained by re-photographing the original negative image. Usually this is on to opaque printing paper, although film can be used when transparent images are required. The negative can simply be placed on top of the printing paper and a diffuse light shone through to produce a contact print. More normally the negative is placed in an enlarger so that the final print is of a convenient size.

A technique that has been much used because of its ability to show full details of images with very wide ranges of contrast is unsharp masking. A contact positive print is made of the original negative on to film, and this is then superimposed on to the negative, but with a small separation. The composite is then printed conventionally with the negative image sharply focused. The positive image is out of focus and acts as a filter to smooth out the density range of the negative.

Hypersensitization

Hypersensitization or hypering includes any process that, when applied to an emulsion, causes it to have a faster response than normal. Its use in astronomical applications is very widespread since even a 10% saving on an exposure of an hour is worthwhile, and the savings in exposure time are frequently very much higher than this. The mechanism that produces the hypersensitization is thought to be the removal of oxygen and water from the emulsion.

Many different approaches to hypersensitization have been tried in the past (for a review of these see earlier editions of this book). Today just heating and gas immersion are employed. The shelf life of hypered plates is often severely reduced, and there is some increase in the fog level.

Heating

The commonest method of hypersensitization in use in astronomy consists simply of baking the emulsion prior to exposure. For blue-sensitive plates, 24 to 72 hours at a temperature of 50 to 75 °C can increase their speed by a factor of two or three. The optimum baking conditions vary between emulsions and even from one batch of a particular emulsion to the next, and so must be determined empirically in each case.

Gas immersion effects

(a) *Nitrogen.* Immersion in an atmosphere of dry nitrogen for several days, or for shorter periods if baking is simultaneously applied, produces speed gains of up to a factor of five with little increase in the fog level.

(b) *Hydrogen* (caution: possible explosion risk). The plates are immersed in an atmosphere of nitrogen mixed with 2–10% hydrogen (known as forming gas) at room temperature for several hours. This is frequently combined with nitrogen baking for some of the commonly used astronomical emulsions, with a resulting gain in sensitivity by about a factor of 25.

(c) *Vacuum.* Lodging the plates in a soft or hard vacuum for up to 24 hours, possibly combined with baking, can give gains of up to a factor of four.

Film types

It is usually unhelpful to include specific commercial products in a book such as this, for they change rapidly. Indeed the first edition of this book discussed 11 different emulsion types, each of which could be wavelength sensitized in eight different ways—today there is just one emulsion to be considered, Kodak's Technical Pan. This is a very fine grain film available in 35 mm and other formats. Its sensitivity extends to 685 nm, making it suitable for solar photography at Hα. Its resolution can reach 300 lines/mm and its speed is variable from ISO 25 to 200 by varying the developing conditions. For long exposures, however, it is essential to hypersensitize it. Many other emulsions are available both from Kodak and other manufacturers as monochrome and colour films. Any of these can be used to obtain images through a telescope (though see the note on colour imaging in the introduction to this section), but their performance will generally be inferior to that of Technical Pan.

Techniques of astronomical photography

The prime difference between astronomical photography and more normal types of photography lies in the exposure length. Because of this, not only must the telescope be driven to follow the motion of the stars, but also it must be continually guided in order to correct any errors in the drive. Usually guiding is by means of a second and smaller telescope attached to the main instrument. The astronomer guides the telescope by keeping cross wires in the eyepiece of the secondary telescope centred on the object, or by the use of automatic systems (section 1.1). For faint and/or diffuse objects, offset guiding may be necessary. That is, the guide telescope is moved from its alignment with the main telescope until it is viewing a nearby bright star while the main telescope continues to point at the required object. Guiding then continues using this bright star. Sometimes the main telescope can act as its own guide by using a guider that inserts a small mirror into the light beam to divert light not producing the main image into the eyepiece.

The determination of the required length of the exposure may be difficult; exposures may be interrupted by cloud, or the star may be a variable whose brightness is not known accurately. Previous experience may provide a guide and records of all exposures, whether successful or not, should always be kept for this reason. Alternatively, several photographs may be taken with differing exposures. The linear part of the characteristic curve is sufficiently long that the exposures can differ by a factor of two or even four if this approach is tried.

Among amateur astronomers, a commonly used technique is to take photographs by eyepiece projection, instead of at prime focus. A camera without its normal lens is positioned a little distance behind the eyepiece, and the focus adjusted to produce a sharp image on the film. This provides much higher plate scales so that even small objects such as the planets can be resolved with a small telescope. A guide to the extra magnification of eyepiece projection over prime focus photography, and to the change in the exposure that may be required, is given by the effective focal ratio

$$\text{EFR} = \frac{fd}{e} \qquad (2.2.9)$$

where f is the focal ratio of the objective, d is the projection distance (i.e. the distance from the optical centre of the eyepiece to the film plane) and e is the focal length of the eyepiece. If T is the time required for an exposure on an extended object at prime focus, then T', the exposure time for the projected image, is given by

$$T' = T\left(\frac{\text{EFR}}{f}\right)^2. \qquad (2.2.10)$$

It is also possible just to point a camera through the eyepiece of a telescope to obtain images. Both the camera and the telescope need to be focused on infinity, and a single-lens reflex (SLR) or a digital camera with a real-time display is essential so that the image can be focused and positioned correctly. This technique is sometimes called magnified imaging and is a very poor second to either direct or eyepiece projection imaging.

The observational limit to the length of exposure times is usually imposed by the background sky brightness. This produces additional fog on the emulsion, and no improvement in the signal-to-noise ratio will occur once the total fog has reached the linear portion of the characteristic curve. Since the brightness of star images is proportional to the square of the objective's diameter, while that of an extended object, including the sky background, is inversely proportional to the square of the focal ratio (section 1.1), it is usually possible to obtain adequate photographs of star fields by using a telescope with a large focal ratio. But for extended objects no such improvement is possible since their brightness scales in the same manner as that of the sky background. Much of the light in the sky background comes from scattered artificial light, so that observatories are normally sited in remote spots well away from built-up areas. Even so, a fast Schmidt camera may have an exposure limit of a few tens of minutes at best. Some further improvement may sometimes then be gained by the use of filters. The scattered artificial light comes largely from sodium and mercury vapour street lighting, so a filter that rejects the most important of the emission lines from these sources can improve the limiting magnitude by two or three stellar magnitudes at badly light-polluted sites. For the study of hot interstellar gas clouds, a narrow band filter centred on the $H\alpha$ line will almost eliminate the sky background while still allowing about half the energy from the source to pass through it.

Analysis of photographic images

Photographs are used by the astronomer to provide two main types of information: the relative positions of objects, and the relative intensities of objects. The analysis for the first of these is straightforward, although care is needed for the highest quality results. The positions of the images are measured by means of a machine that is essentially a highly accurate automated travelling microscope (section 5.1).

To determine the relative intensities of sources from their photographs is more of a problem due to the nonlinear nature of the emulsion's characteristic curve (figure 2.2.5). In order to convert back from the photographic density of the image to the intensity of the source, this curve must be known with some precision. It is obtained by means of a photometric calibration exposure, which is a photograph of a series of sources of differing and known intensities. The characteristic curve is then plotted from the measured densities of the images of these sources. To produce accurate results, great care must be exercised such as:

(a) the photometric calibration film must be from the same batch of film as that used for the main exposure;
(b) the treatment of the two films, including their storage, must be identical;
(c) both films must be processed, hypersensitized etc. together;
(d) the exposure lengths, intermittency effects, temperatures during exposures etc. should be as similar as possible for the two plates;
(e) if the main exposure is for a spectrum, then the calibration exposure should include a range of wavelengths as well as intensities;
(f) both plates must be measured under identical conditions.

With these precautions, a reasonably accurate characteristic curve may be plotted. It is then a simple although tedious procedure to convert the density of the main image back to intensity. The measurement of a plate is undertaken on a microdensitometer. This is a machine that shines a small spot of light through the image and measures the transmitted intensity. Such machines may readily be connected to a small computer, and the characteristic curve also fed into it, so that the conversion of the image to intensity may be undertaken automatically.

Photometry of the highest accuracy is no longer attempted from photographs for stars, but a rough guide to their brightnesses may be obtained from the diameters of their images, as well as by direct microphotometry. Scattering, halation, diffraction rings etc. around star images mean that bright stars have larger images than faint stars. If a sequence of stars of known brightnesses is on the plate, then a calibration curve may be plotted, and estimates of the magnitudes of large numbers of stars made very rapidly.

2.3   Electronic imaging

Introduction

The alternatives to photography for recording images directly are almost all electronic in nature. They have two great advantages over the photographic emulsion. First, the image is usually produced as an electrical signal and can therefore be relayed or transmitted to a remote observer, which is vital for satellite-borne instrumentation and useful in many other circumstances, and also enables the data to be fed directly into a computer. Second, with many of the systems the quantum efficiency is up to a factor of a hundred or so higher than that of the photographic plate. Subsidiary advantages can include intrinsic amplification of the signal, linear response, and long wavelength sensitivity. The major disadvantages include the small number of pixels (1% or less of those available on a large photograph), low reliability, complexity of operation, high cost, short working life, and geometrical distortion of the image.

The most basic form of electronic imaging simply consists of an array of point-source detecting elements. The method is most appropriate for the intrinsically small detectors such as photoconductive cells (section 1.1). Hundreds or thousands or even more of the individual elements (usually termed pixels) can then be used to give high spatial resolution. The array is simply placed at the focus of the telescope or spectroscope etc. in place of the photographic plate. Other arrays such as CCDs, infrared arrays and STJs were reviewed in detail in section 1.1. Millimetre-wave and radio arrays were also considered in section 1.1 and in section 1.2. Here, therefore, we are concerned with other electronic approaches to imaging, most of which have been superseded by array detectors, but which have historical and archival interest.

Television and related systems

Low light level television systems, combined with image intensifiers and perhaps using a slower scan rate than normal, are useful. They are particularly found on the guide systems of large telescopes where they enable the operator at a remote-control console to have a viewing/finding/guiding display.

For some applications the cost of developing a custom-designed and built camera may not be prohibitive. These applications are primarily for satellite instrumentation. Planetary probes make extensive use of television cameras of many varieties for direct imaging. Their scanning rate is slow compared with conventional systems, but this is more often due to the rate of transmission of data back to Earth than to the necessity for intrinsically long integration times. Indirect imaging, such as to record spectra etc., also employs television systems, particularly for ultraviolet work. Secondary electron conduction (SEC) cameras that allow integration for many hours are especially useful in this region when allied to an ultraviolet-to-optical converter.

The SEC and the related EBS (electron bounce silicon) systems are based upon the vidicon television camera. The sensitive element in a basic vidicon is a photoconducting layer (section 1.1). This is deposited on to a

Figure 2.3.1. The vidicon television camera.

transparent conductor which is held at a slight positive potential and which faces the window of the tube. The photoconducting layer changes its conductivity in the pattern of the image that is falling upon it, becoming the most conducting where it is most intensely illuminated. Since the photoconductor is also a semiconductor, the positive charge applied by the transparent electrode will slowly leak through to its inner face. But since the rate of leakage will depend upon the conductivity, the potential on the inner surface will have a pattern similar to that of the original image. This pattern of potential is then scanned by an electron beam (figure 2.3.1) that almost instantaneously raises the inner face potential at the point of impact to that of the cathode. Capacitive coupling across the photoconducting layer to the transparent anode then produces the required video output current. The magnitude of the current is proportional to the original image intensity at the point being scanned by the electron beam. A positively charged field mesh is also necessary to protect the photoconductive layer from positive ions produced in the residual gas inside the tube. Photoconductive materials that may be used include selenium, antimony trisulphide, cadmium selenide, lead oxide and silicon. The latter two substances have low intrinsic resistivities and so must be divided into a grid of separately insulated pixels. The SEC vidicon works on a similar principle except that the image that is scanned is produced by primary electrons, and not directly by the photons. The photons impinge on to a photoemitter (section 1.1) on the window of the tube (figure 2.3.2). The ejected electrons are accelerated

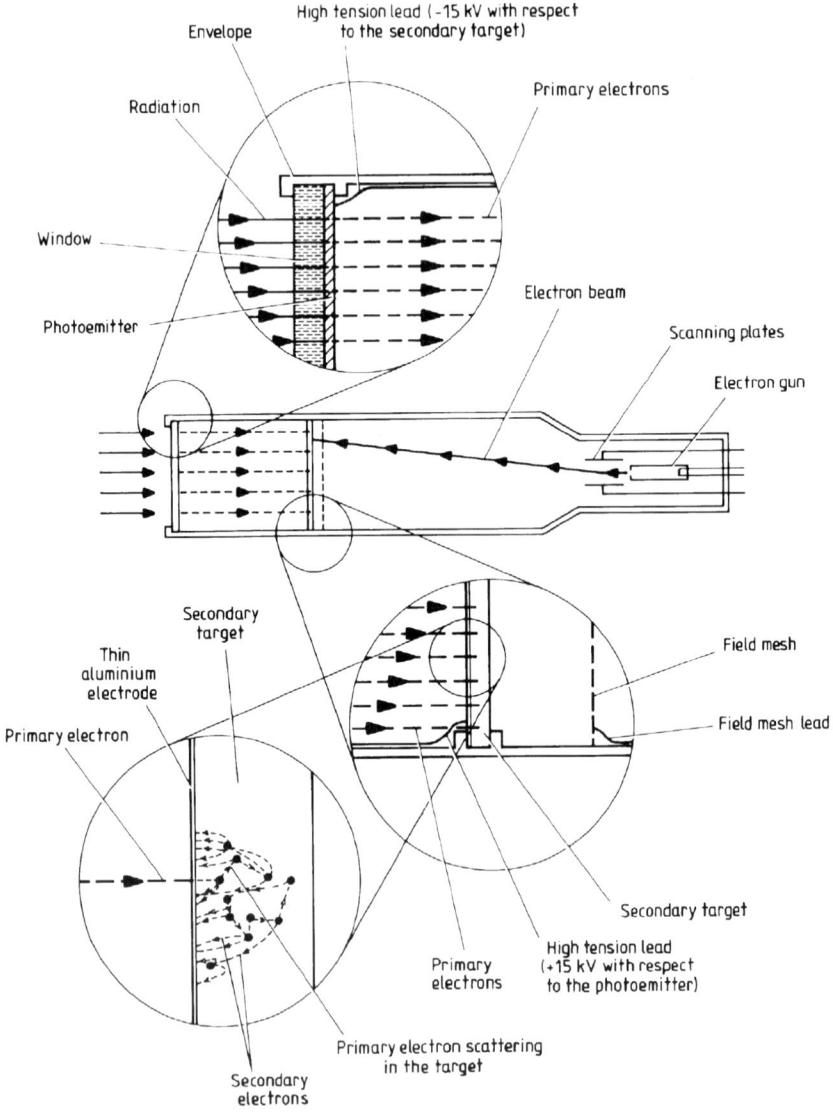

Figure 2.3.2. SEC vidicon television camera.

down the tube by a potential of 10 to 20 kV. On hitting the secondary target, the primary electrons eject many secondary electrons. These are then attracted to a positive electrode and removed, leaving a pattern of potential on the target that matches the original image. This potential image is then read off in a manner identical to that of the normal vidicon. The secondary target is formed of potassium chloride. The primary target can be any photo-emitter, and can be varied to suit the wavelength region being observed (see section 1.1 for a list of suitable photoemitters). The SEC tube has two major advantages. First, the primary electrons produce many times their own number of secondary electrons so that there is an intrinsic amplification of the signal by up to a factor of one hundred. Second, long exposures are possible since the secondary electrons are removed so that the potential pattern may continue to build up in the secondary target without being erased. Integration times of hours are easily achievable. The EBS tube works on the same principle as the SEC tube, but uses silicon for its secondary target. It can have intrinsic gains of up to a factor of one thousand, but requires cooling if integration times of longer than a few seconds are needed.

Image intensifiers

We have already encountered a type of image intensifier in the SEC and EBS television cameras discussed above. They have an intrinsic amplification by a factor of up to 1000 through the production of many secondary electrons by a single primary electron. The conventional use of the term 'image intensifier', however, is usually restricted to devices that produce amplified optical images as their output. The amplification is again achieved through the use of accelerated electrons for all the devices currently in use. There are three main designs for image intensifiers: magnetically focused, electrostatically focused, and proximity focused devices.

Magnetically focused image intensifiers

A standard photoemitter (see section 1.1) for the spectral region to be observed is coated on to the inside of the window of the device (figure 2.3.3). The electrons emitted from this when it is illuminated are then accel-erated down the tube by a potential of up to 40 kV and focused on to the phosphor by the magnetic field of the surrounding solenoid and by the elec-tric fields of the ring electrodes. The phosphor emits light through the rapid recombination of the electron–hole pairs produced in it by the primary electrons. The spectrum of the emitted light corresponds to the energy gap between the valence and the conduction bands of the phosphor. Because of the very high energy of the primary electrons, each primary electron produces many electron–hole pairs, and so the light emitted by the phosphor

Figure 2.3.3. Cross section through a magnetically focused single-stage image intensifier.

is many times brighter than that in the original image. The intensified image can be viewed directly, or recorded by almost any of the imaging systems that we have discussed. The main drawback of the device is the geometrical distortion that it can introduce into the image. Figure 2.3.4 shows a fairly typical example of this distortion for a good quality tube. Some of the other problems of the devices, such as cathode sensitivity variations, are similar to those of photomultipliers (section 1.1). The very high voltages that are used can lead to breakdown of the insulation and/or glow discharges under conditions of high humidity.

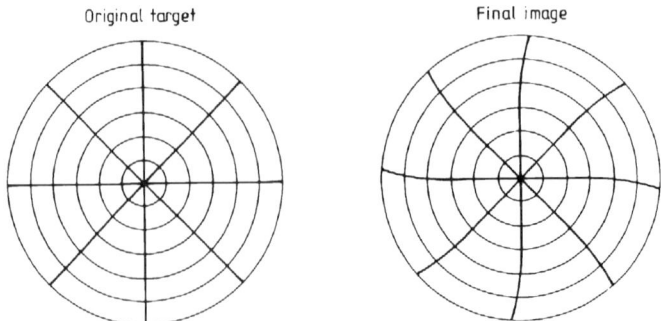

Figure 2.3.4. Rotational and 'S' distortion of the image by an image intensifier.

Electrostatically focused image intensifiers

These are identical to the magnetically focused systems that we have just discussed, except that the focusing of the electron paths is entirely electrostatic. A typical arrangement is shown schematically in figure 2.3.5.

Proximity focused image intensifiers

In these devices the electron paths are kept very short so that they have little opportunity to become defocused, and in addition they may be mechanically restricted by channelling down small tubes. Magnetic and/or electrostatic focusing is not used so that geometrical distortion of the image is avoided; however, the spatial resolution is reduced by the spread of electron paths. The simplest system is just a vacuum tube with the photocathode close to the phosphor. More commonly used is a microchannel plate of the type discussed in section 1.3 for x-ray imaging. The only difference lies in the type of photoemitter being used; one appropriate to the observed wavelength range must be selected instead of those for x-rays. The other details of the system are covered in section 1.3 and so are not repeated here.

Cascades

A single image intensifier of any of the above types amplifies the intensity of the original image by a factor usually between 50 and 100. If larger amplifications are needed, then several of the basic units can be linked together. In this way commercially available devices can achieve amplifications by up to a factor of 10^7. A coherent fibre optic light guide usually

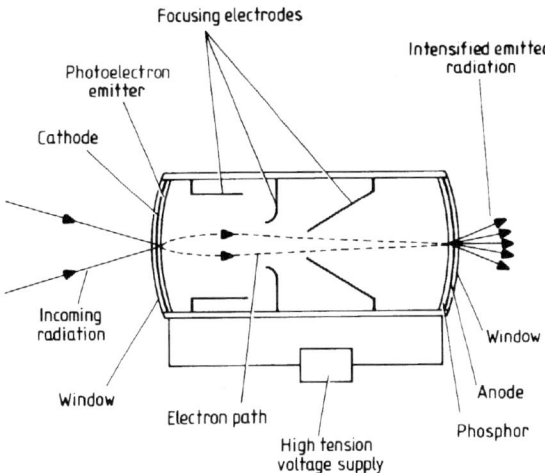

Figure 2.3.5. Cross section through an electrostatically focused single-stage image intensifier.

connects the output from one basic unit to the photocathode of the next. The only modification otherwise required to a basic unit is that the second and subsequent stages must have photoelectron emitters whose response is matched to the emission spectrum of the phosphor, and not to the wavelength region that is being observed.

Recording

The large amplification of the image intensifier is primarily of use in that it allows a detector with a low quantum efficiency such as the eye or a photographic emulsion to work at the quantum efficiency of the photocathode. Since this can reach 50% or more, there is a corresponding gain in the signal-to-noise ratio of the final image. Additionally with photographic plates the higher image intensity may remove or reduce the problem of reciprocity failure. Television cameras and charge-coupled devices are the other methods that are commonly used to detect the final image of an image intensifier. One subsidiary advantage of the image intensifier is that the phosphor of the final stage can be chosen so that its output spectrum matches the response of the detector, irrespective of original wavelengths being observed. The amplification of the image intensity naturally leads to shorter exposure times, but the improvement is only given by the improvement in the effective quantum efficiency of the final detector, if a given signal-to-noise ratio in the image is to be reached. The very large apparent gains of image intensifiers are therefore somewhat misleading as a guide to the savings in telescope time that their use may provide.

Photon counting imaging systems

A combination of a high-gain image intensifier and TV camera led in the 1970s to the development by Alec Boksenberg and others of the IPCS (Image Photon Counting System). In the original IPCS, the image intensifier was placed at the telescope's (or spectroscope's etc) focal plane and produced a blip of some 10^7 photons at its output stage for each incoming photon in the original image. A relatively conventional TV camera then viewed this output and the video signal fed to a computer for storage. In its basic form the IPCS suffered from two major drawbacks. First, there was a large amount of noise from the image intensifier. Second, the image tube spread the cloud of electrons from a single original photon into a blob of significant physical size. The IPCS overcame these problems by using the computer to discriminate between the blips from incoming photons and those from other sources that made up the background noise. The computer also determined the centre of the output 'blob' in order to provide precise positional information. In this way a sharp, almost noise-free, picture could be built up of the original image.

More recently a number of variants on Boksenberg's original IPCS have been developed. These include microchannel plate image intensifiers linked to CCDs, and other varieties of image intensifiers linked to CCDs or other array-type detectors. The principles of the devices remain unchanged, however.

2.4   Scanning

This is such an obvious technique as scarcely to appear to deserve a separate section, and indeed at its simplest it hardly does so. A two-dimensional image may be built up using any point source detector, if that detector is scanned over the image or vice versa. Many images at all wavelengths are obtained in this way. Sometimes the detector moves, sometimes the whole telescope or satellite etc. Occasionally it may be the movement of the secondary mirror of the telescope or of some *ad hoc* component in the apparatus which allows the image to be scanned. Scanning patterns are normally raster or spiral (figure 2.4.1). Other patterns may be encountered, however, such as the continuously nutating roll employed by some of the early artificial satellites. The only precautions advisable are the matching of the scan rate to the response of the detector or to the integration time being used, and the matching of the separation of the scanning lines to the resolution of the system. Scanning by radio telescopes (section 1.2) used to be by observing individual points within the scan pattern for a set interval of time. More recently the scanning has been continuous, or 'on the fly', since this enables slow changes due to the atmosphere or the instrument to be eliminated. Specialized applications of scanning occur in the spectrohelioscope (section 5.3), and the scanning spectrometer (section 4.2). Many Earth observation satellites use push-broom scanning. In this system a linear array of detectors is aligned at right angles to the spacecraft's ground track. The image is then built up as the spacecraft's motion moves the array to look at successive slices of the swathe of ground over which the satellite is passing.

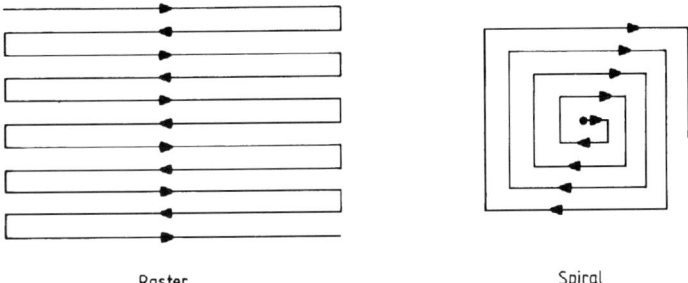

Raster   Spiral

Figure 2.4.1. Scanning patterns.

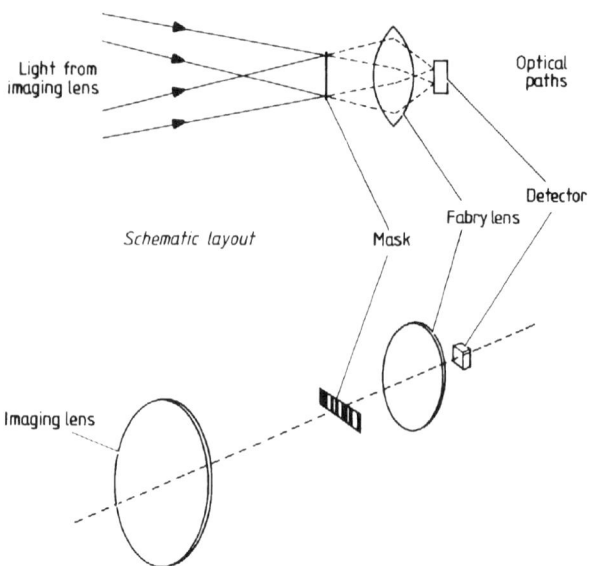

Figure 2.4.2. Schematic arrangement for a Hadamard mask imaging system.

A more sophisticated approach to scanning is to modulate the output of the detector by interposing a mask of some type in the light beam. Examples of this technique are discussed elsewhere and include the modulation collimator and the coded array mask used in x-ray imaging (section 1.3). A very great improvement over either these methods or the basic approach discussed above may be made by using a series of differing masks, or by scanning a single mask through the light beam so that differing portions of it are utilized. This improved method is known as Hadamard mask imaging. We may best illustrate its principles by considering one-dimensional images such as might be required for spectrophotometry. The optical arrangement is shown in figure 2.4.2. The mask is placed in the image plane of the telescope, and the Fabry lens directs all the light that passes through the mask on to the single detector. Thus the output from the system consists of a simple measurement of the intensity passed by the mask. If a different mask is now substituted for the first, then a new and in general different intensity reading will be obtained. If the image is to be resolved into N elements, then N such different masks must be used, and N intensity readings determined. If $\mathbf{D}$ is the vector formed from the detector output readings, $\mathbf{I}$ is the vector of the intensities of the elements of the image, and $\mathbf{M}$ the $N \times N$ matrix whose columns each represent one of the masks, with the individual elements of the matrix corresponding to the transmissions of the individual segments of the mask, i.e.

$$\mathbf{D} = [D_1, D_2, D_3, \ldots, D_N] \tag{2.4.1}$$

$$\mathbf{I} = [I_1, I_2, I_3, \ldots, I_N] \tag{2.4.2}$$

$$\mathbf{M} = \begin{bmatrix} m_{11} & m_{12} & m_{13} & \cdots & m_{1N} \\ m_{21} & & & & \\ m_{31} & & & & \\ \vdots & & & & \\ m_{N1} & & & & m_{NN} \end{bmatrix}. \tag{2.4.3}$$

Then, ignoring any contribution arising from noise, we have

$$\mathbf{D} = \mathbf{IM} \tag{2.4.4}$$

and so

$$\mathbf{I} = \mathbf{DM}^{-1} \tag{2.4.5}$$

Thus the original image is simply obtained by inverting the matrix representing the masks. The improvement of the method over a simple scan lies in its multiplex advantage (cf. the Fourier transform spectrometer, section 4.1). The masks usually comprise segments that either transmit or obscure the radiation completely, that is

$$m_{ij} = 0 \text{ or } 1 \tag{2.4.6}$$

and, on average, about half of the total image is obscured by a mask. Thus $N/2$ image segments contribute to the intensity falling on to the detector at any one time. Hence if a given signal-to-noise ratio is reached in a time T, when the detector observes a single image element, then the total time required to detect the whole image with a simple scan is

$$N \times T \tag{2.4.7}$$

and is approximately

$$\sqrt{2N}T \tag{2.4.8}$$

for the Hadamard masking system. Thus the multiplex advantage is approximately a factor of

$$\sqrt{\frac{N}{2}} \tag{2.4.9}$$

improvement in the exposure length.

In practice, moving a larger mask across the image generates the different masks. The matrix representing the masks must then be cyclic, or in other words, each successive column is related to the previous one by being moved down a row. The use of a single mask in this way can lead to a considerable saving in the construction costs, since $2N - 1$ segments are needed for it compared with N^2 if separate masks are used. An additional constraint on the matrix $\mathbf{M}$ arises from the presence of noise in the output. The errors introduced by this are minimized when

$$\text{Tr}[\mathbf{M}^{-1} \ (\mathbf{M}^{-1})^{\text{T}}] \tag{2.4.10}$$

is minimized. There are many possible matrices that will satisfy these two constraints, but there is no completely general method for their generation. One method of finding suitable matrices and so of specifying the mask is based upon the group of matrices known as the Hadamard matrices (hence the name for this scanning method). These are matrices whose elements are ±1 and which have the property

$$\mathbf{HH}^\mathrm{T} = N\mathbf{I} \tag{2.4.11}$$

where $\mathbf{H}$ is the Hadamard matrix and $\mathbf{I}$ is the identity matrix. A typical example of the mask matrix, $\mathbf{M}$, obtained in this way might be

$$\mathbf{M} = \begin{bmatrix} 0 & 1 & 0 & 1 & 1 & 1 & 0 \\ 0 & 0 & 1 & 0 & 1 & 1 & 1 \\ 1 & 0 & 0 & 1 & 0 & 1 & 1 \\ 1 & 1 & 0 & 0 & 1 & 0 & 1 \\ 1 & 1 & 1 & 0 & 0 & 1 & 0 \\ 0 & 1 & 1 & 1 & 0 & 0 & 1 \\ 1 & 0 & 1 & 1 & 1 & 0 & 0 \end{bmatrix} \tag{2.4.12}$$

for N having a value of 7.

Variations to this scheme can include making the obscuring segments out of mirrors and using the reflected signal as well as the transmitted signal, or arranging for the opaque segments to have some standard intensity rather than zero intensity. The latter device is particularly useful for infrared work. The scheme can also easily be extended to more than one dimension, although it then rapidly becomes very complex. A two-dimensional mask may be used in a straightforward extension of the one-dimensional system to provide two-dimensional imaging. A two-dimensional mask combined suitably with another one-dimensional mask can provide data on three independent variables—for example, spectrophotometry of two-dimensional images. Two two-dimensional masks could then add (for example) polari-metry to this, and so on. Except in the one-dimensional case, the use of a computer to unravel the data is obviously essential, and even in the one-dimensional case it will be helpful as soon as the number of resolution elements rises above five or six.

2.5 Interferometry

Introduction

Interferometry is the technique of using constructive and destructive addition of radiation to determine information about the source of that radiation. Interferometers are also used to obtain high-precision positions for optical and radio sources, and this application is considered in section 5.1. Two principal types of interferometer exist: the Michelson stellar interferometer, first proposed by Armand Fizeau in 1868, and the intensity interferometer proposed by Robert Hanbury Brown in 1949. There are also a number of subsidiary techniques such as amplitude interferometry, nulling interferometry, speckle interferometry, Fourier spectroscopy etc. The intensity interferometer is now largely of historical interest, although a brief description is included at the end of this section for completeness.

Michelson optical stellar interferometer

The Michelson *stellar* interferometer is so called in order to distinguish it from the Michelson interferometer used in the 'Michelson–Morley' experiment and in Fourier spectroscopy. Since the latter type of interferometer is discussed primarily in section 4.1, we shall normally drop the 'stellar' qualification for this section when referring to the first of the two main types of interferometer.

In practice, many interferometers use the outputs from numerous telescopes. However, this is just to reduce the time taken for the observations, and it is actually the outputs from pairs of telescopes that are combined to produce the interference effects. We may start, therefore, by considering the interference effects arising from two beams of radiation.

The complete optical system that we shall study is shown in figure 2.5.1. The telescope objective, which for simplicity is shown as a lens, is covered by an opaque screen with two small apertures, and is illuminated by two monochromatic equally bright point sources at infinity. The objective diameter is D, its focal length F, the separation of the apertures is d, the width of the apertures is Δ, the angular separation of the sources is α, and their emitted wavelength is λ.

Let us first consider the objective by itself, without the screen, and with just one of the sources. The image structure is then just the well-known diffraction pattern of a circular lens (figure 2.5.2). With both sources viewed by the whole objective, two such patterns are superimposed. There are no interference effects between these two images, for their radiation is not mutually coherent. When the main maxima are superimposed upon the first minimum of the other pattern, we have Rayleigh's criterion for the resolution of a lens (figures 1.1.29 and 2.5.3). The Rayleigh criterion for the minimum angle between two separable sources α', as we saw in section 1.1, is

$$\alpha' = \frac{1.22\lambda}{D}. \tag{2.5.1}$$

Now let us consider the situation with the screen in front of the objective, and first consider just one aperture looking at just one of the sources. The

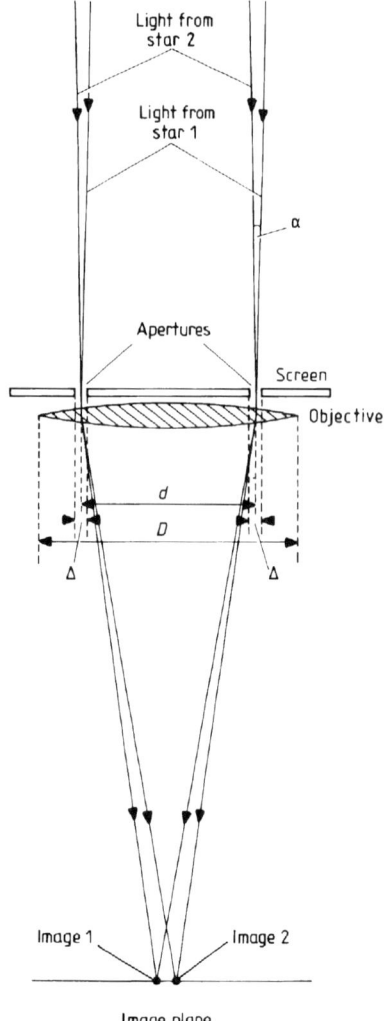

Figure 2.5.1. Optical arrangement of a Michelson interferometer.

situation is then the same as that illustrated in figure 2.5.2, but the total intensity and the resolution are both reduced because of the smaller size of the aperture compared with the objective (figure 2.5.4). Although the image for the whole objective is also shown for comparison in figure 2.5.4, if it were truly to scale for the situation illustrated in figure 2.5.1, then it would be one seventh of the width that is shown and 1800 times higher!

Now consider what happens when one of the sources is viewed simultaneously through both small apertures. If the apertures were infinitely small,

Figure 2.5.2. Image structure for one point source and the whole objective.

Figure 2.5.3. Image structure for two point sources and the whole objective.

then ignoring the inconvenient fact that no light would get through anyway, we would obtain a simple interference pattern (figure 2.5.5). The effect of the finite width of the apertures is to modulate the straightforward variation of figure 2.5.5 by the shape of the image for a single aperture (figure 2.5.4). Since two apertures are now contributing to the intensity and the energy 'lost' at the minima reappears at the maxima, the overall envelope of the image

Figure 2.5.4. Image structure for one point source and one aperture.

Figure 2.5.5. Image structure for a single point source viewed through two infinitely small apertures.

peaks at four times the intensity for a single aperture (figure 2.5.6). Again, for the actual situation shown in figure 2.5.1, there should be a total of 33 fringes inside the major maximum of the envelope.

Finally let us consider the case of two equally bright point sources viewed through two apertures. Each source has an image whose structure is that shown in figure 2.5.6, and these simply add together in the same manner as the case illustrated in figure 2.5.3 to form the combined image. The structure of this combined image will depend upon the separation of the two sources. When the sources are almost superimposed, the image structure will be identical with that shown in figure 2.5.6, except that all the intensities will have doubled. As the sources move apart, the two fringe patterns will also separate, until when the sources are separated by an angle α'', given by

$$\alpha'' = \frac{\lambda}{2d} \tag{2.5.2}$$

the maxima of one fringe pattern will be superimposed upon the minima of the other and vice versa. The fringes will then disappear and the image will be

Figure 2.5.6. Image structure for one point source viewed through two apertures.

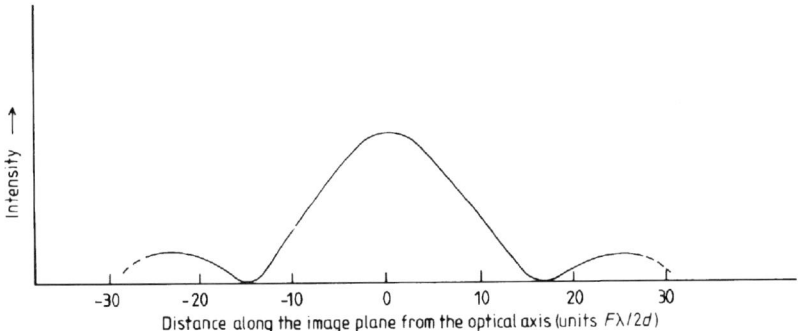

Figure 2.5.7. Image structures for two point sources separated by an angle, $\lambda/2d$, viewed through two apertures.

given by their envelope (figure 2.5.7). There may still be a very slight ripple on this image structure due to the incomplete filling of the minima by the maxima, but it is unlikely to be noticeable. The fringes will reappear as the sources continue to separate until the pattern is almost double that of figure 2.5.6 once again. The image patterns are then separated by a whole fringe width, and the sources by $2\alpha''$. The fringes disappear again for a source separation of $3\alpha''$, and reach yet another maximum for a source separation of $4\alpha''$, and so on. Thus the fringes are most clearly visible when the source's angular separation is given by $2n\alpha''$ (where n is an integer) and they disappear or reach a minimum in clarity for separations of $(2n+1)\alpha''$. Applying the Rayleigh criterion for resolution, we see that the resolution of two apertures is given by the separation of the sources for which the two fringe patterns are mutually displaced by half a fringe width. This as we have seen is simply the angle α'', and so we have

$$\frac{\text{resolution through two apertures}}{\text{resolution of the objective}} = \frac{\alpha''}{\alpha'} \qquad (2.5.3)$$

$$= \frac{2.44d}{D}. \qquad (2.5.4)$$

Imagining the two apertures placed at the edge of the objective (i.e. $d = D$), we see the quite remarkable result that the resolution of an objective may be increased by almost a factor of two and a half by screening it down to two small apertures at opposite ends of one of its diameters. The improvement in the resolution is only along the relevant axis—perpendicular to this, the resolution is just that of one of the apertures. The resolution in both axes may, however, be improved by using a central occulting disc which leaves the rim of the objective clear. This also enables us to get a feeling for the physical basis for this improvement in resolution. We may regard the objective as composed of a series of concentric narrow rings, each of which has a resolution given by equation (2.5.2), with d the diameter of the ring. Since the objective's resolution is the average of all these individual resolutions, it is naturally less than their maximum. In practice some blurring of the image may be detected for separations of the sources that are smaller than the Rayleigh limit (section 1.1). This blurring is easier to detect for the fringes produced by the two apertures than it is for the images through the whole objective. The effective improvement in the resolution may therefore be even larger than that given by equation (2.5.4).

Now for the situation illustrated in figure 2.5.1, the path difference between the two light beams on arriving at the apertures is likely to be small, since the screen will be perpendicular to the line of sight to the objects. In order to produce fringes with the maximum clarity, however, that path difference must be close to zero. This is because radiation is never completely monochromatic; there is always a certain range of wavelengths present known as the bandwidth of the signal. Now for a zero path difference at the apertures, all wavelengths will be in phase when they are combined and will interfere constructively. However, if the path difference is not zero, some wavelengths will be in phase but others will be out of phase to a greater or lesser extent, since the path difference will equal different numbers of cycles or fractions of cycles at the different wavelengths. There will thus be a mix of constructive and destructive interference, and the observed fringes will have reduced contrast. The path difference for which the contrast in the fringes reduces to zero (i.e. no fringes are seen) is called the coherence length, l, and is given by

$$l = \frac{c}{\Delta\nu} = \frac{\lambda^2}{\Delta\lambda} \qquad (2.5.5)$$

where $\Delta\nu$ and $\Delta\lambda$ are the frequency and wavelength bandwidths of the radiation, so that for $\lambda = 500\,\text{nm}$ and $\Delta\lambda = 1\,\text{nm}$, we have a coherence length of 0.25 mm, for white light ($\Delta\lambda \approx 300\,\text{nm}$) this reduces to less than a micron, but in the radio region it can be large: 30 m at $\nu = 1.5\,\text{GHz}$ and $\Delta\nu = 10\,\text{MHz}$, for example. However, as we shall see below, the main output from an interferometer is the fringe contrast, usually known as the fringe visibility, V (equation (2.5.6)), so that in order for this not to be degraded, the path difference to the apertures, or their equivalent, must be kept to a small fraction of the coherence length. For interferometers such as Michelson's 1921 stellar interferometer (figure 2.5.8), the correction to zero path difference is small, and in that case was accomplished through the use of adjustable glass wedges. However, most interferometers, whether operating in the optical or radio regions, now use separate telescopes on the ground (figure 2.5.9 for example), so the path difference to the telescopes can be over 100 m for optical interferometers, and up to thousands of kilometres for very-long-baseline radio interferometry. These path differences have to be corrected either during the observation by hardware, or afterwards during the data processing.

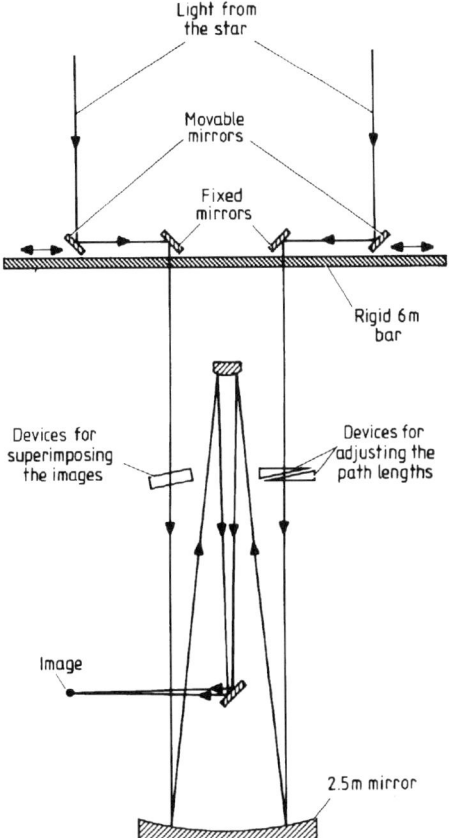

Figure 2.5.8. Schematic optical arrangement of the Michelson stellar interferometer.

Michelson's original interferometer was essentially identical to that shown in figure 2.5.1, except that the apertures were replaced by two movable parallel slits to increase the amount of light available, and a narrow band filter was included to give nearly monochromatic light (earlier we had assumed that the *sources* were monochromatic). To use the instrument to measure the separation of a double star with equally bright components, the slits are aligned perpendicularly to the line joining the stars, and moved close together, so that the image fringe patterns are undisplaced and the image structure is that of figure 2.5.6. The actual appearance of the image in the eyepiece at this stage is shown in figure 2.5.9. The slits are then moved apart until the fringes disappear (figures 2.5.7 and 2.5.10). The distance between the slits is then such that the fringe pattern from one star is filling in the fringe pattern from that other, and their separation is given by α'' (equation (2.5.2), with d given by the distance between the slits). If

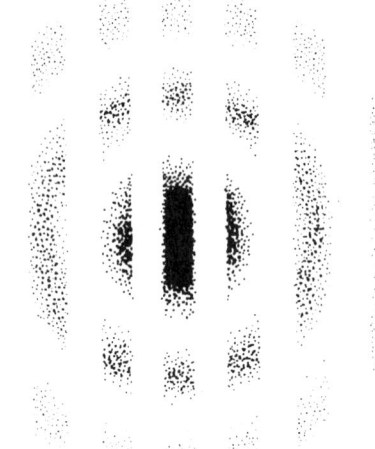

Figure 2.5.9. Appearance of the image of a source seen in a Michelson interferometer at the instant of maximum fringe visibility.

the two stars are of differing brightnesses, then the fringes will not disappear completely, but the fringe visibility, V, given by

$$V = \frac{(I_{\max} - I_{\min})}{(I_{\max} + I_{\min})} \qquad (2.5.6)$$

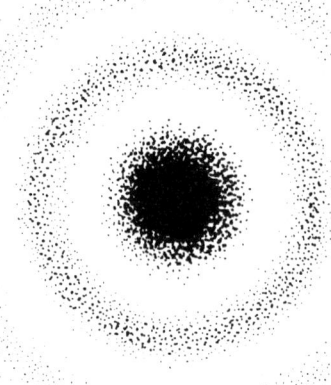

Figure 2.5.10. Appearance of the image of a source seen in a Michelson interferometer at the time of zero fringe visibility.

where I_{max} is the intensity of a fringe maximum and I_{min} is the intensity of a fringe minimum, will reach a minimum at the same slit distance. To measure the diameter of a symmetrical object like a star or the satellite of a planet, the same procedure is used. However, the two sources are now the two halves of the star's disc, represented by point sources at the optical centres of the two semi-circles. The diameter of the star is then given by $2.44\alpha''$, so that in determining the diameters of objects, an interferometer has the same resolution as a conventional telescope with an objective diameter equal to the separation of the slits.

Very stringent requirements on the stability and accuracy of the apparatus are required for the success of this technique. As we have seen, the path difference to the slits must not be greater than a small fraction of the coherence length of the radiation (equation (2.5.5)). Furthermore that path difference must remain constant to considerably better than the wavelength of the radiation that is being used, as the slits are separated. Vibrations and scintillation are additional limiting factors. In general, the paths will not be identical when the slits are close together and the fringes are at their most visible. But the path difference will then be some integral number of wavelengths and provided that it is significantly less than the coherence length the interferometer may still be used. Thus in practice the value that is required for d is the difference between the slit separations at minimum and maximum fringe visibilities. A system such as has just been described was used by Michelson in 1891 to measure the diameters of the Galilean satellites. Their diameters are in the region of $1''$ of arc, so that d has a value of a few tens of millimetres.

Now the angular diameters of stellar discs are very much smaller than those of the Galilean satellites; $0.047''$ for α Orionis (Betelgeuse) for example, one of the largest angular diameter stars in the sky. A separation for the apertures of several metres is thus needed if stellar diameters are to be measured. This led to the improved version of the system that is rather better known than the one used for the Galilean satellites and which is the type of interferometer that is usually intended when the Michelson stellar interferometer is mentioned. It was used by Albert Michelson and Francis Pease in 1921 to measure the diameters of half a dozen or so of the larger stars. The slits were replaced by movable mirrors on a rigid 6 m long bar mounted on top of the 2.5 m Hooker telescope (figure 2.5.8). By this means, d could be increased beyond the diameter of the telescope. A system of mirrors then reflected the light into the telescope. The practical difficulties with the system were very great and although larger systems have since been attempted none has been successful, and so the design is now largely of historical interest only.

Modern optical interferometers are based upon separate telescopes rather than two apertures feeding a single instrument. They are still primarily used to measure stellar diameters, although some aperture synthesis systems (see below) are now working and others are under construction. There is no difference in the principle of operation of these interferometers from that of Michelson, but there are many practical changes. The main change arises because the path difference to the two telescopes can now be many metres. A system for compensating for the path difference is thus required. In most designs the path compensation is via a delay line; the light from one telescope is sent on a longer path than that from the other one before they are mixed together. The extra path length is arranged to be equal to the path difference between the telescopes. Since the path difference will change as the object moves across the sky, the delay must also change, and this is usually accomplished by having one or more of the mirrors reflecting the light on to the longer path mounted on a movable carriage. Even using narrow bandwidths, the small value of the coherence length (equation (2.5.5)) at optical wavelengths means that the carriage must be capable of being positioned to sub-micron accuracy and yet be able to travel tens of metres.

The second major difference is that the telescopes are usually fixed in position, or movable only slowly between observations. It is thus not possible to change their separation in order to observe maximum and minimum fringe visibilities. A change in the effective separation occurs as the angle between the baseline and the source in the sky alters, but this is generally insufficient to range from maximum to minimum fringe visibility. Thus instead a measurement of the fringe visibility is made for just one, or a small number, of separations. This measurement is then fitted to a theoretical plot of the variation of the fringe visibility with the telescope separation and the size of the source in order to determine the latter.

The third difference arises from the atmospheric turbulence. We have seen that there is an upper limit to the size of a telescope aperture, given by Fried's coherence length (equation (1.1.73)). In the visible this is only about 120 mm, though in the infrared it can rise to 300 mm. Telescopes with apertures larger than Fried's limit will thus be receiving radiation through several, perhaps many, atmospheric cells, each of which will have different phase shifts and wavefront distortions. When the beams are combined, the desired fringes will be washed out by all these differing contributions. Furthermore, the atmosphere changes on a time scale of around 5 ms, so that even if small apertures are used, exposures have to be short enough to freeze the atmospheric motion (see also speckle interferometry, section 2.6). Thus many separate observations have to be added together to get a value with a sufficiently high signal-to-noise ratio. In order to use large telescopes as a part of an interferometer, their images have to be corrected through adaptive optics (section 1.1). Even a simple tip–tilt correction will allow the usable aperture to be increased by a factor of three. However, in order to use 8 m and 10 m telescopes, such as ESO's VLT and the Keck telescopes, full atmospheric correction is needed.

When only two telescopes form the interferometer, the interruption of the phase of the signal by the atmosphere results in the loss of that information. Full reconstruction of the morphology of the original object is not then possible, and only the fringe visibility can be determined (see below for a discussion of multi-element interferometers, closure phase, and aperture synthesis). However, the fringe visibility still enables stellar diameters and double star separations to be found, and so several two-element interferometers are currently in operation. Thus ISI[34] based on Mount Wilson, California, and observing at 11 μm, uses two 1.65 m mirrors with a maximum baseline of 35 m. It is planned to add a third mirror and to increase the baseline to 75 m. SUSI[35] in Australia operates in the visible region with two 0.14 m mirrors and a maximum separation of 640 m, while GI2T[36] in France, also operating in the visible, uses two 1.5 m mirrors with a maximum baseline of 65 m. In a few years' time, the LBT[37] on Mount Graham, Arizona, and the Magellan telescopes in the Chilean Andes, with two 8.4 m and two 6.5 m mirrors respectively, will be able to act as two-aperture interferometers with 20 m baselines.

A proposal for the future that would have much in common with stellar coronagraphs (section 5.3) is for a nulling interferometer based in space. This would use destructive interference to suppress the bright central object, thus allowing faint companions (e.g. planets) to be discerned. Earth-like planets around stars 10 or more parsecs away might be detected in observations of a few hours using a four-element interferometer with 50 m baselines. An Earth-based prototype has shown that the principle works. It used two 1.8 m mirrors and operated at 10 μm wavelength. The path differences induced by the atmosphere were in this case utilized to provide the interference effects. Numerous 50 ms exposures were obtained of α Orionis (Betelgeuse), and only those with a 5 μm delay were used. The star's light was largely suppressed, but the radiation from the surrounding dust cloud underwent a different delay, and so could be observed directly.

A variation on the system that also has some analogy with intensity interferometry (see below) is known as amplitude interferometry. This has been used recently in a successful attempt to measure stellar diameters. At its heart is a device known as a Köster prism that splits and combines the two separate light beams (figure 2.5.11). The interfering beams are detected in a straightforward manner using photomultipliers. Their outputs may then be compared. The fringe visibility of the Michelson interferometer appears as the anticorrelated component of the two signals, while the atmospheric scintillation etc. affects the correlated component. Thus the

---

[34] Infrared Spatial Interferometer.
[35] Sydney University Stellar Interferometer.
[36] Grand Interféromètre à 2 Télescopes.
[37] Large Binocular Telescope.

Figure 2.5.11. The Köster prism.

atmospheric effects can be screened out to a large extent, and the stability of the system is vastly improved.

Michelson radio interferometer

Radio antennae are generally only a small number of wavelengths in diameter (section 1.2), and so their intrinsic resolution (equation (1.1.28)) is poor. The use of interferometers to obtain higher resolution therefore dates back to soon after the start of radio astronomy. There is no difference in principle between a radio and an optical interferometer when they are used to measure the separation of double sources, or the diameters of uniform objects. However, the output from a radio telescope contains both the amplitude and phase of the signal, so that complete imaging of the source is possible. Radio interferometers generally use more than two antennae, and these may be in a two-dimensional array. The basic principle, however, is the same as for just two antennae, although the calculations become considerably more involved.

The individual elements of a radio interferometer are usually fairly conventional radio telescopes (section 1.2) and the electrical signal that they output varies in phase with the received signal. In fact with most radio interferometers it is the electrical signals that are mixed to produce the interference effect rather than the radio signals themselves. Their signals may then be combined in two quite different ways to provide the interferometer output. In the simplest version, the signals are simply added together before the square law detector (figure 1.2.2), and the output will then vary with the path difference between the two signals. Such an arrangement, however, suffers from instability problems, particularly due to variations in the voltage gain.

A system that is often preferred to the simple adding interferometer is the correlation or multiplying interferometer. In this, as the name suggests, the IF signals from the receivers are multiplied together. The output of an ideal correlation interferometer will contain only the signals from the source (which are correlated); the other components of the outputs of the telescopes will be zero as shown below.

If we take the output voltages from the two elements of the inter-ferometer to be $(V_1 + V_1')$ and $(V_2 + V_2')$ where V_1 and V_2 are the correlated components (i.e. from the source in the sky) and V_1' and V_2' are the uncorrelated components, the product of the signals is then

$$V = (V_1 + V_1') \times (V_2 + V_2') \tag{2.5.7}$$

$$= V_1 V_2 + V_1' V_2 + V_1 V_2' + V_1' V_2'. \tag{2.5.8}$$

If we average the output over time, then any component of equation (2.5.8) containing an uncorrelated component will tend to zero. Thus

$$\overline{V} = \overline{V_1 V_2}. \tag{2.5.9}$$

In other words, the time-averaged output of a correlation interferometer is the product of the correlated voltages. Since most noise sources contribute primarily to the uncorrelated components, the correlation interferometer is inherently much more stable than the adding interferometer. The phase-switched interferometer (see below) is an early example of a correlation interferometer, though direct multiplying interferometers are now more common.

The systematics of radio astronomy differ from those of optical work (section 1.2), so that some translation is required to relate optical inter-ferometers to radio interferometers. If we take the polar diagram of a single radio antenna (figure 1.2.6 for example), then this is the radio analogue of our image structure for a single source and a single aperture (figures 2.5.2 and 2.5.4). The detector in a radio telescope accepts energy from only a small fraction of this image at any given instant (though a few array detectors with small numbers of pixels are now coming into use—section 1.2). Thus scanning the radio telescope across the sky corresponds to scanning this energy-accepting region through the optical image structure. The main lobe of the polar diagram is thus the equivalent of the central maximum of the optical image, and the side lobes are the equivalent of the diffraction fringes. Since an antenna is normally directed towards a source, the signal from it corresponds to a measurement of the central peak intensity of the optical image (figures 2.5.2 and 2.5.4). If two stationary antennae are now

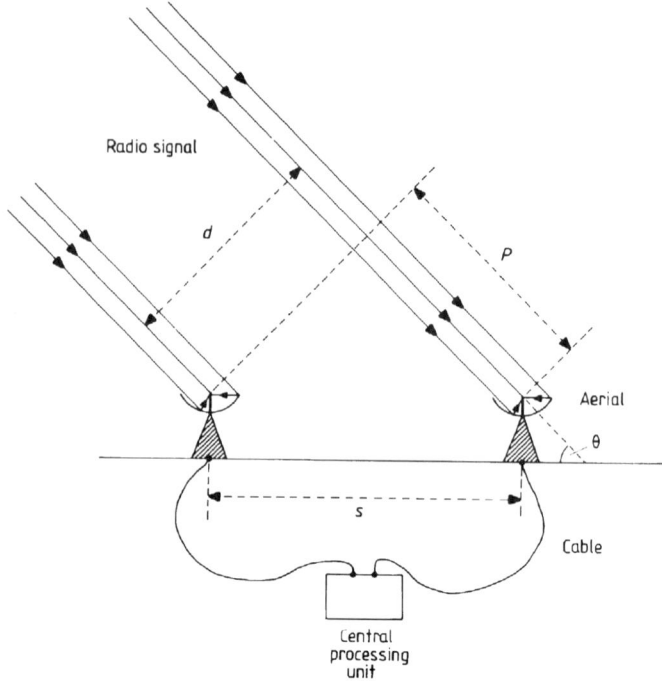

Figure 2.5.12. Schematic arrangement of a radio interferometer with fixed antennae.

considered and their outputs combined, then when the signals arrive without
any path differences, the final output from the radio system as a whole
corresponds to the central peak intensity of figure 2.5.6. If a path difference
does exist, however, then provided that it is less than the coherence length,
the final output will correspond to some other point within that image. In
particular when the path difference is a whole number of wavelengths, the
output will correspond to the peak intensity of one of the fringes, and
when it is a whole number plus half a wavelength, it will correspond to
one of the minima. Now the path differences arise in two main ways: from
the angle of inclination of the source to the line joining the two antennae,
and from delays in the electronics and cables between the antennae and
the central processing station (figure 2.5.12). The latter will normally be
small and constant and may be ignored or corrected. The former will alter
as the rotation of the Earth changes the angle of inclination. Thus the
output of the interferometer will vary with time as the value of P changes.
The output over a period of time, however, will not follow precisely the
shape of figure 2.5.6 because the rate of change of path difference varies
throughout the day, as we may see from the equation for P:

$$P = s \cos \phi \cos(\psi - E) \qquad (2.5.10)$$

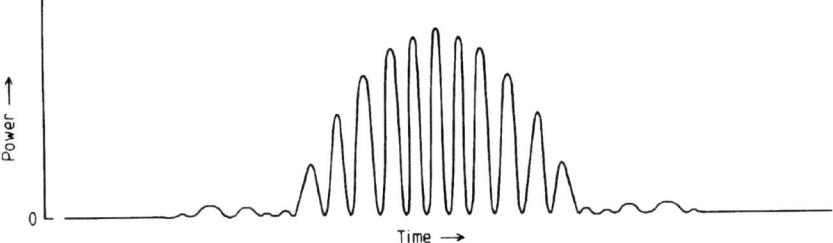

Figure 2.5.13. Output from a radio interferometer with stationary antennae, viewing a single source.

where μ is the altitude of the object and is given by

$$\mu = \sin^{-1}[\sin \delta \sin \phi + \cos \delta \cos \phi \cos(T - \alpha)].\qquad 2.5.11)$$

ψ is the azimuth of the object and is given by

$$\psi = \cot^{-1}[\sin \phi \cot(T - \alpha) - \cos \phi \tan \delta \operatorname{cosec}(T - \alpha)].\qquad (2.5.12)$$

In these equations E is the azimuth of the line joining the antennae, α and δ are the right ascension and declination of the object, T is the local sidereal time at the instant of observation and ϕ is the latitude of the interferometer. The path difference also varies because the effective separation of the antennae, d, where

$$d = s[\sin^2 \mu + \cos^2 \mu \sin^2(\psi - E)]^{1/2}\qquad (2.5.13)$$

also changes with time, thus the resolution and fringe spacing (equation (2.5.2)) are altered. Hence the output over a period of time from a radio interferometer with fixed antennae is a set of fringes whose spacing varies, with maxima and minima corresponding to those of figure 2.5.6 for the instantaneous values of the path difference and effective antenna separation (figure 2.5.13).

An improved type of interferometer, which has increased sensitivity and stability, is the phase-switched interferometer (see earlier discussion). The phase of the signal from one antenna is periodically changed by 180° by, for example, switching an extra piece of cable half a wavelength long into or out of the circuit. This has the effect of oscillating the beam pattern of the interferometer through half a fringe width. The difference in the signal for the two positions is then recorded. The phase switching is generally undertaken in the latter stages of the receiver (section 1.2), so that any transient effects of the switching are not amplified. The output fluctuates either side of the zero position as the object moves across the sky (figure 2.5.14).

When the antennae are driven so that they track the object across the sky (figure 2.5.15), then the output of each antenna corresponds to the central peak intensity of each image (figures 2.5.2 and 2.5.4). The path difference

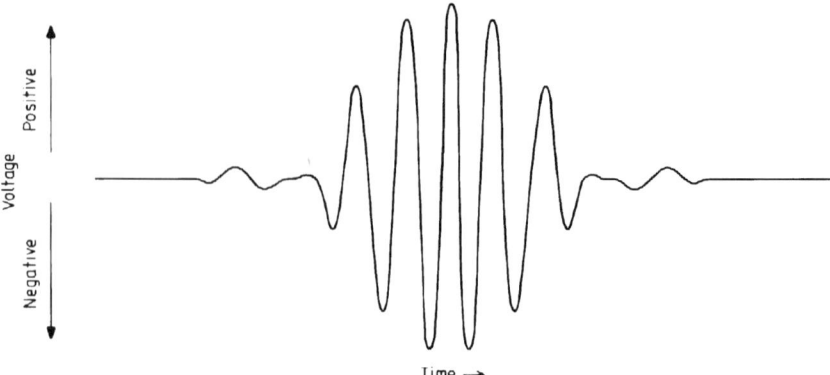

Figure 2.5.14. Output from a radio interferometer with stationary antennae and phase switching, viewing a single point source.

then causes a simple interference pattern (figure 2.5.16 and cf. figure 2.5.5) whose fringe spacing alters due to the varying rate of change of path difference and antenna effective spacing as in the previous case. The maxima are now of constant amplitude since the antenna's projected effective areas are constant. In reality, many more fringes would occur than are shown in figure 2.5.16. As with the optical interferometer, however, the path

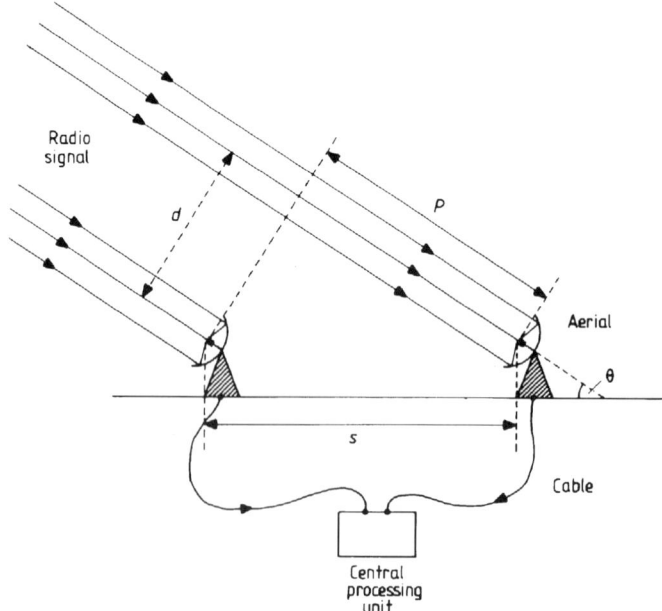

Figure 2.5.15. Schematic arrangement of a radio interferometer with tracking antennae.

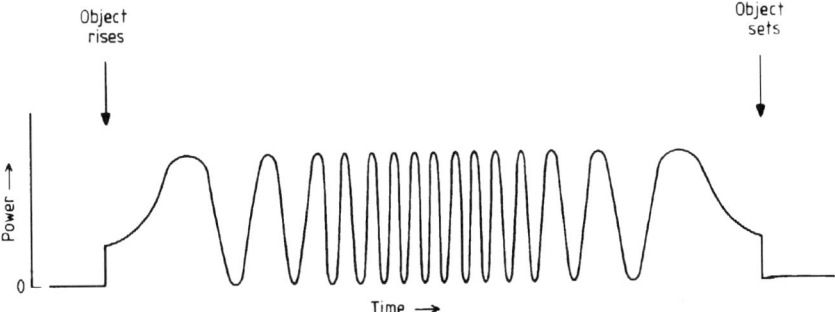

Figure 2.5.16. Output from a radio interferometer with tracking antennae viewing a single point source.

differences at the antennae arising from the inclination of the source to the baseline have to be compensated to a fraction of the coherence length. However, this is much easier at the longer wavelengths since the coherence length is large—30 m for a 10 MHz bandwidth, and 300 m for a 1 MHz bandwidth. The path difference can therefore be corrected by switching in extra lengths of cable or by shifting the recorded signals with respect to each other during data processing.

Reflection interferometers

These are not an important class of interferometer, but they are included for the sake of completeness and for their historical interest. They are closely related to the phased array type of radio telescope discussed in section 1.2. The first astronomical radio interferometer measurements were made with this type of instrument and were of the Sun. Only one antenna is required, and it receives both the direct and the reflected rays (figure 2.5.17). These interfere, and the changing output as the object moves around the sky can be interpreted in terms of the structure of the object.

Aperture synthesis

An interferometer works most efficiently, in the sense of returning the most information, for sources whose separation is comparable with its resolution. For objects substantially larger than the resolution little or no useful information may be obtained. We may, however, obtain information about a larger source by using an interferometer with a *smaller* separation of its elements. The resolution is thereby degraded until it is comparable with the angular size of the source. By combining the results of two interferometers of differing separations, one might thus obtain information on both the large- and small-scale structure of the source. Following this idea

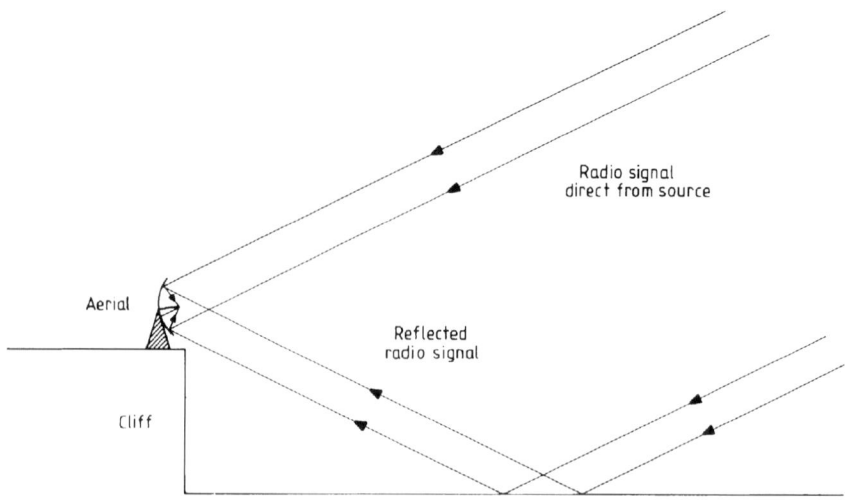

Figure 2.5.17. Schematic possible arrangement for reflection interferometry.

to its logical conclusion led Sir Martin Ryle in the early 1960s to the invention and development of the technique of aperture synthesis. He was awarded the Nobel prize for physics in 1974 for this work.

By this technique, which also goes under the name of Earth-rotation synthesis and is closely related to synthetic aperture radar (section 2.8), observations of a stable source by a number of interferometers are combined to give the effect of an observation using a single very large telescope. The simplest way to understand how aperture synthesis works is to take an alternative view of the operation of an interferometer. The output from a two-aperture interferometer viewing a monochromatic point source is shown in figure 2.5.5, and is a simple sine wave. This output function is just the Fourier transform (equation (2.1.3)) of the source. Such a relationship between source and interferometer output is no coincidence, but is just a special case of the van Cittert–Zernicke theorem:

> The instantaneous output of a two-element interferometer is a measure of one component of the two-dimensional Fourier transform (equation (2.1.10)) of the objects in the field of view of the telescopes.

Thus, if a large number of two-element interferometers were available, so that all the components of the Fourier transform could be measured, then the inverse two-dimensional Fourier transform (equation (2.1.11)) would immediately give a map of the portion of the sky under observation.

Now the complete determination of the Fourier transform of even a small field of view would require an infinite number of interferometers. In practice, therefore, the technique of aperture synthesis is modified in several ways. The most important of these is relaxing the requirement to measure all the Fourier components *at the same instant*. However, once the measurements are spread over time, the source(s) being observed must remain unvarying over the length of time required for those measurements.

Given, then, a source that is stable at least over the measurement time, we may use one or more interferometer pairs to measure the Fourier components using different separations and angles. Of course, it is still not possible to make an infinite number of such measurements, but we may use the discrete versions of equations (2.1.10) and (2.1.11) to bring the required measurements down to a finite number (cf. the one-dimensional analogues, equations (2.1.3), (2.1.4), (2.1.8) and (2.1.9)) though at the expense of losing the high-frequency components of the transform, and hence the finer details of the image (section 2.1). The problem of observing with many different baselines is eased because we are observing from the rotating Earth. Thus, if a single pair of telescopes observes a source over 24 hours, the orientation of the baseline revolves through 360° (figure 2.5.18). If the object is not at the north (or south) pole then the projected spacing of the interferometer will also vary and they will seem to trace out an ellipse. The continuous output of such an interferometer is a complex function whose amplitude is proportional to the amplitude of the Fourier transform and whose phase is the phase shift in the fringe pattern.

The requirement for 24 hours of observation would limit the technique to circumpolar objects. Fortunately, however, only 12 hours are actually

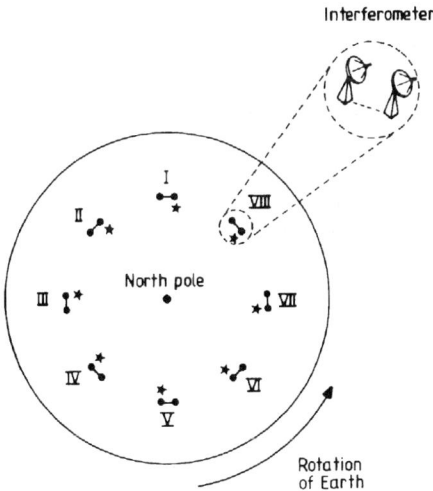

Figure 2.5.18. Changing orientation of an interferometer. The Earth is viewed from above the North Pole and successive positions of the interferometer at three-hour intervals are shown. Notice how the orientation of the starred antenna changes through 360° with respect to the other antenna during a day.

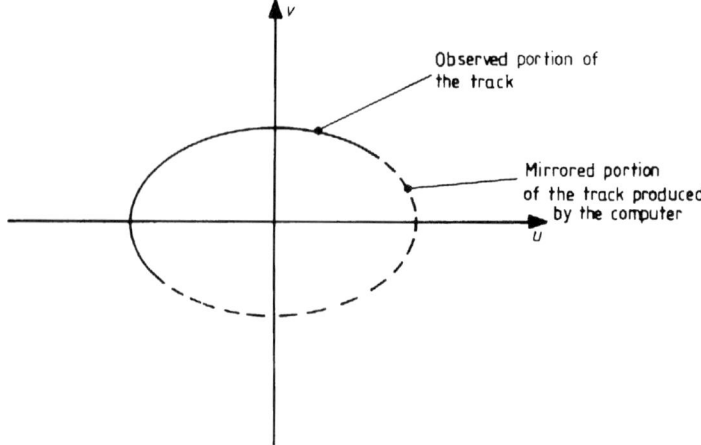

Figure 2.5.19. Track of one element of an interferometer with respect to the other in the u–v plane for an object at a declination of $\pm 35°$.

required, and the other 12 hours can then be calculated by the computer from the conjugates of the first set of observations. Hence, aperture synthesis can be applied to any object in the same hemisphere as the interferometer.

Two elements arranged upon an east–west line follow a circular track perpendicular to the Earth's rotational axis. It is thus convenient to choose the plane perpendicular to the Earth's axis to work in, and this is usually termed the u–v plane. If the interferometer is not aligned east–west, then the paths of its elements will occupy a volume in the u–v–w space, and additional delays will have to be incorporated into the signals to reduce them to the u–v plane. The paths of the elements of an interferometer in the u–v plane range from circles for an object at a declination of $\pm 90°$ through increasingly narrower ellipses to a straight line for an object with a declination of $0°$ (figure 2.5.19).

A single 12 hour observation by a two-element interferometer thus samples all the components of the Fourier transform of the field of view covered by its track in the u–v plane. We require, however, the whole of the u–v plane to be sampled. Thus a series of 12 hour observations must be made, with the interferometer baseline changed by the diameter of one of its elements each time (figure 2.5.20). The u–v plane is then sampled completely out to the maximum baseline possible for the interferometer, and the inverse Fourier transform will give an image equivalent to that from a single telescope with a diameter equal to the maximum baseline.

A two-element radio interferometer with (say) 20 m diameter antennae and a maximum baseline of 1 km would need to make fifty 12 hour observations in order to synthesize a 1 km diameter telescope. By using more than two antennae, however, the time required for the observations can be

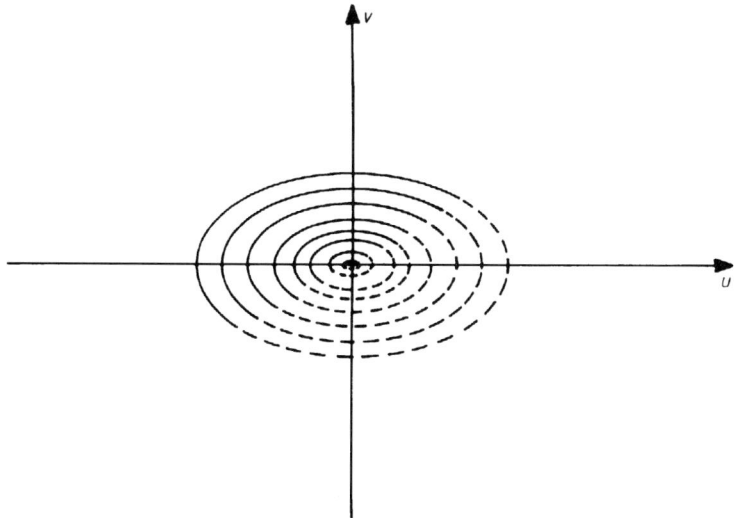

Figure 2.5.20. Successive tracks in the u–v plane of a two-element interferometer as its baseline is varied.

much reduced. Thus with six elements, there are 15 different pairs. If the spacings of these pairs of elements are all different (non-redundant spacing, something that it is not always possible to achieve) then the 50 visibility functions required by the previous example can be obtained in just four 12 hour observing sessions. If the source has a reasonably smooth spectrum, the u–v plane may be sampled even more rapidly by observing simultaneously at several nearby frequencies (multi-frequency synthesis). Since the path differences between the antennae are dependent upon the operating wavelength, this effectively multiplies the number of interferometer pairs with different separations by the number of frequencies being observed. The analysis of the results from such observations, however, will be considerably complicated by any variability of the source over the frequencies being used.

A radio aperture synthesis system such as we have just been considering is a filled-aperture system. That is, the observations synthesize the sensitivity and resolution of a telescope with a diameter equal to the maximum available baseline. Although many aperture synthesis systems are of this type, it becomes increasingly impractical and expensive to sample the whole u–v plane as the baseline extends beyond 10 km or so. The largest such system is the VLA[38] in New Mexico. It uses twenty-seven 25 m dishes arranged in a 'Y' pattern with a maximum baseline of 36 km.

---

[38] Very Large Array.

Aperture synthesis systems larger than the VLA have unfilled apertures. This also applies to some smaller radio and to optical aperture synthesis systems. The Fourier transform of the source is thus not fully sampled. Special techniques, known as hybrid mapping (see below), are then required to produce the final maps of radio sources. The MERLIN[39] system based at Jodrell Bank in the UK is typical of these larger unfilled systems. MERLIN has seven fixed antennae with baselines ranging from 6 to 230 km, and can reach resolutions of $0.008''$ at a wavelength of 10 mm. In the Netherlands, the WSRT[40] uses ten fixed and four movable 25 m dishes over a maximum baseline of 3 km. For the future, the SKA is a proposal for some thirty 200 m diameter radio dishes that will provide a collecting area of a square kilometre with a maximum baseline of 1000 km; an alternative design for this project is based upon thousands of smaller Luneburg lens-based receivers.

Beyond MERLIN and similar radio systems, we have very-long-base-line interferometry (VLBI). For VLBI, the elements of the interferometer may be separated by thousands of kilometres, over several continents, and provide resolutions of $0.001''$ or better. In VLBI, the signals from each element are separately recorded along with timing pulses from an atomic clock. The recordings are then physically brought together and processed by a computer that uses the time signals to ensure the correct registration of one radio signal with respect to another. The VLBA,[41] for example, uses ten 25 m telescopes spread out over the USA from Hawaii to the US Virgin islands. Its maximum baseline is 10 000 km, and at a wavelength of 7 mm its resolution can be $0.00015''$. Recently the Japanese Halca spacecraft carrying an 8 m radio telescope has been linked with up to 40 ground-based dishes, to give baselines up to 30 000 km in length.

At millimetre and sub-millimetre wavelengths the Owens valley radio observatory in California is operating an interferometer with six 10 m tele-scopes and can achieve a resolution of $0.5''$, while on the Plateau de Bure in France, IRAM[42] operates a system with six 15 m telescopes providing imaging to better than one second of arc. The Atacama Large Millimetre Array (ALMA) is due to be completed by 2009. It will have sixty-four 12 m antennas and operate at wavelengths from 10 mm to 0.35 mm and have angular resolutions down to $0.01''$ at its maximum antenna separation of 10 km.

Aperture synthesis at visible and near infrared wavelengths has only recently been successfully attempted, because of the very stringent require-ments on the stability and accuracy of construction of the instruments

[39] Multi-Element Radio Linked INterferometer.
[40] Westerbork Synthesis Radio Telescope.
[41] Very Long Baseline Array.
[42] Institut de Radio Astronomie Millimétrique.

imposed by the small wavelengths involved. COAST[43] based at Cambridge (UK) produced its first images in 1995, and now uses five 0.4 m telescopes with a maximum baseline of 48 m. The CHARA[44] array on Mount Wilson can reach a resolution of $0.0002''$ using six 1 m telescopes with a maximum baseline of 350 m. The four 8 m telescopes of ESO's VLTI[45] have recently produced their first fringes and it is planned to add four 1.8 m telescopes that will be dedicated to aperture synthesis and be able to be moved to 30 different positions offering a maximum baseline of 200 m. It is also planned that the two 10 m Keck telescopes will shortly combine with four 1.8 m telescopes to act as an interferometer. There are several projects currently being planned for space-based interferometers. GAIA and SIM are intended for astrometry and are discussed in section 5.1. TPF[46] will use five spacecraft and four 3.5 m telescopes in an infrared nulling interferometer specifically designed to detect Earth-like planets.

The extraction of the image of the sky from an aperture synthesis system is complicated by the presence of noise and errors. Overcoming the effects of these adds additional stages to the data reduction process. In summary it becomes:

(a) data calibration,
(b) inverse Fourier transform,
(c) deconvolution of instrumental effects,
(d) self-calibration.

Data calibration is required in order to compensate for problems such as errors in the locations of the elements of the interferometer, and variations in the atmosphere. It is carried out by comparing the theoretical instrumental response function with the observed response to an isolated stable point source. These responses should be the same and if there is any difference then the data calibration process attempts to correct it by adjusting the amplitudes and phases of the Fourier components. Often ideal calibration sources cannot be found close to the observed field. Then, calibration can be attempted using special calibration signals, but these are insensitive to the atmospheric variations, and the result is much inferior to that obtained using a celestial source.

Applying the inverse Fourier transform to the data has already been discussed, and further general details are given in section 2.1.

After the inverse Fourier transformation has been completed, we are left with the 'dirty' map of the sky. This is the map contaminated by artefacts introduced by the PSF of the interferometer. The primary components of

[43] Cambridge Optical Aperture Synthesis Telescope.
[44] Centre for High Angular Resolution Astronomy.
[45] Very Large Telescope Interferometer.
[46] Terrestrial Planet Finder.

the PSF, apart from the central response, are the side lobes. These appear on the dirty map as a series of rings that extend widely over the sky, and which are centred on the central response of the PSF. The PSF can be calculated from interferometry theory to a high degree of precision. The deconvolution can then proceed as outlined in section 2.1.

The maximum entropy method discussed in section 2.1 has recently become widely used for determining the best source function to fit the dirty map. Another method of deconvolving the PSF, however, has long been in use, and is still used by many workers, and that is the method known as CLEAN.

The CLEAN algorithm was introduced by Jan Högbom in 1974. It involves the following stages.

(a) Normalize the PSF (instrumental profile or 'dirty' beam) to (gI_{max}), where I_{max} is the intensity of the point of maximum intensity in the dirty map and g is the 'loop gain' with a value between 0 and 1.
(b) Subtract the normalized PSF from the dirty map.
(c) Find the point of maximum intensity in the new map—this may or may not be the same point as before—and repeat the first two steps.
(d) Continue the process iteratively until I_{max} is comparable with the noise level.
(e) Produce a final clear map by returning all the components removed in the previous stages in the form of 'clean beams' with appropriate positions and amplitudes. The clean beams are typically chosen to be Gaussian with similar widths to the central response of the dirty beam.

CLEAN has proved to be a useful method, despite its lack of a substantial theoretical basis, for images made up of point sources. For extended sources MEMs are better because CLEAN may then require thousands of iterations, though as pointed out in section 2.1, MEMs also suffer from problems, especially the variation of resolution over the image.

Many packages for the reduction of radio interferometric data are available within the Astronomical Image Processing System (AIPS and AIPS++) produced by the National Radio Astronomy Observatory (NRAO). This is the most widely used set of software for radio work, and is available on Starlink for UK professional astronomers. It may be downloaded from the web free of charge for use on personal computers (see http://www.aoc.nrao.edu/aips/). You will also need your computer to have a UNIX or LINUX operating system. The latter is available, also without charge for downloading on to PCs, at http://www.linux.org/. You are warned, however, that though available free of charge, these packages are not easy for an inexperienced computer user to implement successfully. However, with the increasing speed and power of computers, real-time processing of the data from radio telescopes and aperture synthesis systems is becoming more common.

The final stage of self-calibration is required for optical systems and for radio systems when the baselines become more than a few kilometres in length, since the atmospheric effects then differ from one telescope to another. Under such circumstances with three or more elements we may use the closure phase that is independent of the atmospheric phase delays. The closure phase is defined as the sum of the observed phases for the three baselines made by three elements of the interferometer. It is independent of the atmospheric phase delays as we may see by defining the phases for the three baselines in the absence of an atmosphere to be ϕ_{12}, ϕ_{23} and ϕ_{31}, and the atmospheric phase delays at each element as a_1, a_2 and a_3. The observed phases are then

$$\phi_{12} + a_1 - a_2$$

$$\phi_{23} + a_2 - a_3$$

$$\phi_{31} + a_3 - a_1.$$

The closure phase is then given by the sum of the phases around the triangle of the baselines:

$$\phi_{123} = \phi_{12} + a_1 - a_2 + \phi_{23} + a_2 - a_3 + \phi_{31} + a_3 - a_1 \qquad (2.5.14)$$

$$= \phi_{12} + \phi_{23} + \phi_{31}. \qquad (2.5.15)$$

From equation (2.5.15) we may see that the closure phase is independent of the atmospheric effects and is equal to the sum of the phases in the absence of an atmosphere. In a similar way an atmosphere-independent closure amplitude can be defined for four elements:

$$G_{1234} = \frac{A_{12}A_{34}}{A_{13}A_{24}} \qquad (2.5.16)$$

where A_{12} is the amplitude for the baseline between elements 1 and 2 etc. Neither of these closure quantities are actually used to form the image, but they are used to reduce the number of unknowns in the procedure. At millimetre wavelengths, the principal atmospheric effects are due to water vapour. Since this absorbs the radiation, as well as leading to the phase delays, monitoring the sky brightness can provide information on the amount of water vapour along the line of sight, and so provide additional information for correcting the phase delays.

For VLBI, 'hybrid mapping' is required since there is insufficient information in the visibility functions to produce a map directly. Hybrid mapping is an iterative technique that uses a mixture of measurements and guesswork. An initial guess is made at the form of the required map. This may well actually be a lower resolution map from a smaller interferometer. From this map the visibility functions are predicted, and the true phases estimated. A new map is then generated by Fourier inversion. This map is then improved and used to provide a new start to the cycle. The iteration is continued until the hybrid map is in satisfactory agreement with the observations.

Intensity interferometer

Until the last decade, the technical difficulties of an optical Michelson interferometer severely limited its usefulness. Most of these problems, however, may be reduced in a device that correlates intensity fluctuations. Robert Hanbury-Brown originally invented the device in 1949 as a radio interferometer, but it has found its main application in the optical region. The disadvantage of the system compared with the Michelson interferometer is that phase information is lost, and so the structure of a complex source cannot be reconstituted. The far greater ease of operation of the intensity interferometer led to it being able to measure some hundred stellar diameters between 1965 and 1972. Although it has been long closed down, the instrument is briefly discussed here for its historical interest, and because its principle of operation differs markedly from that of the Michelson interferometers.

The principle of the operation of the interferometer relies upon phase differences in the low-frequency beat signals from different mutually incoherent sources at each antenna, combined with electrical filters to reject the high frequency components of the signals. The schematic arrangement of the system is shown in figure 2.5.21. We may imagine the signal from a source resolved into its Fourier components and consider the interaction of one such component from one source with another component from the other source. Let the frequencies of these two components be ν_1 and ν_2, then when they are mixed, there will be two additional frequencies—the upper and lower beat frequencies, $(\nu_1 + \nu_2)$ and $(\nu_1 - \nu_2)$—involved. For light waves, the lower beat frequency will be in the radio region, and this component may easily be filtered out within the electronics from the much higher original and upper beat frequencies. The low-frequency outputs from the two telescopes are multiplied and integrated over a short time interval and the signal bandwidth to produce the correlation function, K. Hanbury-Brown and Richard Twiss were able to show that K was simply the square of the fringe visibility (equation 2.5.6) for the Michelson interferometer.

The intensity interferometer was used to measure stellar diameters and $K(d)$ reaches its first zero when the angular stellar diameter, θ', is given by

$$\theta' = \frac{1.22\lambda}{d} \tag{2.5.17}$$

where d is the separation of the receivers. So the resolution of an intensity interferometer (and also of a Michelson interferometer) for stellar discs is

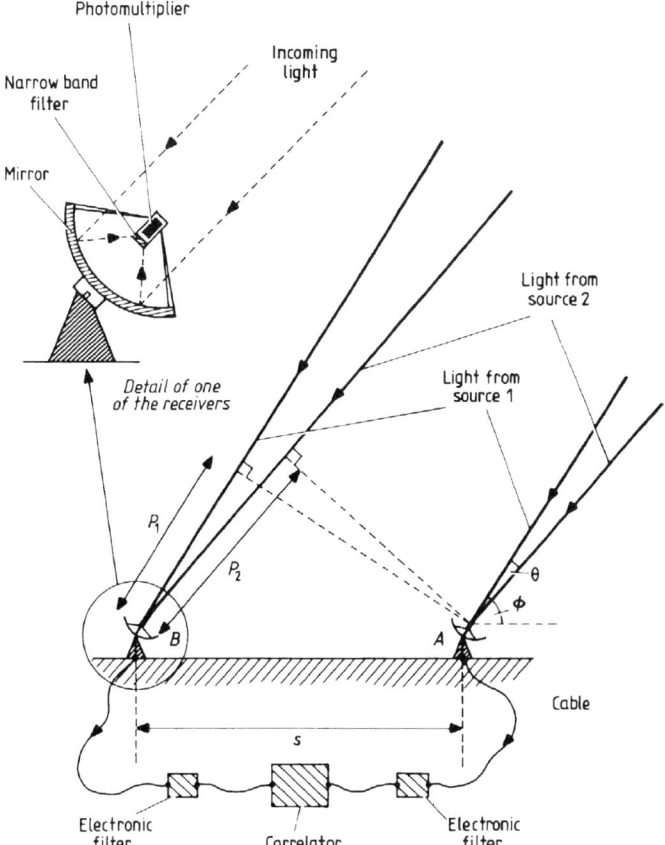

Figure 2.5.21. Schematic arrangement of an intensity interferometer.

the same as that of a telescope (equation (2.5.1)) whose diameter is equal to the separation of the receivers (figure 2.5.22).

The greater ease of construction and operation of the intensity interferometer over the Michelson interferometer arises from its dependence upon the beat frequency of two light beams of similar wavelengths, rather than upon the actual frequency of the light. A typical value of the lower beat frequency is 100 MHz, which corresponds to a wavelength of 3 m. Thus the path differences for the two receivers may vary by up to about 0.3 m during an observing sequence without ill effects. Scintillation, by the same argument, is also negligible.

Only one working intensity interferometer has been constructed. It was built by Hanbury-Brown at Narrabri in Australia. It has now been decommissioned. It used two 6.5 m reflectors that were formed from several hundred smaller mirrors. There was no need for very high optical

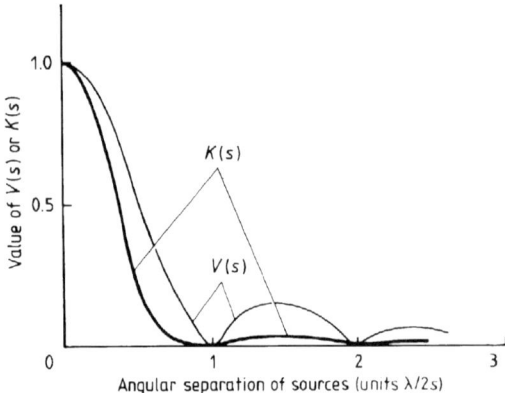

Figure 2.5.22. Comparison of the fringe visibility of a Michelson interferometer with the correlation function of an intensity interferometer.

quality since the reflectors simply acted as light buckets and only the brightest stars could be observed. The reflectors were mounted on trolleys on a circular track 94 m in radius. The line between the reflectors could therefore always be kept perpendicular to the line of sight to the source, and their separation could be varied from 0 to 196 m (figure 2.5.23). It operated at a wavelength of 433 nm, giving it a maximum resolution of about $0.0005'$ of arc.

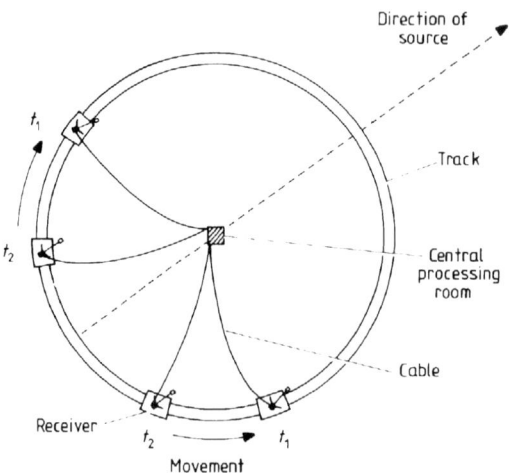

Figure 2.5.23. Schematic layout of the intensity interferometer at Narrabri, showing the positions of the receivers for observing in the same direction with different baselines. Note there were only *two* receivers; the diagram shows the arrangement at two separate baselines, superimposed.

Exercises

2.5.1 Calculate the separation of the slits over a telescope objective in the first type of Michelson stellar interferometer, which would be required to measure the diameter of Ganymede (a) at opposition and (b) near conjunction of Jupiter. Take Ganymede's diameter to be 5000 km and the eye's sensitivity to peak at 550 nm.

2.5.2 Observations of a quasar that is thought to be at a distance of 1500 Mpc just reveal structure to a VLBI working at 50 GHz. If its maximum baseline is 9000 km, what is the linear scale of this structure?

2.5.3 Calculate the maximum distance at which the Narrabri intensity interferometer would be able to measure the diameter of a solar-type star.

2.6 Speckle interferometry

This technique is a kind of poor man's space telescope since it provides near diffraction-limited performance from Earth-based telescopes. It works by obtaining images of the object sufficiently rapidly to freeze the blurring of the image that arises from atmospheric scintillation. The total image then consists of a large number of small dots or speckles, each of which is a diffraction-limited image for some objective diameter up to and including the diameter of the actual objective (figure 2.6.1). An alternative to this technique is the adaptive optics telescope (section 1.1), where adjusting the telescope optics to compensate for the atmospheric effects recombines the speckles.

We may see how this speckled image structure arises by considering the effect of scintillation upon the incoming wavefront from the object. If we

Figure 2.6.1. Schematic image of a point source obtained using a very short exposure time.

assume that above the atmosphere the wavefront is planar and coherent, then the main effect of scintillation is to introduce differential phase delays across it. The delays arise because the atmosphere is non-uniform with different cells within it having slightly different refractive indices. A typical cell size is 0.1 m, and the scintillation frequencies usually lie in the range 1 to 100 Hz. Thus some 100 atmospheric cells will affect an average image from a 1 m telescope at any given instant. These will be rapidly changing, and over a normal exposure time, which can range from seconds to hours, they will form an integrated image that will be large and blurred compared with the diffraction-limited image. Even under the best seeing conditions, the image is rarely less than one second of arc across. An exposure of a few milliseconds, however, is sufficiently rapid to freeze the image motion, and the observed image is then just the resultant of the contributions from the atmospheric cells across the telescope objective at that moment. Now the large number of these cells renders it highly probable that some of the phase delays will be similar to each other, and so some of the contributions to the image will be in phase with each other. These particular contributions will have been distributed over the objective in a random manner. Considering two such contributions, we have in fact a simple interferometer, and the two beams of radiation will combine in the image plane to produce results identical with those of an interferometer whose baseline is equal to the separation of the contributions on the objective. We have already seen what this image structure might be (figure 2.5.6). If several collinear contributions are in phase, then the image structure will approach that shown in figure 4.1.2 modulated by the intensity variation due to the aperture of a single cell. The resolution of the images is then given by the maximum separation of the cells. When the in-phase cells are distributed in two dimensions over the objective, the images have resolutions in both axes given by the maximum separations along those axes at the objective. The smallest speckles in the total image therefore have the diffraction-limited resolution of the whole objective, assuming always of course that the optical quality of the telescope is sufficient to reach this limit. Similar results will be obtained for those contributions to the image that are delayed by an integral number of wavelengths with respect to each other. Intermediate phase delays will cause destructive interference to a greater or lesser extent at the point in the image plane that is symmetrical between the two contributing beams of radiation, but will again interfere constructively at other points, to produce an interference pattern that has been shifted with respect to its normal position. Thus all pairs of contributions to the final image interfere with each other to produce one or more speckles.

In its most straightforward application, speckle interferometry can produce an actual image of the object. Each speckle is a photon noise limited image of the object, so that in some cases the structure can even be seen by eye on the original image. A schematic instantaneous image of a

Figure 2.6.2. Schematic image of a just-resolved double source obtained with a very short exposure time.

double star that is just at the diffraction limit of the objective is shown in figure 2.6.2. Normally, however, the details of the image are completely lost in the noise. Then, many images, perhaps several thousand, must be averaged before any improvement in the resolution is gained.

More commonly, the data are Fourier analysed, and the power spectrum obtained. This is the square of the modulus of the Fourier transform of the image intensity (equation (2.1.3)). The Fourier transform can be obtained directly using a large computer, or optical means can be used. In the latter case, the photograph is illuminated by collimated coherent light (figure 2.6.3). It is placed at one focus of an objective and its Fourier transform is imaged at the back focus of the objective. A spatial filter at this point can be used to remove unwanted frequencies if required, and

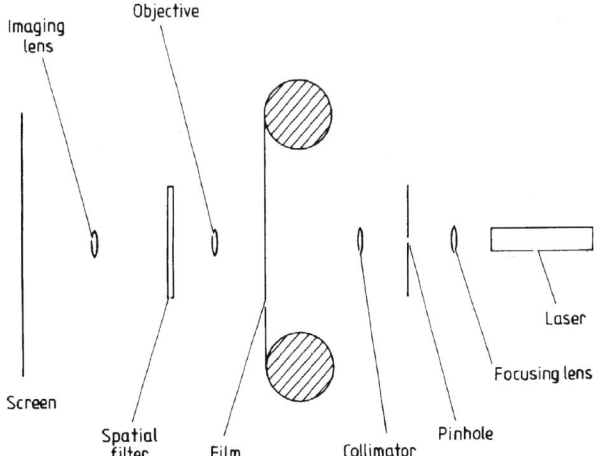

Figure 2.6.3. Arrangement for obtaining the Fourier transform of a speckle photograph by optical means.

then the Fourier transform re-imaged. However it may be obtained, the power spectrum can then be inverted to give centrisymmetric information such as diameters, limb darkening, oblateness and binarity. Non-centrisymmetric structure can only be obtained if there is a point source close enough for its image to be obtained simultaneously with that of the object and for it to be affected by the atmosphere in the same manner as the image of the object (i.e. the point source is within the same isoplanatic patch of sky as the object of interest). Then deconvolution (section 2.1) can be used to retrieve the image structure.

The practical application of the technique requires the use of large telescopes and high-gain image intensifiers. Not only are very short exposures required (0.001 to 0.1 s), but very large plate scales (0.1 to 1 seconds of arc per millimetre) are also needed in order to separate the individual speckles. Furthermore, the wavelength range must be restricted by a narrow-band filter to 20 to 30 nm. Even with such a restricted wavelength range, it may still be necessary to correct any remaining atmospheric dispersion using a low-power direct vision spectroscope. Thus the limiting magnitude of the technique is currently about $+18^m$, and it has found its major applications in the study of red supergiants, the orbits of close binary stars, Seyfert galaxies, asteroids and fine details of solar structure. Speckle interferometry, however, is a recent invention and it is still developing, with CCD detectors now being used in place of photographic emulsions.

2.7   Occultations

Background

An occultation is the more general version of an eclipse. It is the obscuration of a distant astronomical object by the passage in front of it of a closer object. Normally the term implies occultation by the Moon, or more rarely by a planet, since other types of occultation are almost unknown. It is one of the most ancient forms of astronomical observation, with records of lunar occultations stretching back some two and a half millennia. More recently interest in the events has been revived for their ability to give precise positions for objects observed at low angular resolution in, say, the x-ray region, and for their ability to give structural information about objects at better than the normal diffraction-limited resolution of a telescope. To see how the latter effect is possible, we must consider what happens during an occultation.

First let us consider Fresnel diffraction at a knife-edge (figure 2.7.1) for radiation from a monochromatic point source. The phase difference, δ, at a point P between the direct and the diffracted rays is then

$$\delta = \frac{2\pi}{\lambda}\{d_1 + [d_2^2 + (d_2 \tan \theta)^2]^{1/2} - [(d_1 + d_2)^2 + (d_2 \tan \theta)^2]^{1/2}\} \quad (2.7.1)$$

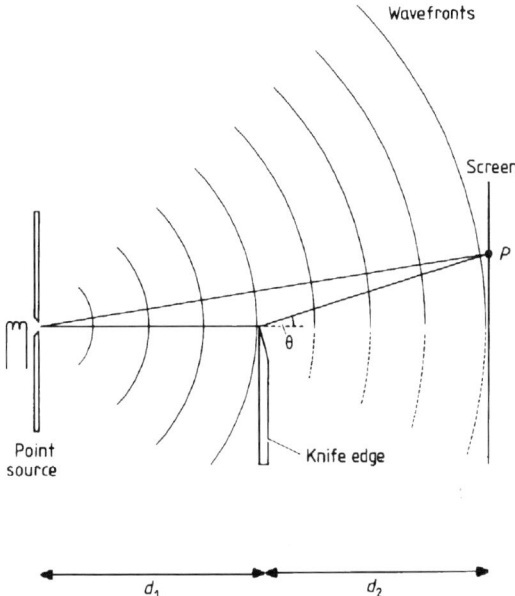

Figure 2.7.1. Fresnel diffraction at a knife-edge.

which, since θ is very small, simplifies to

$$\delta = \frac{\pi d_1 d_2}{\lambda(d_1 + d_2)}\, \theta^2. \tag{2.7.2}$$

The intensity at a point in a diffraction pattern is obtainable from the Cornu spiral (figure 2.7.2), by the square of the length of the vector, $\mathbf{A}$. $\mathbf{P'}$ is the point whose distance along the curve from the origin, l, is given by

$$l = \left(\frac{2 d_1 d_2}{\lambda(d_1 + d_2)}\right)^{1/2}\theta \tag{2.7.3}$$

and the phase difference at $\mathbf{P'}$ is the angle that the tangent to the curve at that point makes with the x axis, or from equation (2.7.2)

$$\delta = \tfrac{1}{2}\pi l^2. \tag{2.7.4}$$

The coordinates of $\mathbf{P'}$, which is the point on the Cornu spiral giving the intensity at $\mathbf{P}$, x and y, are obtainable from the Fresnel integrals

$$x = \frac{1}{\sqrt{2}} \int_0^l \cos(\tfrac{1}{2}\pi l^2)\, \mathrm{d}l \tag{2.7.5}$$

$$y = \frac{1}{\sqrt{2}} \int_0^l \sin(\tfrac{1}{2}\pi l^2)\, \mathrm{d}l \tag{2.7.6}$$

whose pattern of behaviour is shown in figure 2.7.3.

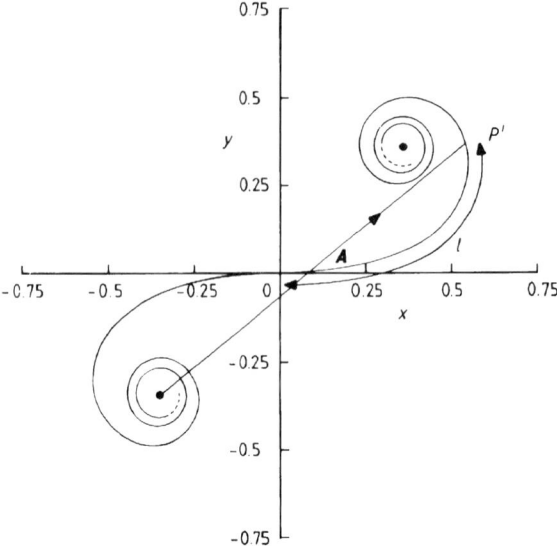

Figure 2.7.2. The Cornu spiral.

If we now consider a star occulted by the Moon, then we have

$$d_1 \gg d_2 \tag{2.7.7}$$

so that

$$l \approx \left(\frac{2d_2}{\lambda} \right) \theta \tag{2.7.8}$$

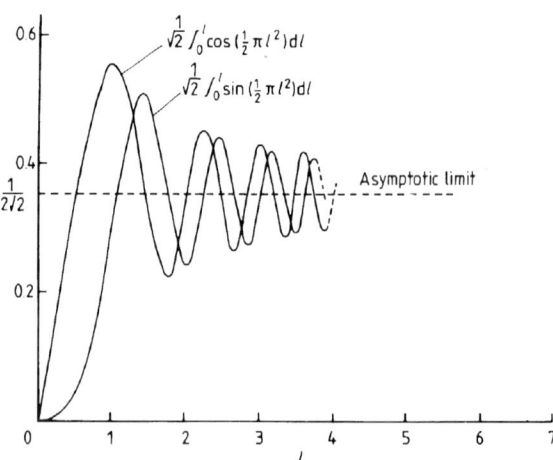

Figure 2.7.3. The Fresnel integrals.

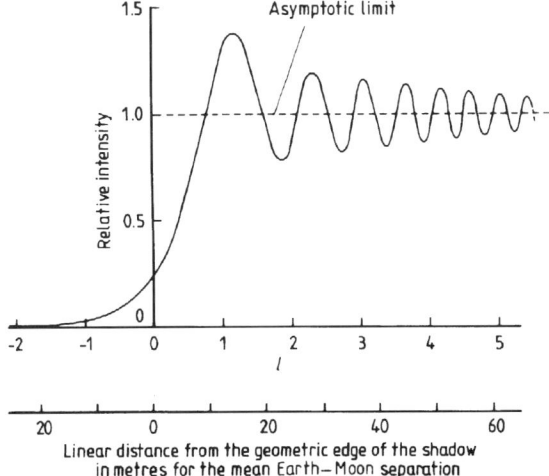

Figure 2.7.4. Fringes at the edge of the lunar shadow.

but otherwise the situation is unchanged from the one we have just discussed.
The edge of the Moon, of course, is not a sharp knife-edge, but since even a
sharp knife-edge is many wavelengths thick, the two situations are not in
practice any different. The shadow of the Moon cast by the star on to the
Earth therefore has a standard set of diffraction fringes around its edge.
The intensities of the fringes are obtainable from the Cornu spiral and
equations (2.7.5) and (2.7.6), and are shown in figure 2.7.4. The first
minimum occurs for

$$l = 1.22 \tag{2.7.9}$$

so that for the mean Earth–Moon distance

$$d_2 = 3.84 \times 10^8 \, \mathrm{m} \tag{2.7.10}$$

we obtain

$$\theta = 0.0064'' \tag{2.7.11}$$

at a wavelength of 500 nm. The fringes have a linear width of about 12 m
therefore, when the shadow is projected on to a part of the Earth's surface
that is perpendicular to the line of sight. The mean rate of angular motion
of the Moon, $\dot{\theta}$, is

$$\dot{\theta} = 0.55'' \, \mathrm{s}^{-1} \tag{2.7.12}$$

so that at a given spot on the Earth, the fringes will be observed as intensity
variations of the star as it is occulted, with a basic frequency of up to 85 Hz.
Usually the basic frequency is lower than this since the Earth's rotation can
partially off-set the Moon's motion, and because the section of the lunar

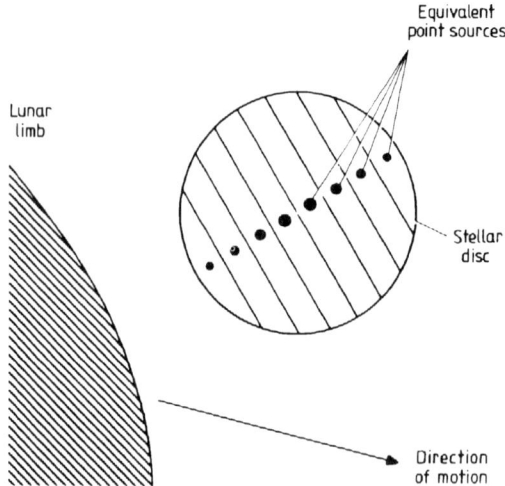

Figure 2.7.5. Schematic decomposition of a non-point source into equivalent point sources for occultation purposes.

limb that occults the star will generally be inclined to the direction of the motion.

If the star is not a point source, then the fringe pattern shown in figure 2.7.4 becomes modified. We may see how this happens by imagining two point sources separated by an angle of about $0.0037''$ in a direction perpendicular to the limb of the Moon. At a wavelength of 500 nm the first maximum of one star is then superimposed upon the first minimum of the other star, and the amplitude of the resultant fringes is much reduced compared with the single-point-source case. The separation of the sources parallel to the lunar limb is, within reason, unimportant in terms of its effect on the fringes. Thus an extended source can be divided into strips parallel to the lunar limb (figure 2.7.5), and each strip then behaves during the occultation as though it were a centred point source of the relevant intensity. The fringe patterns from all these point sources are then superimposed in the final tracing of the intensity variations. The precise nature of the alteration from the point source case will depend upon the size of the star and upon its surface intensity variations through such factors as limb darkening and gravity darkening etc. as discussed later in this section. A rough guide, however, is that changes from the point source pattern are just detectable for stellar diameters of $0.002''$, while the fringes disappear completely for object diameters of about $0.2''$. In the latter situation, the diameter may be recoverable from the total length of time that is required for the object to fade from sight. Double stars may also be distinguished from point sources when their separation exceeds about $0.002''$.

Techniques

In complete contrast to almost all other areas of astronomy, the detection of occultations in the optical region is best undertaken using comparatively small telescopes. This arises from the linear size of the fringes, which we have seen is about 12 m. Telescopes larger than about 1 m will therefore simultaneously sample widely differing parts of the fringe so that the detected luminosity variations become smeared. Since the signal-to-noise ratio decreases with decreasing size of telescope, the optimum size for observing occultations usually lies between 0.5 and 2 m.

Most of the photometers described in section 3.2 can be used to detect the star. However, we have seen that the observed intensity variations can have basic frequencies of 85 Hz with harmonics of several hundred hertz. The photometer and its associated electronics must therefore be capable of responding sufficiently rapidly to pick up these frequencies. Thus long integration times are not possible. Also chopping and phase-sensitive detection are ruled out unless they are operating at 10 kHz or more. The photometers used to detect occultations thus require response times of one to a few milliseconds.

Fringes are also blurred by the waveband over which the photometer is operating. We have seen from equation (2.7.8) that the monochromatic fringe pattern is wavelength dependent. A bichromatic source with wavelengths differing by a factor of 2.37 would have the first maximum for one wavelength superimposed upon the first minimum for the other, so that the fringes would almost disappear. Smaller bandwidths will still degrade the visibility of the fringes although to a lesser extent. Using the standard UBV filters (section 3.1) only five or six fringes will be detectable even in the absence of all other noise sources. A bandwidth of about 20 nm must be used if the fringes found using a medium-sized telescope are not to deteriorate through this effect.

The main difference between a photometer used for occultation observations and one used for other purposes lies in the recording technique. In normal photometry the data rate is usually sufficiently slow that a chart recorder can be used, or the readings can even be read off a meter and noted down by hand. An occultation by contrast only lasts for a second or two at most, and upwards of a thousand data points must be recorded in that time. Although ultraviolet chart recorders can operate at chart speeds of metres per second and so could record this quantity of data, electronic methods are generally preferred for the ease with which the data may then be fed into a computer. Since standard commercial tape recorders have responses ranging to over 10 kHz, they can be used to record the occultation if allied to a voltage-to-frequency converter. Alternatively, that data can be fed straight into a computer memory via an analogue-to-digital converter. In the latter case a completely automatic system can be arranged by using a recycling store. Some few kilobits of information are stored and the latest data overwrite the oldest. A simple detection system to stop the recording a second or two after the occultation will then leave the desired data in the computer's memory ready for processing.

CCDs can also be used to determine the diffraction pattern of an occulted star. The diffraction pattern moves over the CCD at a calculable velocity, and the charges in the pixels are moved through the device at the same rate (cf. image tracking for liquid mirrors, section 1.1). By this process of time delay and integration (TDI), high signal-to-noise ratios can be reached because each portion of the diffraction pattern is observed for a longer time than when a single element detector is used.

Since the observations are obviously always carried out within a few minutes of arc of the brightly illuminated portion of the Moon, the scattered background light is a serious problem. It can be minimized by using clean, dust free optics, precisely made light baffles, and a very small entrance aperture for the photometer, but it can never be completely eliminated. The total intensity of the background can easily be many times that of the star even when all these precautions have been taken. Thus electronic methods must be used to compensate for its effects. The simplest system is to use a differential amplifier that subtracts a preset voltage from the photometer's signal before amplification. It must be set manually a few seconds before the occultation since the background is continually changing. A more sophisticated approach uses a feedback to the differential amplifier that continually adjusts the preset voltage so that a zero output results. The time constant of this adjustment, however, is arranged to be long, so that while the slowly changing background is counteracted, the rapid changes during the occultation are unaffected. The latter system is particularly important for observing disappearances, since it does not require last-minute movements of the telescope on to and off the star.

An occultation can be observed at the leading edge of the Moon (a disappearance event) or at the trailing edge (when it is a reappearance) providing that the limb near the star is not illuminated. Disappearances are by far the easiest events to observe since the star is visible and can be accurately tracked right up to the instant of the occultation. Reappearances can be observed only if a very precise off-set is available for the photometer or telescope, or if the telescope can be set on to the star before its disappearance and then kept tracking precisely on its position for tens of minutes until it reappears. The same information content is available, however, from either event. The Moon often has quite a large motion in declination, so that it is occasionally possible for both a disappearance and the following reappearance to occur at a dark limb. This is a particularly important situation since the limb angles will be different between the two events, and a complete two-dimensional map of the object can be found.

Scintillation is another major problem in observing an occultation. The frequencies involved in scintillation are similar to those of the occultation so that they cannot be filtered out. The noise level is therefore largely due to the effects of scintillation, and it is rare for more than four fringes to be detectable in practice. Little can be done to reduce the problems caused by scintillation since the usual precautions of observing only near the zenith and on the steadiest nights cannot be followed; occultations have to be observed when and where they occur.

Occultations can be used to provide information other than whether the star is a double or not, or to determine its diameter. Precise timing of the event can give the lunar position to $0.05''$ or better, and can be used to calibrate ephemeris time. Until recently, when navigation satellites became available, lunar occultations were used by surveyors to determine the positions on the Earth of isolated remote sites, such as oceanic islands. Occultations of stars by planets are quite rare but are of considerable interest. Their main use is as a probe for the upper atmosphere of the planet, since it may be observed through the upper layers for some time before it is completely obscured. There has also been the recent serendipitous discovery of the rings of Uranus through their occultation of a star. In spectral regions other than the visible, lunar occultations have in the past found application in the determination of precise positions of objects.

Analysis

The analysis of the observations to determine the diameters and/or duplicity of the sources is simple to describe, but very time-consuming to carry out. It involves the comparison of the observed curve with synthesized curves for various models. The synthetic curves are obtained in the manner indicated in figure 2.7.5, taking account also of the bandwidth and the size of the telescope. One additional factor to be taken into account is the inclination of the lunar limb to the direction of motion of the Moon. If the Moon were a smooth sphere this would be a simple calculation, but the smooth limb is distorted by lunar surface features. Sometimes the actual inclination at the point of contact may be determinable from artificial satellite photographs of the Moon. On other occasions, the lunar slope must appear as an additional unknown to be determined from the analysis. Very rarely, a steep slope may be encountered and the star may reappear briefly before its final extinction. More often the surface may be too rough on a scale of tens of metres to provide a good 'straight-edge'. Both these cases can usually be recognized from the data, and have to be discarded, at least for the purposes of determination of diameters or duplicity.

The analysis for positions of objects simply comprises determining the precise position of the edge of the Moon at the instant of occultation. One observation limits the position of the source to a semicircle corresponding to the leading (or trailing) edge of the Moon. Two observations from separate sites, or from the same site at different lunations, fix the object's position to the one or two intersections of two such semicircles. This will usually be sufficient to allow an unambiguous optical identification for the object, but if not, further observations can continue to be added to reduce the ambiguity in its position and to increase the accuracy, for as long as the occultations continue. Generally a source that is occulted one month as observed from the Earth or from a close Earth-orbiting satellite will continue to be occulted for several succeeding lunations. Eventually, however, the rotation of the lunar orbit in the Saros cycle of $18\frac{2}{3}$ years will move the Moon away from the object's position, and no further occultations will occur for several years.

2.8 Radar

Introduction

Radar astronomy and radio astronomy are very closely linked because the same equipment is often used for both purposes. Radar astronomy is less familiar to the astrophysicist, however, for only one star, the Sun, has ever been studied with it, and this is likely to remain the case until other stars are visited by spacecraft, which will be a while yet. Other aspects of astronomy benefit from the use of radar to a much greater extent, so the technique is included for completeness, despite being almost outside the limits of this book, and because its results may find applications in some areas of astrophysics.

Radar, as used in astronomy, can be fairly conventional equipment for use on board spacecraft, or for studying meteor trails, or it can be of highly specialized design and construction for use in detecting the Moon, planets, Sun etc. from the Earth. Its results potentially contain information on three aspects of the object being observed—distance, surface fine scale structure, and relative velocity.

Theoretical principles

Consider a radar system as shown in figure 2.8.1. The transmitter has a power P and is an isotropic emitter. The radar cross section of the object is α, defined as the cross sectional area of a perfectly isotropically scattering sphere which would return the same amount of energy to the receiver as the object. Then the flux at the receiver, f, is given by

$$f = \frac{P\alpha}{4\pi R_1^2 4\pi R_2^2}.$$ (2.8.1)

Normally we have the transmitter and receiver close together, if they are not actually the same antenna, so that

$$R_1 = R_2 = R.$$ (2.8.2)

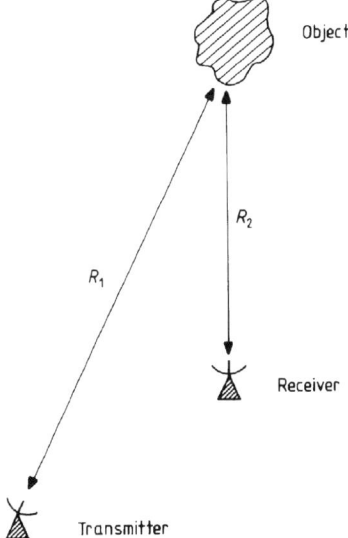

Figure 2.8.1. Schematic radar system.

Furthermore the transmitter would not in practice be isotropic, but would have a gain, g (see section 1.2). If the receiver has an effective collecting area of A_e then the received signal, F, is given by

$$F = \frac{A_e \alpha P g}{16\pi^2 R^4}.$$ (2.8.3)

For an antenna, we have the gain from equation (1.2.10)

$$g = \frac{4\pi \nu^2 A'_e}{c^2}$$ (2.8.4)

where A'_e is the effective area of the transmitting antenna. Much of the time in radar astronomy, the transmitting and receiving dishes are the same, so that

$$A'_e = A_e$$ (2.8.5)

and

$$F = \frac{P \alpha A_e^2 \nu^2}{4\pi c^2 R^4}.$$ (2.8.6)

This last equation is valid for objects that are not angularly resolved by the radar. For targets that are comparable with the beam width or larger, the radar cross section, α, must be replaced by an appropriate integral. Thus for a spherical target which has the radar beam directed towards its centre, we have

$$F \approx \frac{P A_{\mathrm{e}}^2 \nu^2}{4\pi c^2 R^4} \int_0^{\pi/2} 2\pi r \alpha(\phi) \sin \phi \left[s\left(\frac{r \sin \phi}{R} \right) \right]^2 \mathrm{d}\phi \qquad (2.8.7)$$

where r is the radius of the target and is assumed to be small compared with the distance R, ϕ is the angle at the centre of the target to a point on its surface illuminated by the radar beam and $\alpha(\phi)$ is the radar cross section for the surface of the target when the incident and returned beams make an angle ϕ to the normal to the surface. The function is normalized so that the integral tends to α as the beam width increases. $s(\theta)$ is the sensitivity of the transmitter/receiver at an angle θ to its optical axis (cf. equation (1.2.11)).

The amount of flux that is received is not the only criterion for the detection of objects, as we saw in section 1.2 (equation (1.2.4)). The flux must also be sufficiently stronger than the noise level of the whole system if the returned pulse is to be distinguishable. Now we have seen in section 1.2 that the receiver noise may be characterized by comparing it with the noise generated in a resistor at some temperature. We may similarly characterize all the other noise sources, and obtain a temperature, T_{s}, for the whole system, which includes the effects of the target, and of the transmission paths as well as the transmitter and receiver. Then from equation (1.2.5), we have the noise in power terms, N

$$N = 4k T_{\mathrm{s}} \Delta\nu \qquad (2.8.8)$$

where $\Delta\nu$ is the bandwidth of the receiver in frequency terms. The signal-to-noise ratio is therefore

$$\frac{F}{N} = \frac{P \alpha A_{\mathrm{e}}^2 \nu^2}{16\pi k c^2 R^4 T_{\mathrm{s}} \Delta\nu} \qquad (2.8.9)$$

and must be unity or larger if a single pulse is to be detected. Since the target that is selected fixes α and R, the signal-to-noise ratio may be increased by increasing the power of the transmitter, the effective area of the antenna, or the frequency, or by decreasing the system temperature or the bandwidth. Only over the last of these items does the experimenter have any real control—all the others will be fixed by the choice of antenna. However, the bandwidth is related to the length of the radar pulse. Even if the transmitter emits monochromatic radiation, the pulse will have a spread of frequencies given by the Fourier transform of the pulse shape. To a first approximation

$$\Delta\nu \approx \frac{1}{\tau} \quad \mathrm{Hz} \qquad (2.8.10)$$

where τ is the length of the transmitted pulse in seconds. Thus increasing the pulse length can increase the signal-to-noise ratio. Unfortunately for accurate ranging, the pulse needs to be as short as possible, and so an optimum value which minimizes the conflict between these two requirements must be sought, and this will vary from radar system to radar system, and from target to target.

When ranging is not required, so that information is only being sought on the surface structure and the velocity, the pulse length may be increased very considerably. Such a system is then known as a continuous wave (CW) radar, and its useful pulse length is only limited by the stability of the transmitter frequency, and by the spread in frequencies introduced into the returned pulse through Doppler shifts arising from the movement and rotation of the target. In practice CW radars work continuously even though the signal-to-noise ratio remains that for the optimum pulse. The alternative to CW radar is pulsed radar. Astronomical pulsed radar systems have a pulse length typically between 10 µs and 10 ms, with peak powers of several tens of megawatts. For both CW and pulsed systems, the receiver pass band must be matched to the returned pulse in both frequency and bandwidth. The receiver must therefore be tunable over a short range to allow for the Doppler shifting of the emitted frequency.

The signal-to-noise ratio may be improved by integration. With CW radar, samples are taken at intervals given by the optimum pulse length. For pulsed radar, the pulses are simply combined. In either case the noise is reduced by a factor of $N^{1/2}$, where N is the number of samples or pulses that are combined. If the radiation in separate pulses from a pulsed radar system is coherent, then the signal-to-noise ratio improves directly with N. Coherence, however, may only be retained over the time interval given by the optimum pulse length for the system operating in the CW mode. Thus if integration is continued over several times the coherence interval, the signal-to-noise ratio for a pulsed system decreases as $(NN'^{1/2})$, where N is the number of pulses in a coherence interval, and N' the number of coherence intervals over which the integration extends.

For most astronomical targets, whether they are angularly resolved or not, there is some depth resolution with pulsed systems. The physical length of a pulse is $(c\tau)$, and this ranges from about 3 to 3000 km for most practical astronomical radar systems. Thus if the depth of the target, in the sense of the distance along the line of sight between the nearest and farthest points returning detectable echoes, is larger than the physical pulse length, the echo will be spread over a greater time interval than the original pulse. This has both advantages and disadvantages. First, it provides information on the structure of the target with depth. If the depth information can be combined with Doppler shifts arising through the target's rotation, then the whole surface may be mappable, although there may still remain a twofold ambiguity about the equator. Second, and on the debit side, the signal-to-noise ratio is reduced in proportion approximately to the ratio of the lengths of the emitted and returned pulses.

Equation (2.8.9) and its variants that take account of resolved targets etc. is often called the radar equation. Of great practical importance is its R^{-4}

dependence. Thus if other things are equal, the power required to detect an object increases by a factor of 16 when the distance to the object doubles. Thus Venus at greatest elongation needs 37 times the power required for its detection at inferior conjunction. The high dependence upon R can be made to work in our favour, however, when using satellite-borne radar systems. The value of R can then be reduced from tens of millions of kilometres to a few hundred. The enormous gain in the sensitivity that results means that the small low-power radar systems that are suitable for satellite use are sufficient to map the target in detail. The prime example of this to date is, of course, Venus. Most of the data we have on the surface features have come from the radar carried by Magellan and the various Mariner and Pioneer spacecraft.

The radar on board Magellan is of a different type from those so far considered. It is known as Synthetic Aperture Radar (SAR) and has much in common with the technique of aperture synthesis (section 2.5). Such radars have also been used on board several remote sensing satellites, such as Seasat and ERS 1 for studying the Earth.

Synthetic aperture radar uses a single antenna that, since it is on board a spacecraft, is moving with respect to the surface of the planet below. The single antenna therefore successively occupies the positions of the multiple antennae of an array (figure 2.8.2). To simulate the observation of such an array, all that is necessary, therefore, is to record the output from the radar when it occupies the position of one of the elements of the array, and then to add the successive recordings with appropriate shifts to counteract their various time delays. In this way the radar can build up an image whose resolution, along the line of flight of the spacecraft, is many times better than that which the simple antenna would provide by itself. In practice, of course, the SAR operates continually, not just at the positions of the elements of the synthesized array.

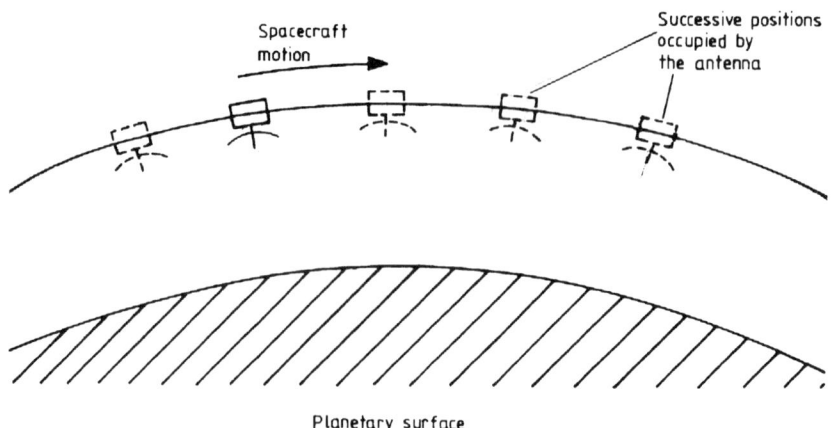

Figure 2.8.2. Synthesis of an interferometer array by a moving single antenna.

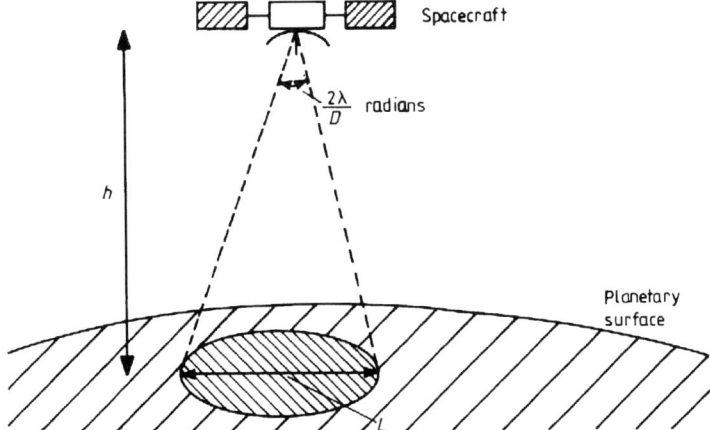

Figure 2.8.3. Radar footprint.

The maximum array length that can be synthesized in this way is limited by the period over which a single point on the planet's surface can be kept under observation. The resolution of a parabolic dish used as a radar is approximately λ/D. If the spacecraft is at a height, h, the radar 'footprint' of the planetary surface is thus about

$$L = \frac{2\lambda h}{D} \tag{2.8.11}$$

in diameter (figure 2.8.3). Since h, in general, will be small compared with the size of the planet, we may ignore the curvature of the orbit. We may then easily see (figure 2.8.4) that a given point on the surface only remains observable while the spacecraft moves a distance L.

Now, the resolution of an interferometer (section 2.5) is given by equation (2.5.2). Thus, the angular resolution of an SAR of length L is

$$\text{resolution} = \frac{\lambda}{2L} = \frac{D}{4H} \quad \text{radians.} \tag{2.8.12}$$

The diameter of the synthesized radar footprint is then

$$\text{footprint diameter} = 2\left(\frac{D}{4h}\right)h = \frac{D}{2}. \tag{2.8.13}$$

Equation (2.8.13) shows the remarkable result that the linear resolution of an SAR is improved by decreasing the size of the radar dish! The linear resolution is just half the dish diameter. Other considerations, such as signal-to-noise ratio (equation (2.8.9)) require the dish to be as large as possible. Thus, an actual SAR system has a dish size that is a compromise, and existing systems typically have 5 to 10 m diameter dishes.

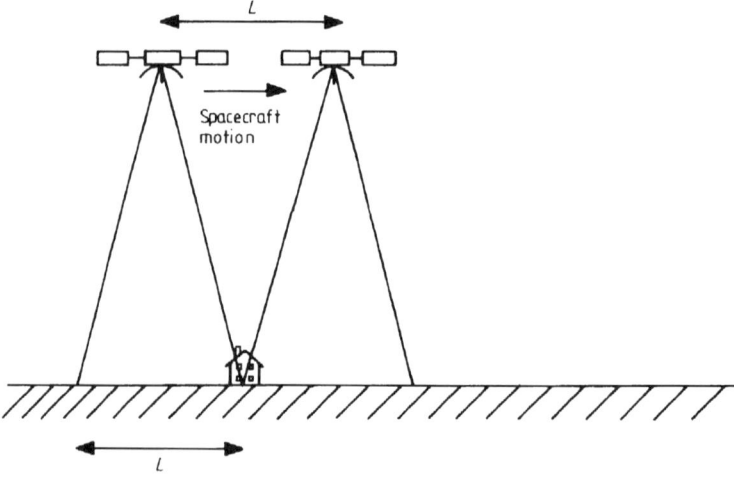

Figure 2.8.4. Maximum length of an SAR.

The above analysis applies to a focused SAR, that is to say, an SAR in which the changing distance of the spacecraft from the point of observation is compensated by phase-shifting the echoes appropriately. In an unfocused SAR the point at which the uncorrelated phase shifts due to the changing distance to the observed object reach about $\lambda/4$ limits the synthesized array length. The diameter of the array footprint is then about

$$\text{unfocused SAR footprint} = \sqrt{2\lambda h}. \qquad (2.8.14)$$

Equipment

Radar systems for use in studying planets etc. have much in common with radio telescopes. Almost any filled aperture radio antenna is usable as a radar antenna. Steerable paraboloids are particularly favoured because of their convenience in use. Unfilled apertures, such as collinear arrays (section 1.2), are not used since most of the transmitted power is wasted.

The receiver is also generally similar to those in use in radio astronomy, except that it must be very accurately tunable to provide compensation for Doppler shifts, and the stability of its frequency must be very high. Its bandpass should also be adjustable so that it may be matched accurately to the expected profile of the returned pulse in order to minimize noise.

Radar systems, however, do require three additional components that are not found in radio telescopes. First, and obviously, there must be a transmitter; second, a very-high-stability master oscillator; third, an accurate timing system. The transmitter is usually second only to the antenna as a proportion of the cost of the whole system. Different targets or purposes will generally require different powers, pulse lengths, frequencies etc. from the transmitter. Thus it must be sufficiently flexible to cope with demands that might typically range from pulse emission at several megawatts for a millisecond burst to continuous wave generation at several hundred kilowatts. Separate pulses must be coherent, at least over a time interval equal to the optimum pulse length for the CW mode, if full advantage of integration to improve signal-to-noise ratio is to be gained. The master oscillator is in many ways the heart of the system. In a typical observing situation, the returned pulse must be resolved to 0.1 Hz for a frequency in the region of 1 GHz, and this must be accomplished after a delay between the emitted pulse and the echo that can range from minutes to hours. Thus the frequency must be stable to one part in 10^{12} or 10^{13}, and it is the master oscillator that provides this stability. It may be a temperature-controlled crystal oscillator, or an atomic clock. In the former case, corrections for ageing will be necessary. The frequency of the master oscillator is then stepped up or down as required and used to drive the exciter for the transmitter, the local oscillators for the heterodyne receivers, the Doppler compensation system, and the timing system. The final item, the timing system, is essential if the range of the target is to be found. It is also essential to know with quite a high accuracy when to expect the echo. This arises because the pulses are not broadcast at constant intervals. If this were to be the case then there would be no certain way of relating a given echo to a given emitted pulse. Thus instead, the emitted pulses are sent out at varying and coded intervals, and the same pattern sought among the echoes. Since the distance between the Earth and the target is also varying, an accurate timing system is vital to allow the resulting echoes with their changing intervals to be found among all the noise.

Data analysis

The analysis of the data in order to obtain the distance to the target is simple in principle, being merely half the delay time multiplied by the speed of light. In practice the process is much more complicated. Corrections for the atmospheres of the Earth and the target (should it have one) are of particular importance. The radar pulse will be refracted and delayed by the Earth's atmosphere, and there will be similar effects in the target's atmosphere with in addition the possibility of reflection from an ionosphere as well as, or instead of, reflection from the surface. This applies especially to solar detection. Then reflection can often occur from layers high above the visible surface: 10^3 km above the photosphere, for example, for a radar frequency of 1 GHz. Furthermore if the target is a deep one, i.e. it is resolved in depth, then the returned pulse will be spread over a greater time interval than the emitted one so that the delay length becomes ambiguous. Other effects such as refraction or delay in the interplanetary medium may also need to be taken into account.

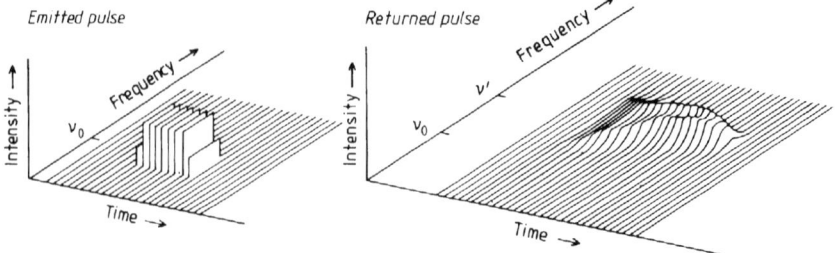

Figure 2.8.5. Schematic change in pulse shape for radar pulses reflected from a deep rotating target. The emitted pulse would normally have a Gaussian profile in frequency, but it is shown here as a square wave in order to enhance the clarity of the changes between it and the echo.

For a deep, rotating target, the pulse is spread out in time and frequency, and this can be used to plot maps of the surface of the object (figure 2.8.5). A cross section through the returned pulse at a given instant of time will be formed from echoes from all the points over an annular region such as that shown in figure 2.8.6. For a rotating target, which for simplicity we assume has its rotational axis perpendicular to the line of sight, points such as A and A′ will have the same approach velocity, and their echoes will be shifted to the same higher frequency. Similarly the echoes from points such as B and B′ will be shifted to the same lower frequency. Thus the intensity of the echo at a particular instant and frequency is related to the radar properties of two

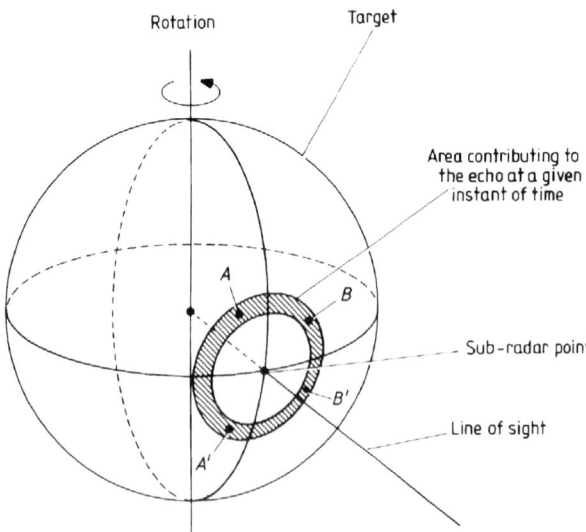

Figure 2.8.6. Region of a deep radar target contributing to the echo at a particular instant of time.

points that are equidistant from the equator and are north and south of it. Hence a map of the radar reflectivity of the target's surface may be produced, though it will have a two-fold ambiguity about the equator. If the rotational axis is not perpendicular to the line of sight, then the principle is similar, but the calculations are more complex. However, if such a target can be observed at two different angles of inclination of its rotational axis, then the two-fold ambiguity may be removed and a genuine map of the surface produced. Again in practice the process needs many corrections, and must have the frequency and time profiles of the emitted pulse removed from the echo by deconvolution (section 2.1). The features on radar maps are often difficult to relate to more normal surface features, since they are a conflation of the effects of surface reflectivity, small scale structure, height, gradient etc.

The relative velocity of the Earth and the target is much simpler to obtain. Provided that the transmitter is sufficiently stable in frequency (see the earlier discussion), then it may be found from the shift in the mean frequency of the echo compared with the mean frequency of the emitted pulse, by the application of the Doppler formula. Corrections for the Earth's orbital and rotational velocities are required. The former is given by equation (4.1.40), while the latter may be obtained from

$$\nu_c = \nu_o + 462\cos\delta\sin(\alpha - \text{LST})\sin(\text{LAT}) \quad \text{m s}^{-1} \qquad (2.8.15)$$

where ν_c is the corrected velocity (m s^{-1}), ν_o is the observed velocity (m s^{-1}), α and δ are the right ascension and declination of the target, LST is the local sidereal time at the radar station when the observation is made and LAT is the latitude of the radar station.

The combination of accurate knowledge of both distance and velocity for a planet on several occasions enables its orbit to be calculated to a very high degree of precision. This then provides information on basic quantities such as the astronomical unit, and enables tests of metrical gravitational theories such as general relativity to be made, and so is of considerable interest and importance even outside the area of astronomy. Producing a precise orbit for a planet by this method requires, however, that the Earth's and the planet's orbits both be known precisely so that the various corrections etc. can be applied. Thus the procedure is a 'bootstrap' one, with the initial data leading to improved orbits, and these improved orbits then allowing better interpretation of the data and so on over many iterations.

The radar reflection contains information on the surface structure of the target. Changes in the phase, polarization and of the manner in which the radar reflectivity changes with angle, can in principle be used to determine rock types, roughness of the surface and so on. In practice, the data are rarely sufficient to allow anything more than vague possibilities to be indicated. The range of unfixed parameters describing the surface is so large that very many different models for the surface can fit the same data. This aspect of radar astronomy has therefore been of little use to date.

Meteors

The ionized vapour along the track of a meteor reflects radar transmissions admirably. Over the past few decades, therefore, many stations have been set up to detect meteors by such means. Since they can be observed equally well during the day and at night, many previously unknown daytime meteor showers have been discovered. Either pulsed or CW radar is suitable, and a wavelength of a few metres is usually used.

The echoes from a meteor trail are modulated into a Fresnel diffraction pattern (figure 2.7.4), with the echoed power expressible in terms of the Fresnel integrals (equations (2.7.5) and (2.7.6)) as

$$F(t) = F(x^2 + y^2) \qquad (2.8.16)$$

where F is the echo strength for the fully formed trail and $F(t)$ is the echo strength at time t during the formation of the trail. The parameter l of the Fresnel integrals in this case is the distance of the meteor from the minimum range point. The modulation of the echo arises because of the phase differences between echoes returning from different parts of the trail. The distance of the meteor, and hence its height, may be found from the delay time as usual. In addition to the distance, however, the cross range velocity can also be obtained from the echo modulation. Several stations can combine, or a single transmitter can have several well-separated receivers, in order to provide further information such as the position of the radiant, atmospheric deceleration of the meteor etc.

Exercise

2.8.1   If a 10 MW peak power radar is used to observe Venus from the Earth at inferior conjunction, using a 300 m dish antenna, calculate the peak power required for a satellite-borne radar situated 1000 km above Venus's surface which has a 1 m dish antenna if the same signal-to-noise ratio is to be achieved. Assume that all the other parameters of the radar systems are similar, and take the distance of Venus from the Earth at inferior conjunction to be 4.14×10^7 km.

If the linear resolution of the Earth-based radar at Venus is 10^4 km, what is that of the satellite-based system?

2.9   Electronic images

Image formats

There are various ways of storing electronic images. One currently in wide-spread use for astronomical images is the Flexible Image Transport System, or FITS. There are various versions of FITS, but all have a header, the image in binary form, and an end section. The header must be a multiple of 2880 bytes (this number originated in the days of punched card input to computers and represents 36 cards each of 80 bytes). The header contains information on the number of bits representing the data in the image, the number of dimensions of the image (e.g. one for a spectrum, two for a monochromatic image, three for a colour image, etc.), the number of elements along each dimension, the object observed, the telescope used, details of the observation and other comments. The header is then padded out to a multiple of 2880 with zeros. The image section of a FITS file is simply the image data in the form specified by the header. The end section then pads the image section with zeros until it too is an integer multiple of 2880 bytes.

Other formats may be encountered such as JPEG, GIF and TIFF that are more useful for processed images than raw data.

Image compression

Most images, howsoever they may be produced and at whatever wavelength they were obtained, are nowadays stored in electronic form. This is even true of many photographic images that may be scanned using micro-densitometers (section 2.2) and fed into a computer. Electronic images may be stored by simply recording the precise value of the intensity at each pixel in an appropriate two-dimensional matrix. However, this uses large amounts of memory; for example 2 megabytes for a 16-bit image from a 1000×1000 pixel CCD (and 32-bit, $10\,000 \times 10\,000$ CCDs, requiring 400 megabytes per image, are being envisaged). Means whereby storage space for electronic images may be utilized most efficiently are therefore at a premium. The most widely used approach to the problem is to compress the images.

Image compression relies for its success upon some of the information in the image being redundant. That information may therefore be omitted or stored in shortened form without losing information from the image as a whole. Image compression that conserves all the information in the original image is known as loss-less compression. For some purposes a reduction in the amount of information in the image may be acceptable (for example, if one is only interested in certain objects or intensity ranges within the image), and this is the basis of lossy compression (see below).

There are three main approaches to loss-less image compression. The first is differential compression. This stores not the intensity values for each pixel, but the differences between the values for adjacent pixels. For an image with smooth variations, the differences will normally require fewer bits for their storage than the absolute intensity values. The second approach, known as run compression, may be used when many sequential pixels have the same intensity. This is often true for the sky background in astronomical images. Each run of constant intensity values may be stored as a single intensity combined with the number of pixels in the run, rather than storing the value for every pixel. The third approach, which is widely used on the Internet and elsewhere, where it goes by various proprietary names such as ZIP and DoubleSpace, is similar to run compression except that it stores repeated sequences whatever their nature as a single code, not just constant runs. For some images, loss-less compressions by a factor of five or ten times are possible.

Sometimes not all the information in the original images is required. Lossy compression can then be by factors of 20, 50 or more. For example, a 32-bit image to be displayed on a computer monitor which has 256 bright-ness levels need only be stored to 8-bit accuracy, or only the (say) 64 × 64 pixels covering the image of a planet, galaxy etc. need be stored from a 1000 × 1000 pixel image if that is all that is of interest within the image. Most workers, however, will prefer to store the original data in a loss-less form.

Compression techniques do not work effectively on noisy images since the noise is incompressible. Then a lossy compression may be of use since it will primarily be the noise that is lost. An increasingly widely used lossy compression for such images is based on the wavelet transform

$$T = \frac{1}{\sqrt{|a|}} \int f(t) \psi \left(\frac{t-b}{a} \right) dt. \tag{2.9.1}$$

This can give high compression ratios for noisy images. The details of the technique are, however, beyond the scope of this book.

Image processing

Image processing is the means whereby images are produced in an optimum form to provide the information that is required from them. Photographic image-processing techniques are possible, such as unsharp masking (section 2.2), but most image-processing techniques are applied to electronic images.

Image processing divides into data reduction and image optimization. Data reduction is the relatively mechanical process of correcting for known problems and faults. Many aspects of it are discussed elsewhere, such as dark frame subtraction, flat fielding, and cosmic ray spike elimination on CCD images (section 1.1), deconvolution and maximum entropy processing (section 2.1) and CLEANing aperture synthesis images (section 2.5) etc.

Image optimization is more of an art than a science. It is the further processing of an image with the aim of displaying those parts of the image that are of interest to their best effect for the objective for which they are required. The same image may therefore be optimized in quite different ways if it is to be used for different purposes. For example, an image of Jupiter and its satellites would be processed quite differently if the surface features on Jupiter were to be studied, compared with if the positions of the satellites with respect to the planet were of interest. There are dozens of techniques used in this aspect of image processing, and there are few rules other than experience to suggest which will work best for a particular application. The main techniques are outlined below, but the interested reader will need to look to other sources and to obtain practical experience before being able to apply them with consistent success.

Grey scaling

This is probably the technique in most widespread use. It arises because many detectors have dynamic ranges far greater than those of the computer monitors or hard copy devices that are used to display their images. In many cases the interesting part of the image will cover only a restricted part of the dynamic range of the detector. Grey scaling then consists of stretching that section of the dynamic range over the available intensity levels of the display device.

For example, the spiral arms of a galaxy on a 16-bit CCD image (with a dynamic range from 0 to 65 535) might have CCD level values ranging from 20 100 to 20 862. On a computer monitor with 256 intensity levels, this would be grey scaled so that CCD levels 0 to 20 099 corresponded to intensity 0 (i.e. black) on the monitor, levels 20 100 to 20 103 to intensity 1, levels 20 104 to 20 106 to intensity 2, and so on up to levels 20 859 to 20 862 corresponding to intensity 254. CCD levels 20 863 to 65 535 would then all be lumped together into intensity 255 (i.e. white) on the monitor.

Other types of grey scaling may aim at producing the most visually acceptable version of an image. There are many variations of this technique whose operation may best be envisaged by its effect on a histogram of the display pixel intensities. In histogram equalization, for example, the image pixels are mapped to the display pixels in such a way that there are equal numbers of image pixels in each of the display levels. For many astronomical images, a mapping which results in an exponentially decreasing histogram for the numbers of image pixels in each display level often gives a pleasing image since it produces a good dark background for the image. The number of other mappings available is limited only by the imagination of the operator.

Image combination

Several images of the same object may be added together to provide a resultant image with lower noise. More powerful combination techniques, such as the subtraction of two images to highlight changes, or false-colour displays, are, however, probably in more widespread use.

Spatial filtering

This technique has many variations that can have quite different effects upon the image. The basic process is to combine the intensity of one pixel with those of its surrounding pixels in some manner. The filter may best be displayed by a 3×3 matrix (or 5×5, 7×7 matrices etc.) with the pixel of interest in the centre. The elements of the matrix are the weightings given to each pixel intensity when they are combined and used to replace the original value for the central pixel (figure 2.9.1). Some commonly used filters are shown in figure 2.9.1.

Various commercial image processing packages are available and with today's (and tomorrow's) personal computers enable anyone to undertake extensive image processing. Much of this, though, is biased towards commercial artwork and the production of images for advertising etc. There are, however, packages specifically designed for astronomical image processing, and many are available free of charge. The most widespread of these packages for optical images is IRAF (Image Reduction and Analysis Facility), produced and maintained by NOAO (National Optical Astronomy Observatories). IRAF is available free for downloading via the web (see http://iraf.noao.edu/) or at a cost of a few tens of pounds/dollars on CDROM (see previous web site for further details and also for details of IRAF's image processing packages). You will also need your computer to have a UNIX or LINUX operating system. The latter is available, also without charge for downloading on to PCs, at http://www.linux.org/. You are warned, however, that though available free of charge, these packages are not easy for an inexperienced computer user to implement successfully. IRAF and many other software packages are available to UK professional

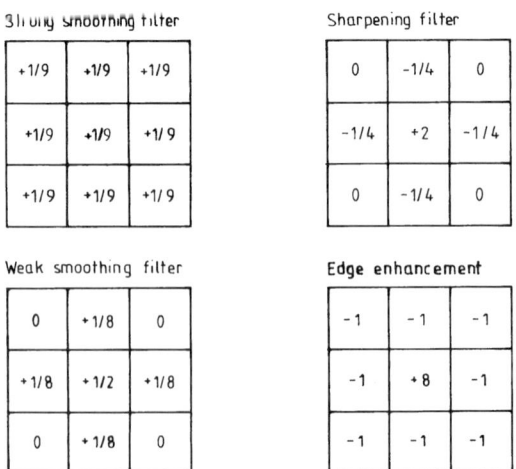

Figure 2.9.1. Examples of commonly used spatial filters.

astronomers via the Starlink computer network. If you have purchased a CCD camera designed for amateur astronomical use, then the manufacturer will almost certainly have provided at least a basic image processing package as a part of the deal. Most such manufacturers also have web sites where other software produced by themselves and by users of the cameras may be found. For radio astronomical data, AIPS and AIPS++ are the most widely used data reduction packages (section 2.5).

Chapter 3

Photometry

3.1 Photometry

Background

Magnitudes

The system used by astronomers to measure the brightnesses of stars is a very ancient one, and for that reason is a very awkward one. It originated with the earliest of the astronomical catalogues, that due to Hipparchos in the late second century BC. The stars in this catalogue were divided into six classes of brightness, the brightest being of the first class, and the faintest of the sixth class. With the invention and development of the telescope such a system proved to be inadequate, and it had to be refined and extended. William Herschel initially undertook this, but our present-day system is based upon the work of Norman Pogson in 1856. He suggested a logarithmic scale that approximately agreed with the earlier measures. As we have seen in section 1.1, the response of the eye is nearly logarithmic, so that the ancient system due to Hipparchos was logarithmic in terms of intensity. Hipparchos' class five stars were about $2\frac{1}{2}$ times brighter than his class six stars, and were about $2\frac{1}{2}$ times fainter than class four stars and so on. Pogson expressed this as a precise mathematical law in which the magnitude was related to the logarithm of the diameter of the minimum size of telescope required to see the star (i.e. the limiting magnitude, equation (3.1.3)). Soon after this, the proposal was adapted to express the magnitude directly in terms of the energy coming from the star, and the equation is now usually used in the form:

$$m_1 - m_2 = -2.5 \log_{10}\left(\frac{E_1}{E_2}\right) \tag{3.1.1}$$

where m_1 and m_2 are the magnitudes of stars 1 and 2 and E_1 and E_2 are the energies per unit area at the surface of the Earth for stars 1 and 2. The scale is awkward to use because of its peculiar base, leading to the ratio of the energies of two stars whose brightnesses differ by one magnitude being

276

2.512.[47] Also the brighter stars have magnitudes whose absolute values are smaller than those of the fainter stars. However, it seems unlikely that astronomers will change the system, so the student must perforce become familiar with it as it stands. One aspect of the scale is immediately apparent from equation (3.1.1), and that is that the scale is a relative one—the magnitude of one star is expressed in terms of that of another; there is no direct reference to absolute energy or intensity units. The zero of the scale is therefore fixed by reference to standard stars. On Hipparchos' scale, those stars that were just visible to the naked eye were of class six. Pogson's scale is thus arranged so that stars of magnitude six are just visible to a normally sensitive eye from a good observing site on a clear, moonless night. The standard stars are chosen to be non-variables, and their magnitudes are assigned so that the above criterion is fulfilled. The primary stars are known as the North Polar Sequence, and comprise stars within $2°$ of the pole star, so that they may be observed from almost all northern hemisphere observatories throughout the year. They are listed in appendix 1. Secondary standards may be set up from these to allow observations to be made in any part of the sky.

The faintest stars visible to the naked eye, from the definition of the scale, are of magnitude six; this is termed the limiting magnitude of the eye. For point sources the brightnesses are increased by the use of a telescope by a factor, G, called the light grasp of the telescope (section 1.1). Since the dark-adapted human eye has a pupil diameter of about $7\,\mathrm{mm}$, G is given by

$$G \approx 2 \times 10^4 d^2 \tag{3.1.2}$$

where d is the telescope's objective diameter in metres. Thus the limiting magnitude through a visually used telescope, m_L, is

$$m_L = 16.8 + 5\log_{10} d. \tag{3.1.3}$$

CCDs and other detection techniques will generally improve on this by some two to five stellar magnitudes.

The magnitudes so far discussed are all apparent magnitudes and so result from a combination of the intrinsic brightness of the object and its distance. This is obviously an essential measurement for an observer, since it is the criterion determining exposure times etc.; however, it is of little intrinsic significance for the object in question. A second type of magnitude is therefore defined that is related to the actual brightness of the object. This is called the absolute magnitude and it is 'the apparent magnitude of the object if its distance from the Earth were ten parsecs'. It is usually denoted by M, while apparent magnitude uses the lower case, m. The relation between apparent and absolute magnitudes may easily be obtained from

[47] $2.512 = 10^{0.4}$. For a difference of two magnitudes, the energies differ by $\times 6.31$ (2.512^2), for three magnitudes by $\times 15.85$ (2.512^3), four magnitudes by $\times 39.81$ (2.512^4) and by five magnitudes by exactly $\times 100$ (2.512^5).

equation (3.1.1). Imagine the object moved from its real distance to ten parsecs from the Earth. Its energy per unit area at the surface of the Earth will then change by a factor $(D/10)^2$, where D is the object's actual distance in parsecs. Thus

$$M - m = -2.5 \log_{10} \left(\frac{D}{10} \right)^2 \qquad (3.1.4)$$

$$M = m + 5 - 5 \log_{10} D. \qquad (3.1.5)$$

The difference between apparent and absolute magnitudes is called the distance modulus and is occasionally used in place of the distance itself:

$$\text{distance modulus} = m - M = 5 \log_{10} D - 5. \qquad (3.1.6)$$

Equations (3.1.5) and (3.1.6) are valid so long as the only factors affecting the apparent magnitude are distance and intrinsic brightness. However, light from the object often has to pass through interstellar gas and dust clouds, where it may be absorbed. A more complete form of equation (3.1.5) is therefore

$$M = m + 5 - 5 \log_{10} D - AD \qquad (3.1.7)$$

where A is the interstellar absorption in magnitudes per parsec. A typical value for A for lines of sight within the galactic plane is $0.002 \, \text{mag pc}^{-1}$. Thus we may determine the absolute magnitude of an object via equations (3.1.5) or (3.1.7) once its distance is known. More frequently, however, the equations are used in the reverse sense in order to determine the distance. Often the absolute magnitude may be estimated by some independent method, and then

$$D = 10^{[(m-M+5)/5]} \quad \text{pc.} \qquad (3.1.8)$$

Such methods for determining distance are known as standard candle methods, since the object is in effect acting as a standard of some known luminosity. The best known examples are the classical Cepheids with their period–luminosity relationship

$$M = -1.9 - 2.8 \log_{10} P \qquad (3.1.9)$$

where P is the period of the variable in days. Many other types of star such as dwarf Cepheids, RR Lyrae stars, W Virginis stars and β Cepheids are also suitable. Also the brightest novae or supernovae, or the brightest globular clusters around a galaxy, or even the brightest galaxy in a cluster of galaxies can provide rough standard candles. Yet another method is due to Wilson and Bappu who found a relationship between the width of the emission core of the ionized calcium line at 393.3 nm (the Ca(II) K line) in late-type stars, and their absolute magnitudes. The luminosity is then proportional to the sixth power of the line width.

Both absolute and apparent magnitudes are normally measured over some well-defined spectral region. While the above discussion is quite general, the equations only have validity within a given spectral region. Since the spectra of two objects may be quite different, their relationship at one wavelength may be very different from that at another. The next section discusses the definitions of these spectral regions and their interrelationships.

Filter systems

A very wide range of filter systems and filter–detector combinations has been devised. They may be grouped into wide-, intermediate- and narrow-band systems according to the widths of their transmission curves. Wide-band filters typically have bandwidths of around 100 nm, intermediate-band filters range from 10 to 50 nm, while narrow-band filters range from 0.05 to 10 nm. The division is convenient for the purposes of discussion here, but is not of any real physical significance.

The filters used in photometry are of two main types based upon either absorption/transmission or on interference. The absorption/transmission filters use salts such as nickel or cobalt oxides dissolved in glass or gelatine, or a suspension of colloid particles. These filters typically transmit over a 100 nm wide region. They are thus mostly used for the wide-band photometric systems. Many of these filters will also transmit in the red and infrared, and so must be used with a red-blocking filter made with copper sulphate. Short-wave blocking filters may be made using cadmium sulphide or selenide, sulphur or gold. Although not normally used for photometry, two other types of filter may usefully be mentioned here: dichroic mirrors and neutral density filters. Neutral density filters absorb by a constant amount over a wide range of wavelengths and may be needed when observing very bright objects like the Sun (section 5.3). They are normally a thin deposit of a reflecting metal such as aluminium or stainless steel on a glass or very thin plastic substrate. Dichroic mirrors reflect over one wavelength range and transmit over another. For example a mirror thinly plated with a gold coating is transparent to visual radiation, but reflects the infrared. They may also be produced using multi-layer interference coatings. Dichroic mirrors may be used in spectroscopy to feed two spectroscopes operating over different wavelength ranges (section 4.2).

Interference filters are mostly Fabry–Pérot etalons (section 4.1) with a very small separation of the mirrors. The transmission wavelength and the bandwidth of the filter can be tuned by changing the mirror separation and/ or their reflectivities (see also tunable filters, section 4.1 and solar Hα filters, section 5.3). Combining several Fabry–Pérot cavities can alter the shape of the transmission band. Interference filters are usually used for the narrower filters of a photometric system since they can be made with bandwidths ranging from a few tens of nanometres to a hundredth of a nanometre. The recently developed rugate filter can transmit or reflect simultaneously at several wavelengths. It uses interference coatings in which the refractive index varies continuously throughout the depth of the layer.

The earliest 'filter' system was given by the response of the human eye (figure 1.1.3) and peaks around 510 nm with a bandwidth of 200 nm or so. Since all eyes have roughly comparable responses, the importance of the spectral region within which a magnitude was measured was not realized until the application of the photographic plate to astronomical detection. Then it was discovered that the magnitudes of stars that had been determined from such plates often differed significantly from those found visually. The discrepancy arose from the differing sensitivities of the eye and the photographic emulsion. Early emulsions and today's unsensitized emulsions (section 2.2) have a response that peaks near 450 nm, and little contribution is made to the image by radiation of a wavelength longer than about 500 nm. The magnitudes determined with the two methods became known as visual and photographic magnitudes respectively, and are denoted by m_v and m_p or M_v and M_p. This system is still used, for it allows the brightnesses of many stars to be found rapidly from plates taken by Schmidt cameras. The visual magnitude, however, is now found from the images on an orthochromatic plate that mimics the eye's sensitivity closely when it is combined with a yellow filter, and is then usually known as the photovisual magnitude (m_{pv} or M_{pv}).

With the development of photoelectric methods of detection of light, and their application to astronomy by Joel Stebbins and others earlier this century, other spectral regions became available, and more precise definitions of them were required. There are now a very great number of different photometric systems. While it is probably not quite true that there is one system for every observer, there is certainly at least one separate system for every slightly differing purpose to which photometric measurements may be put. Many of these are developed for highly specialized purposes, and so are unlikely to be encountered outside their particular area of application. Other systems have a wide application and so are worth knowing in detail.

For a long while, the most widespread of these more general systems was the UBV system, defined by Harold Johnson and William Morgan in 1953. This is a wide band system with the B and V regions corresponding approximately to the photographic and visual responses, and with the U region in the violet and ultraviolet. The precise definition requires a reflecting telescope with aluminized mirrors and uses an RCA 1P21 photomultiplier. The filters are listed in table 3.1.1.

The response curves for the filter–photomultiplier combination are shown in figure 3.1.1. The absorption in the Earth's atmosphere normally cuts off the short-wavelength edge of the U response. At sea level roughly half of the incident energy at a wavelength of 360 nm is absorbed, even for small zenith angles. More importantly, the absorption can vary with

Table 3.1.1.

	U	B	V
Corning 3384	for the V region		
Corning 5030 plus Schott GG 13	for the B region		
Corning 9863	for the U region		
Filter name	U	B	V
Central wavelength (nm)	365	440	550
Bandwidth (nm)	70	100	90

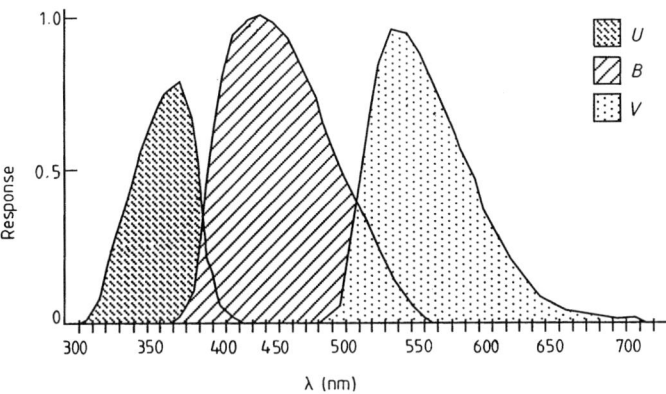

Figure 3.1.1. UBV response curves, excluding the effects of atmospheric absorption.

changes in the atmosphere. Ideal U response curves can therefore only be approached at very high-altitude, high-quality observing sites. The effect of the atmosphere on the U response curve is shown in figure 3.1.2. The B and V response curves are comparatively unaffected by differential atmospheric absorption across their wavebands; there is only a total reduction of the flux through these filters (see section 3.2). The scales are arranged so that the magnitudes through all three filters are equal to each other for A0 V

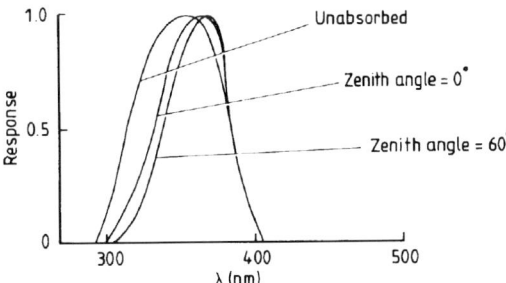

Figure 3.1.2. Effect of atmospheric absorption at sea level upon the U response curve (normalized).

Table 3.1.2.

Filter name	R	I	J	K	L	M	N	Q
Central wavelength (nm)	700	900	1250	2200	3400	4900	10 200	20 000
Bandwidth (nm)	220	240	380	480	700	300	5 000	5 000

stars such as α Lyrae (Vega). The choice of filters for the UBV system was made on the basis of what was easily available at the time. The advent of CCDs has led to several new photometric systems being devised (see below), with filters more related to the requirements of astrophysics, but the UBV system and its extensions are still in use, if only to enable comparisons to be made with measurements taken some time ago.

There are several extensions to the UBV system for use at longer wavelengths. A widely used system was introduced by Johnson in the 1960s for lead sulphide infrared detectors. With the photographic R and I bands this adds another eight regions, listed in table 3.1.2.

The longer wavelength regions in this system are defined by atmospheric transmission and so are very variable. Recently therefore filters have been used to define all the pass bands and the system has been adapted to make better use of the CCDs' response and the improvements in infrared detectors. This has resulted in slight differences to the transmission regions in what is sometimes called the JCG system,[48] shown in table 3.1.3.

Two other wide-band filter systems are now used, because of the large data sets that are available. These are from the Hubble space telescope (HST) and the Sloan Digital Sky Survey (SDSS). The HST system has six filters (table 3.1.4).

The SDSS is based upon the 2.5 m telescope at the Apache Point Observatory in New Mexico and aims to measure 100 million stars. It uses five filters that cover the whole range of sensitivity of CCDs (table 3.1.5).

Table 3.1.3.

Filter name	U	B	V	R	I	Z[49]	J	H	K	L	M
Central wavelength (nm)[50]	367	436	545	638	797	908	1220	1630	2190	3450	4750
Bandwidth (nm)	66	94	85	160	149	96	213	307	390	472	460

[48] For Johnson, Alan Cousins and Ian Glass.
[49] Added for use with CCDs.
[50] Data from *The Encyclopedia of Astronomy and Astrophysics*, ed P Murdin, IoP Press 2001, page 1642.

Table 3.1.4.

Central wavelength (nm)	336	439	450	555	675	814
Bandwidth (nm)	47	71	107	147	127	147

Table 3.1.5.

Filter name	u'	g'	r'	i'	z'
Central wavelength (nm)	358	490	626	767	907
Bandwidth (nm)	64	135	137	154	147

Table 3.1.6.

Filter name	u	v	b	y	β_n	β_w
Central wavelength (nm)	349	411	467	547	486	486
Bandwidth (nm)	30	19	18	23	3	15

Among the intermediate pass band systems, the most widespread is the uvby or Strömgren system (table 3.1.6). This was proposed by Bengt Strömgren in the late 1960s and is now in fairly common use. Its transmission curves are shown in figure 3.1.3. It is often used with two additional filters centred on the Hβ line, 3 nm and 15 nm wide respectively, to provide better temperature discrimination for the hot stars.

Narrow-band work concentrates on isolating spectral lines. Hα and Hβ are common choices, and their variations can be determined by measurements through a pair of filters which are centred on the line but have different bandwidths. No single system is in general use, so that a more detailed discussion is not very profitable. Other spectral features can

Figure 3.1.3. Normalized transmission curves for the uvby system, not including the effects of atmospheric absorption.

be studied with narrow-band systems where one of the filters isolates the spectral feature, and the other is centred on a nearby section of the continuum. The reason for the invention of these and other photometric systems lies in the information that may be obtained from the comparison of the brightness of the object at different wavelengths. Many of the processes contributing to the final spectrum of an object as it is received on Earth preferentially affect one or more spectral regions. Thus estimates of their importance may be obtained simply and rapidly by measurements through a few filters. Some of these processes are discussed in more detail in the next subsection and in section 3.2. The most usual features studied in this way are the Balmer discontinuity and other ionization edges, interstellar absorption, and strong emission or absorption lines or bands.

There is one additional photometric 'system' that has not yet been mentioned. This is the bolometric system, or rather bolometric magnitude, since it has only one pass band. The bolometric magnitude is based upon the total energy emitted by the object at all wavelengths. Since it is not possible to observe this in practice, its value is determined by modelling calculations based upon the object's intensity through one or more of the filters of a photometric system. Although x-ray, ultraviolet, infrared and radio data are now available to assist these calculations, many uncertainties remain, and the bolometric magnitude is still rather imprecise especially for high temperature stars. The calculations are expressed as the difference between the bolometric magnitude and an observed magnitude. Any filter of any photometric system could be chosen as the observational basis, but in practice the V filter of the standard UBV system is normally used. The difference is then known as the bolometric correction, BC,

$$BC = m_{bol} - V \tag{3.1.10}$$

$$= M_{bol} - M_V \tag{3.1.11}$$

and its scale is chosen so that it is zero for main sequence stars with a temperature of about 6500 K, i.e. about spectral class F5 V. The luminosity of a star is directly related to its absolute bolometric magnitude:

$$L_* = (3 \times 10^{28}) \times 10^{-0.4 M_{bol}} \quad \text{W.} \tag{3.1.12}$$

Similarly, the flux of the stellar radiation just above the Earth's atmosphere, f_*, is related to the apparent bolometric magnitude

$$f_* = (2.5 \times 10^{-8}) \times 10^{-0.4 m_{bol}} \quad \text{W m}^{-2}. \tag{3.1.13}$$

The bolometric corrections are plotted in figure 3.1.4.

Measurements in one photometric system can sometimes be converted to another system. This must be based upon extensive observational calibrations or upon detailed calculations using specimen spectral distributions. In

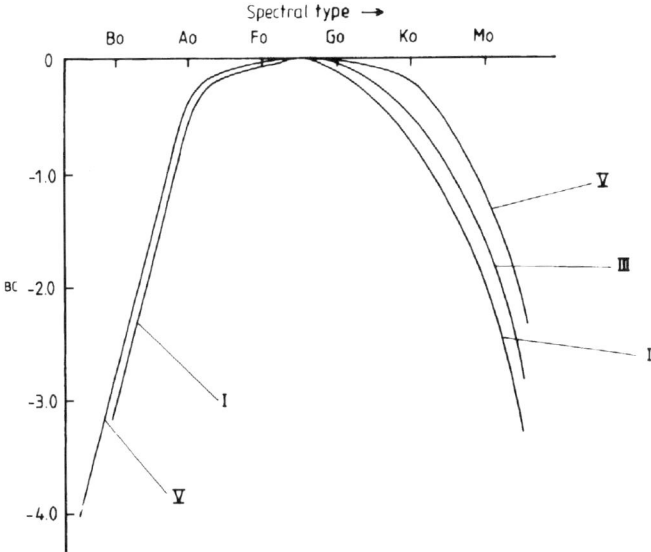

Figure 3.1.4. Bolometric corrections for main sequence stars (type V), giants (type III) and supergiants (type I).

either case, the procedure is very much second best to obtaining the data directly in the required photometric system, and requires great care in its application. Its commonest occurrence is for the conversion of data obtained with a slightly non-standard UBV system to the standard system.

Stellar parameters

Filters may be used when observing gravitational lensing events to separate them from changes arising from the star being a variable. Gravitational lensing affects all wavelengths equally, whereas variables almost always have different amplitudes at different wavelengths. However, the usual purpose of making measurements of stars in the various photometric systems is to determine some aspect of the star's spectral behaviour by a simpler and more rapid method than that of actually obtaining the spectrum. The simplest approach to highlight the desired information is to calculate one or more colour indices. The colour index is just the difference between the star's magnitudes through two different filters, such as the B and V filters of the standard UBV system. The colour index, C, is then just

$$C = B - V \qquad (3.1.14)$$

where B and V are the magnitudes through the B and V filters respectively. Similar colour indices may be obtained for other photometric systems such as 439–555 for the HST system and g'–r' for the SDSS. Colour indices for the

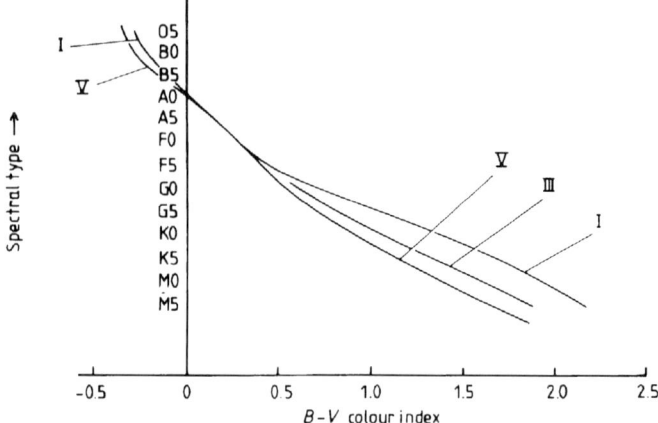

Figure 3.1.5. Relationship between spectral type and $B - V$ colour index.

uvby intermediate band system are discussed below. It should be noted, though, that a fairly common alternative usage is for C to denote the so-called international colour index, which is based upon the photographic and photovisual magnitudes. The interrelationship is

$$m_p - m_{pv} = C = B - V - 0.11. \tag{3.1.15}$$

The $B - V$ colour index is closely related to the spectral type (figure 3.1.5) with an almost linear relationship for main sequence stars. This arises from the dependence of both spectral type and colour index upon temperature. For most stars, the B and V regions are located on the long wavelength side of the maximum spectral intensity. In this part of the spectrum the intensity then varies approximately in a black body fashion for many stars. If we assume that the B and V filters effectively transmit at 440 and 550 nm respectively, then using the Planck equation

$$F_\lambda = \frac{2\pi hc^2}{\lambda^5 [e^{hc/\lambda kT} - 1]} \tag{3.1.16}$$

we obtain

$$B - V = -2.5$$

$$\times \log_{10} \left\{ \frac{(5.5 \times 10^{-7})^5 \left[\exp\left(\dfrac{(6.62 \times 10^{-34}) \times (3 \times 10^8)}{(5.5 \times 10^{-7}) \times (1.38 \times 10^{-23}) \times T} \right) - 1 \right]}{(4.4 \times 10^{-7})^5 \left[\exp\left(\dfrac{(6.62 \times 10^{-34}) \times (3 \times 10^8)}{(4.4 \times 10^{-7}) \times (1.38 \times 10^{-23}) \times T} \right) - 1 \right]} \right\}$$

$$\tag{3.1.17}$$

which simplifies to

$$B - V = -2.5 \log_{10} \left(3.05 \frac{\{\exp[(2.617 \times 10^4)/T] - 1\}}{\{\exp[(3.27 \times 10^4)/T] - 1\}} \right) \qquad (3.1.18)$$

which for $T < 10\,000\,K$ is approximately

$$B - V \approx -2.5 \log_{10} \left(3.05 \frac{\exp[(2.617 \times 10^4)/T]}{\exp[(3.27 \times 10^4)/T]} \right) \qquad (3.1.19)$$

$$= -1.21 + \frac{7090}{T}. \qquad (3.1.20)$$

Now the magnitude scale is an arbitrary one as we have seen, and it is defined in terms of standard stars, with the relationship between the B and the V magnitude scales such that $B - V$ is zero for stars of spectral type A0 (figure 3.1.5). Such stars have a surface temperature of about $10\,000\,K$, and so a correction term of $+0.5$ must be added to equation (3.1.20) to bring it into line with the observed relationship. Thus we get

$$B - V = -0.71 + \frac{7090}{T} \qquad (3.1.21)$$

giving

$$T = \frac{7090}{(B - V) + 0.71} \quad K. \qquad (3.1.22)$$

Equations (3.1.21) and (3.1.22) are still only poor approximations because the filters are very broad, and so the monochromatic approximation used in obtaining the equations is a very crude one. Furthermore, the effective wavelengths (i.e. the 'average' wavelength of the filter, taking account of the energy distribution within the spectrum) of the filters change with the stellar temperature (figure 3.1.6). However, an equation of a similar nature may be fitted empirically to the observed variation (figure 3.1.7), with great success for the temperature range 4000 to $10\,000\,K$, and this is

$$B - V = -0.865 + \frac{8540}{T} \qquad (3.1.23)$$

$$T = \frac{8540}{(B - V) + 0.865} \quad K. \qquad (3.1.24)$$

At higher temperatures, the relationship breaks down, as would be expected from the approximations made to obtain equation (3.1.19). The complete observed relationship is shown in figure 3.1.8. Similar relationships may be found for other photometric systems that have filters near the wavelengths of the standard B and V filters. For example, the relationship between

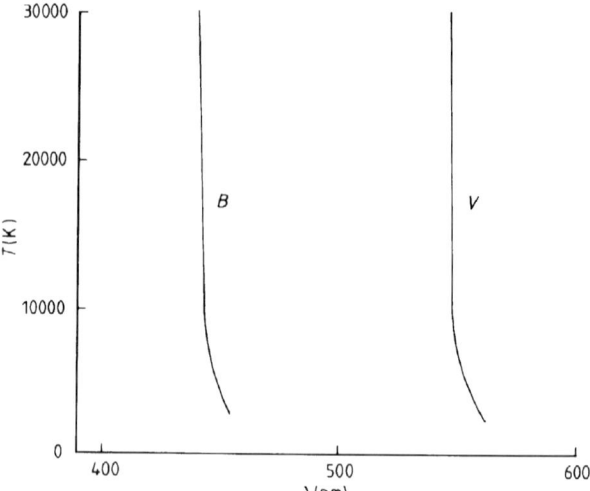

Figure 3.1.6. Change in the effective wavelengths of the standard B and V filters with changing black body temperature.

spectral type and the $b - y$ colour index of the uvby system is shown in figure 3.1.9.

For filters that differ from the B and V filters, the relationship of colour index with spectral type or temperature may be much more complex. In

Figure 3.1.7. Observed and semi-empirical $(B - V)/T$ relationships over the lower part of the main sequence.

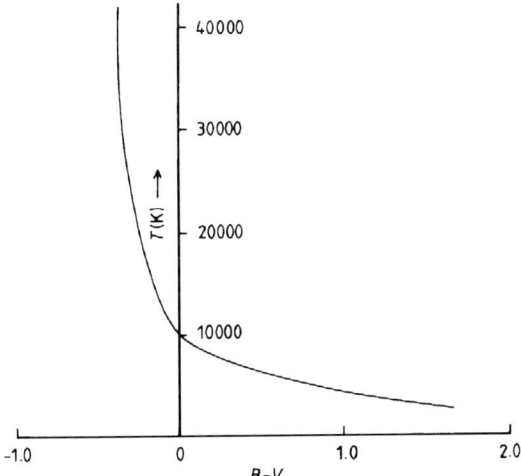

Figure 3.1.8. Observed $B - V$ versus T relationship for the whole of the main sequence.

many cases the colour index is a measure of some other feature of the spectrum, and the temperature relation is of little use or interest. The $U - B$ index for the standard system is an example of such a case, since the U and B responses bracket the Balmer discontinuity (figure 3.1.10). The extent of

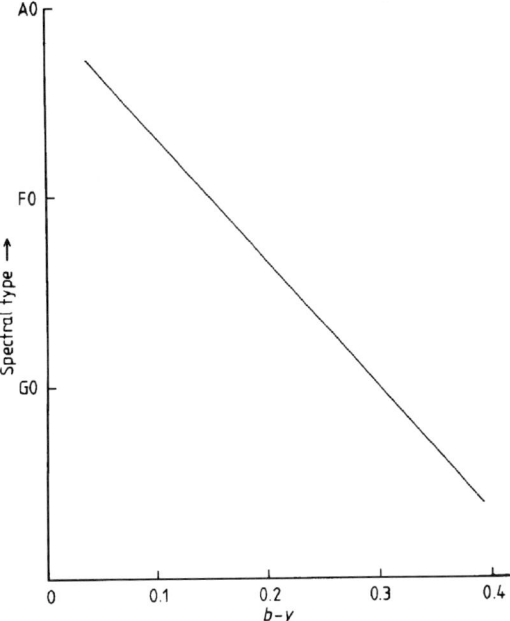

Figure 3.1.9. Relationship between spectral type and $b - y$ colour index.

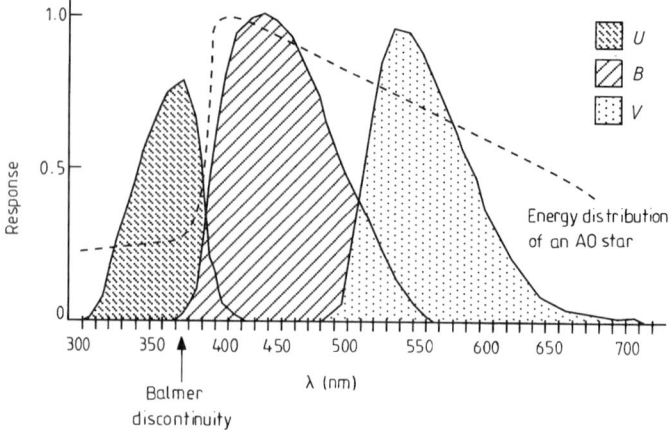

Figure 3.1.10. Position of the standard UBV filters with respect to the Balmer discontinuity.

the Balmer discontinuity is measured by a parameter, D, defined by

$$D = \log_{10}\left(\frac{I_{364+}}{I_{364-}}\right) \tag{3.1.25}$$

where I_{364+} is the spectral intensity at wavelengths just longer than the Balmer discontinuity (which is at or near 364 nm) and I_{364-} is the spectral intensity at wavelengths just shorter than the Balmer discontinuity. The variation of D with the $U - B$ colour index is shown in figure 3.1.11. The relationship is complicated by the overlap of the B filter with the Balmer discontinuity and the variation of the effective position of that discontinuity with spectral and luminosity class. The discontinuity reaches a maximum at about A0 for main sequence stars and at about F0 for supergiants. It almost

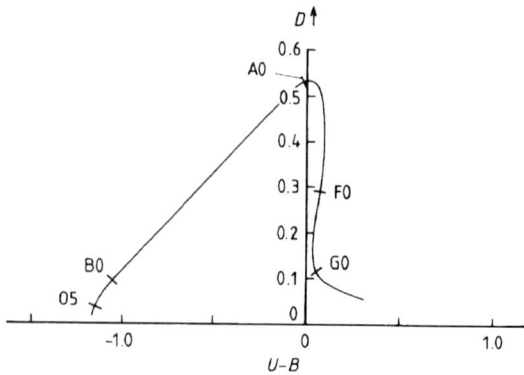

Figure 3.1.11. Variation of D with $U - B$ for main sequence stars.

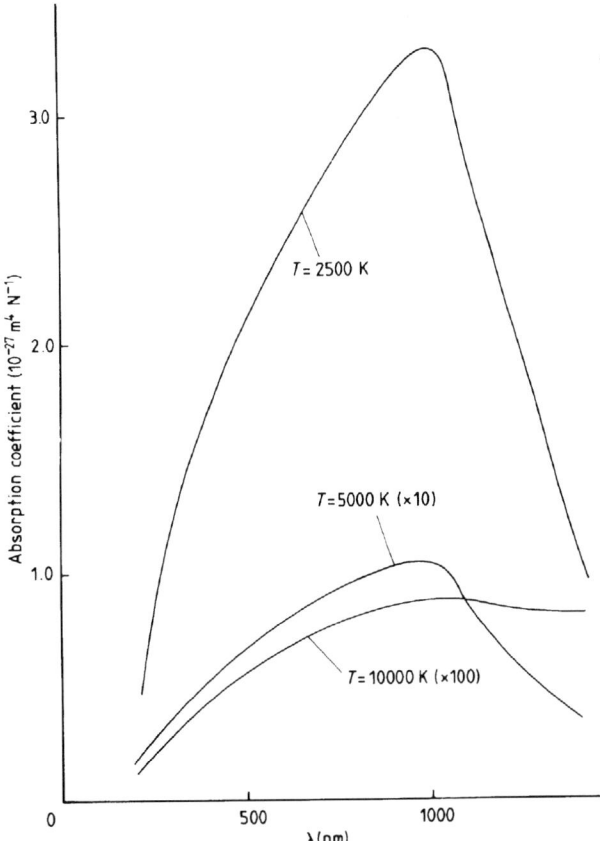

Figure 3.1.12. Change in the absorption coefficient of the negative hydrogen ion with temperature.

vanishes for very hot stars, and for stars cooler than about G0, corresponding to the cases of too-high and too-low temperatures for the existence of
significant populations in the hydrogen $n = 2$ level. The colour index is
also affected by the changing absorption coefficient of the negative hydrogen
ion (figure 3.1.12), and by line blanketing in the later spectral types. A similar
relationship may be obtained for any pair of filters that bracket the Balmer
discontinuity, and another commonly used index is the c_1 index of the
uvby system

$$c_1 = u + b - 2v. \tag{3.1.26}$$

The filters are narrower and do not overlap the discontinuity, leading to a
simpler relationship (figure 3.1.13), but the effects of line blanketing must
still be taken into account.

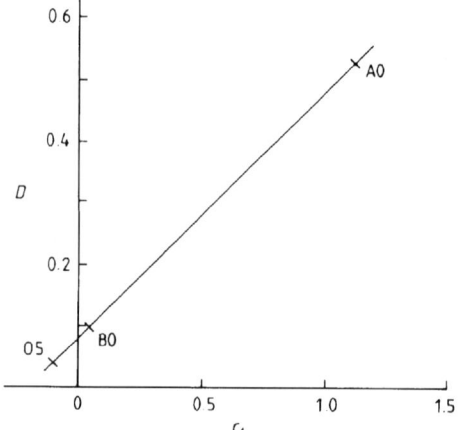

Figure 3.1.13. Variation of D with the c_1 index of the uvby system for main sequence stars.

Thus the $B - V$ colour index is primarily a measure of stellar tempera-ture, while the $U - B$ index is a more complex function of both luminosity and temperature. A plot of one against the other provides a useful classifica-tion system analogous to the Hertzsprung–Russell diagram. It is commonly called the colour–colour diagram and is shown in figure 3.1.14. The deviations of the curves from that of a black body arise from the effects just mentioned: Balmer discontinuity, line blanketing, negative hydrogen ion absorption coefficient variation, and also because the radiation originates over a region of the stellar atmosphere rather than in a single layer charac-terized by a single temperature. The latter effect is due to limb darkening, and because the thermalization length (the distance between successive absorptions and re-emissions) normally corresponds to a scattering optical depth many times unity. The final spectral distribution therefore contains contributions from regions ranging from perhaps many thousands of optical depths below the visible surface to well above the photosphere and so cannot be assigned a single temperature.

Figure 3.1.14 is based upon measurements of nearby stars. More distant stars are affected by interstellar absorption, and since this is strongly inversely dependent upon wavelength (figure 3.1.15), the $U - B$ and the $B - V$ colour indices are altered. The star's spectrum is progressively weakened at shorter wavelengths and so the process is often called interstellar reddening.[51] The colour excesses measure the degree to

---

[51] This is not the same as the redshift observed for distant galaxies. That results from the galaxies' velocities away from us and changes the observed wavelengths of the spectrum lines. With interstellar reddening the spectrum lines' wavelengths are unchanged.

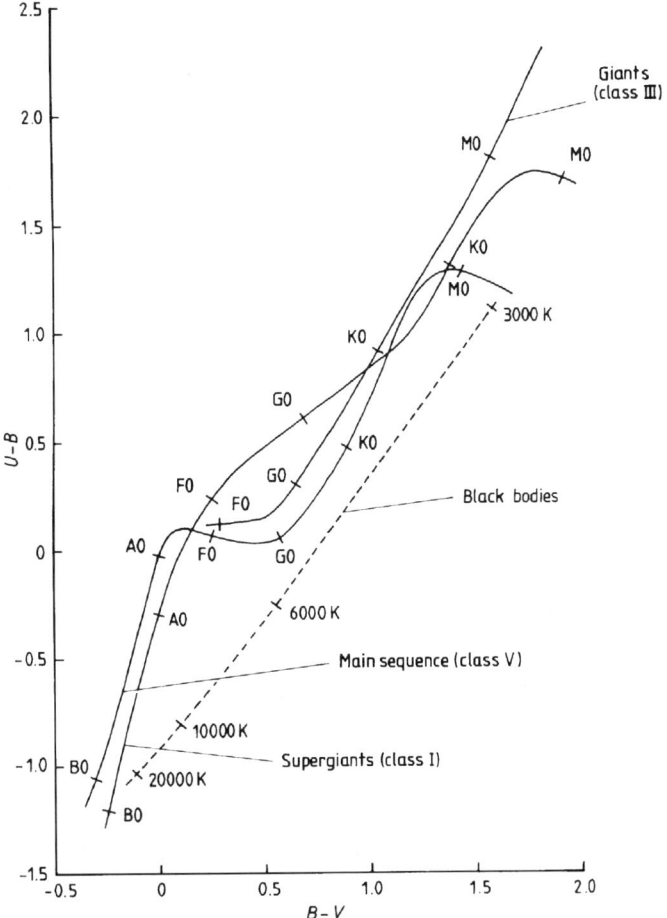

Figure 3.1.14. $(U - B)$ versus $(B - V)$ colour–colour diagram.

which the spectrum is reddened

$$E_{U-B} = (U - B) - (U - B)_0 \qquad (3.1.27)$$
$$E_{B-V} = (B - V) - (B - V)_0 \qquad (3.1.28)$$

where the subscript 0 denotes the unreddened quantities. These are called the intrinsic colour indices and may be obtained from the spectral type and figures 3.1.5 and 3.1.16. The interstellar absorption (figure 3.1.15) varies over most of the optical spectrum in a manner that may be described by the semi-empirical relationship

$$A_\lambda = \frac{6.5 \times 10^{-10}}{\lambda} - 2.0 \times 10^{-4} \quad \text{mag pc}^{-1}. \qquad (3.1.29)$$

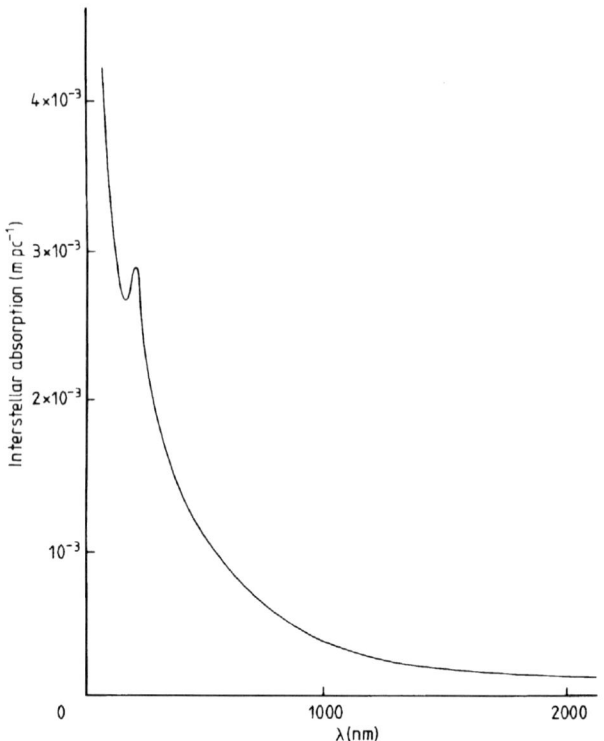

Figure 3.1.15. Average interstellar absorption.

Hence, approximating U, B and V filters by monochromatic responses at their central wavelengths, we have, for a distance of D parsecs,

$$\frac{E_{U-B}}{E_{B-V}} = \frac{(U - U_0) - (B - B_0)}{(B - B_0) - (V - V_0)} \tag{3.1.30}$$

$$= \frac{\left[\dfrac{6.5 \times 10^{-10}}{3.65 \times 10^{-7}} - 2.0 \times 10^{-4}\right]D - \left[\dfrac{6.5 \times 10^{-10}}{4.4 \times 10^{-7}} - 2.0 \times 10^{-4}\right]D}{\left[\dfrac{6.5 \times 10^{-10}}{4.4 \times 10^{-7}} - 2.0 \times 10^{-4}\right]D - \left[\dfrac{6.5 \times 10^{-10}}{5.5 \times 10^{-7}} - 2.0 \times 10^{-4}\right]D} \tag{3.1.31}$$

$$= \frac{\left[\dfrac{1}{365} - \dfrac{1}{440}\right]}{\left[\dfrac{1}{440} - \dfrac{1}{550}\right]} \tag{3.1.32}$$

$$= 1.027. \tag{3.1.33}$$

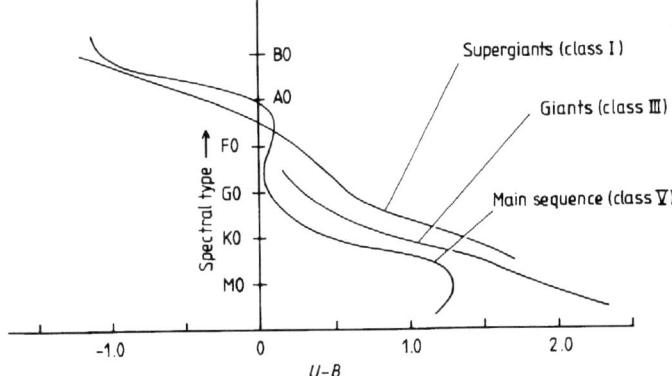

Figure 3.1.16. Relationship between spectral type and $U - B$ colour index.

Thus the ratio of the colour excesses is independent of the reddening. This is an important result and the ratio is called the reddening ratio. Its actual value is somewhat different from that given in equation (3.1.33), because of our mono-chromatic approximation, and furthermore it is slightly temperature dependent and not quite independent of the reddening. Its best empirical values are

$$\frac{E_{U-B}}{E_{B-V}} = (0.70 \pm 0.10) + (0.045 \pm 0.015)E_{B-V} \qquad \text{at } 30\,000 \text{ K} \qquad (3.1.34)$$

$$\frac{E_{U-B}}{E_{B-V}} = (0.72 \pm 0.06) + (0.05 \pm 0.01)E_{B-V} \qquad \text{at } 10\,000 \text{ K} \qquad (3.1.35)$$

$$\frac{E_{U-B}}{E_{B-V}} = (0.82 \pm 0.12) + (0.065 \pm 0.015)E_{B-V} \qquad \text{at } 5000 \text{ K}. \qquad (3.1.36)$$

The dependence upon temperature and reddening is weak, so that for many purposes we may use a weighted average reddening ratio

$$\left(\frac{E_{U-B}}{E_{B-V}}\right) = 0.72 \pm 0.03. \qquad (3.1.37)$$

The colour factor, Q, defined by

$$Q = (U - B) - \left(\frac{E_{U-B}}{E_{B-V}}\right)(B - V) \qquad (3.1.38)$$

is then independent of reddening, as we may see from its expansion:

$$Q = (U - B)_0 + E_{U-B} - \left(\frac{E_{U-B}}{E_{B-V}}\right)[(B - V)_0 + E_{B-V}] \qquad (3.1.39)$$

$$= (U - B)_0 - \left(\frac{E_{U-B}}{E_{B-V}}\right)(B - V)_0 \qquad (3.1.40)$$

$$= (U - B)_0 - 0.72(B - V)_0 \qquad (3.1.41)$$

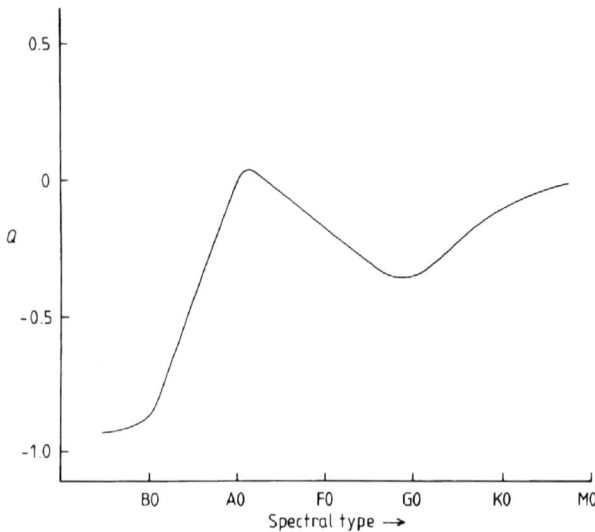

Figure 3.1.17. Variation of colour factor with spectral type.

and provides a precise measure of the spectral class of the B type stars (figure 3.1.17). Since the intrinsic $B - V$ colour index is closely related to the spectral type (figure 3.1.5), we therefore also have the empirical relation for the early spectral type

$$(B - V)_0 = 0.332Q \tag{3.1.42}$$

shown in figure 3.1.18. Hence we have

$$E_{B-V} = (B - V) - 0.332Q = 1.4E_{U-B}. \tag{3.1.43}$$

Thus simple UBV photometry for hot stars results in determinations of temperature, Balmer discontinuity, spectral type and reddening. The latter may also be translated into distance when the interstellar absorption in the star's direction is known. Thus we have potentially a very high return of information for a small amount of observational effort, and the reader may see why the relatively crude methods of UBV and other wide band photometry still remain popular.

Similar or analogous relationships may be set up for other filter systems. Only one further example is briefly discussed here. The reader is referred to the bibliography for further information on these and other systems. The uvby system plus filters centred on Hβ (often called the uvbyβ system) has several indices. We have already seen that $b - y$ is a good temperature indicator (figure 3.1.9), and that c_1 is a measure of the Balmer discontinuity for hot stars (figure 3.1.13). A third index is labelled m_1 and is given by

$$m_1 = v + y - 2b. \tag{3.1.44}$$

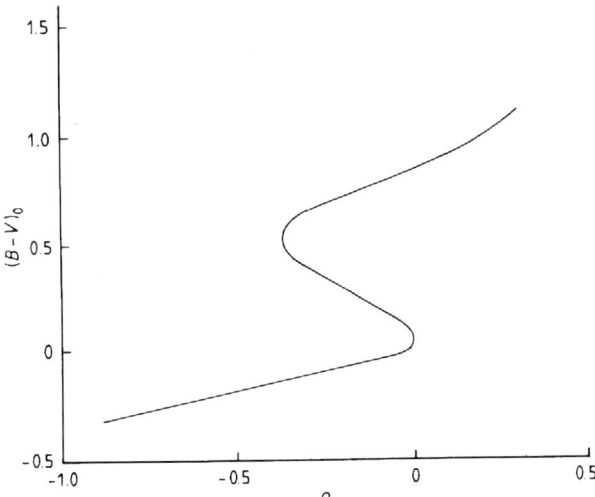

Figure 3.1.18. Relationship between colour factor and the $B - V$ intrinsic colour index.

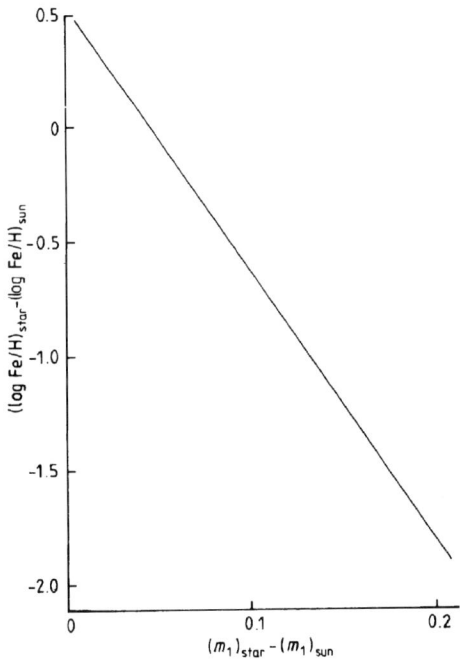

Figure 3.1.19. Relationship between metallicity index and iron abundance for solar type stars.

It is sometimes called the metallicity index, for it provides a measure of the number of absorption lines in the spectrum and hence, for a given temperature the abundance of the metals (figure 3.1.19). The fourth commonly used index of the system uses the wide and narrow Hβ filters, and is given by

$$\beta = m_n - m_w \tag{3.1.45}$$

where m_n and m_w are the magnitudes through the narrow and wide Hβ filters respectively. β is directly proportional to the equivalent width of Hβ (figure 3.1.20), and therefore acts as a guide to luminosity and temperature among the hotter stars (figure 3.1.21). Colour excesses may be defined as before

$$E_{b-y} = (b-y) - (b-y)_0 \tag{3.1.46}$$

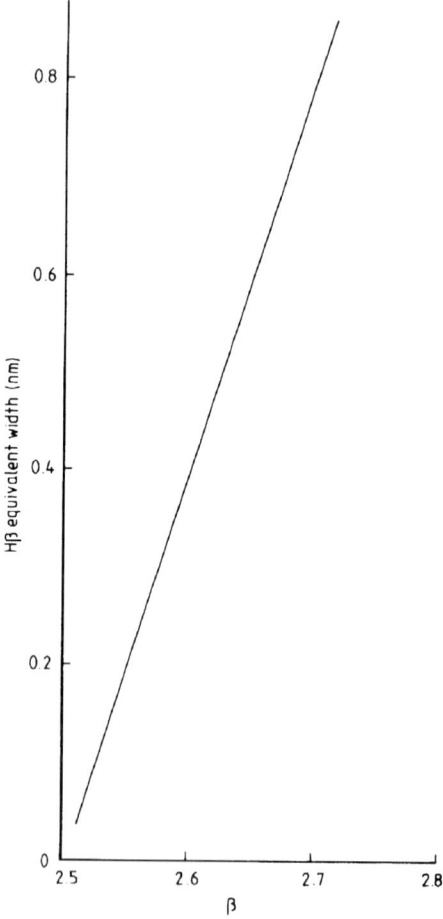

Figure 3.1.20. Relationship between β and the equivalent width of Hβ.

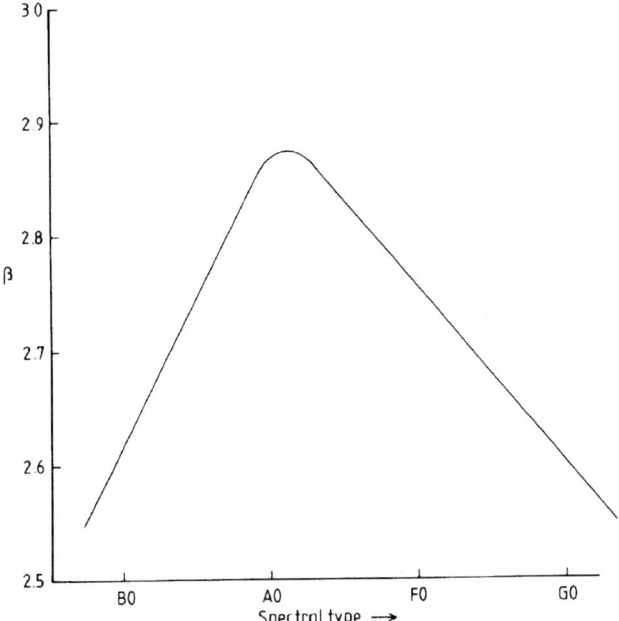

Figure 3.1.21. Relationship of β with spectral type.

and so on. Correction for interstellar absorption may also be accomplished in a similar manner to the UBV case and results in

$$(c_1)_0 = c_1 - 0.20(b - y) \qquad (3.1.47)$$

$$(m_1)_0 = m_1 + 0.18(b - y) \qquad (3.1.48)$$

$$(u - b)_0 = (c_1)_0 + 2(m_1)_0. \qquad (3.1.49)$$

By combining the photometry with model atmosphere work, much additional information such as effective temperatures and surface gravities may be deduced.

Exercises

3.1.1 Calculate the absolute magnitude of Jupiter from its apparent magnitude at opposition, -2.6, and the absolute magnitude of M31 from its apparent magnitude, $+3.5$ (assuming it to be a point source). Their distances are respectively 4.2 AU and 670 kpc.

3.1.2 Calculate the distance of a Cepheid whose apparent magnitude is $+13^m$ on average, and whose absolute magnitude is -4^m on average, assuming that it is affected by interstellar absorption at a rate of $1.5 \times 10^{-3} \, \text{mag pc}^{-1}$.

3.1.3 Standard UBV measures of a main sequence star give the following results: $U = 3.19$; $B = 4.03$; $V = 4.19$. Hence calculate or find: $(U - B)$, $(B - V)$; Q; $(B - V)_0$, E_{B-V}; E_{U-B}, $(U - B)_0$; spectral type; temperature; distance (assuming that the interstellar absorption shown in figure 3.1.15 may be applied); U_0, B_0, V_0 and M_V.

3.2 Photometers

Instruments

Introduction

We may distinguish three basic methods of measuring stellar brightness, based upon the eye, the photographic plate, or upon a variety of photo-electric devices. All three are still in use today. Visual photometry is the almost exclusive prerogative of the amateur astronomer. Although visual photometry suffers from large errors and lacks consistency, it is one of the few areas of astronomy wherein the amateur can still make a genuine contribution. Monitoring and long-term schemes, that could not be justified on a large telescope because of the pressure of higher-priority programmes, can easily be undertaken visually using small telescopes. The light curves of many stars are known only through this long-term work, and one need only remember that amateurs discover numerous novae, supernovae etc. to realize the value of the monitoring undertaken this way. The reduction in the cost of CCDs means that many amateur variable star observers now use them. However, if high accuracy is not needed, visual estimation is far quicker when many stars distributed over the whole sky are to be monitored. Photographic techniques have also been superseded for precise work by the photoelectric methods, but they are still valuable for survey work—a Schmidt camera combined with an automatic measuring machine provides a system that can determine magnitudes at a rate of many thousands of stars per hour. The recently produced Hubble space telescope second Guide Star Catalogue (GSC 2) containing details of 500 million objects, for example, is based upon photographic plates obtained by the 1.2 m Schmidt cameras at Mount Palomar and the Anglo-Australian Observatory. Precision photometry, however, is now exclusively undertaken by means of photoelectric devices, with the CCD predominating in the optical region, and solid-state devices in the infrared. We now look at some of the principles and designs for instruments for each of these three techniques in more detail.

Visual photometers

The eye is comparatively sensitive to small differences in intensity between two sources, but performs poorly when judging the absolute intensity of a single source. Eye estimates of a star's brightness to accuracy of a tenth of a stellar magnitude can thus be achieved if the variable star has two comparison stars, one brighter and one fainter, within the field of view. The variable's brightness is estimated as a fraction of the difference between the brightnesses of the comparison stars. This approach is known as Argelander's step method or the fractional estimate method. The results are usually recorded at the telescope as A-n-B, where A is the fainter star and B the brighter, and n is the number of tenths of the difference between the two that the variable is brighter than A. The variable's magnitude is then calculated from the magnitudes of A and B.

For (slightly) greater precision, or if suitable comparison stars are unavailable, various instruments can be used to improve the visual estimate. However, these now have little advantage over a CCD, and have more-or-less fallen out of use. They are briefly mentioned for their historical interest. The simplest, the extinction photometer, uses the threshold detection limit for the eye itself as the comparison. The image of the source is progressively reduced in brightness by some method such as a variable diaphragm over the telescope's objective, two polarizing filters with a variable mutual angle etc. The setting of the device that is reducing the star's intensity at the instant when the star disappears from view may then be calibrated in terms of the original magnitude of the star. Other designs use a comparison source that may be a second star or an artificial source (figure 3.2.1), and which is arranged to be visible in the field of view simultaneously with the star or source of interest. Either one or the other of the two sources is then adjusted in brightness until both are judged to be equally bright. The highest precision for visual work is reached with the flicker photometer (figure 3.2.2) where the device superimposes the variable and its comparison and then chops between them at a few hertz. Adjusting the comparison's brightness until the flickering disappears can give the brightness of the variable to within a few hundredths of a stellar magnitude.

Photographic photometry

Apart from being used with Schmidt cameras and automatic plate-measuring machines to produce relatively low accuracy ($\pm 0.01^m$) measurements of the brightnesses of large numbers of stars and other objects, photography has been replaced by the CCD, even for most amateur work. Only a brief overview of the method is therefore included here. The emulsion's characteristic curve (figure 2.2.5) must first be found. This may be via a separate calibration exposure on the edge of the image, or if sufficient standard stars of known magnitudes are on the image, as is likely to be the case with large area Schmidt plates, then they may be measured and used instead. The images of stars of different magnitudes will be of different sizes. This arises through the image of the star not being an ideal point. It is spread out by

Figure 3.2.1. Visual photometers. (a) Schematic plan of a visual comparison photometer using an artificial source. (b) Schematic plan of a visual comparison photometer using a second star for comparison.

diffraction, scintillation, scattering within the emulsion, and by reflection off the emulsion's support (halation). The actual image structure will therefore have an intensity distribution that probably approaches a Gaussian pattern. However, the recorded image will have a size governed by the point at which the illumination increases the photographic density above the gross fog level. Images of stars of differing magnitudes will then be recorded as shown schematically in figure 3.2.3.

The measurement that is made of the stellar images may simply be that of its diameter, when a semi-empirical relationship of the form

$$D = A + B\log_{10} I \tag{3.2.1}$$

(where D is the image diameter, I the intensity of the star, and A and B are constants to be determined from the data) can often be fitted to the

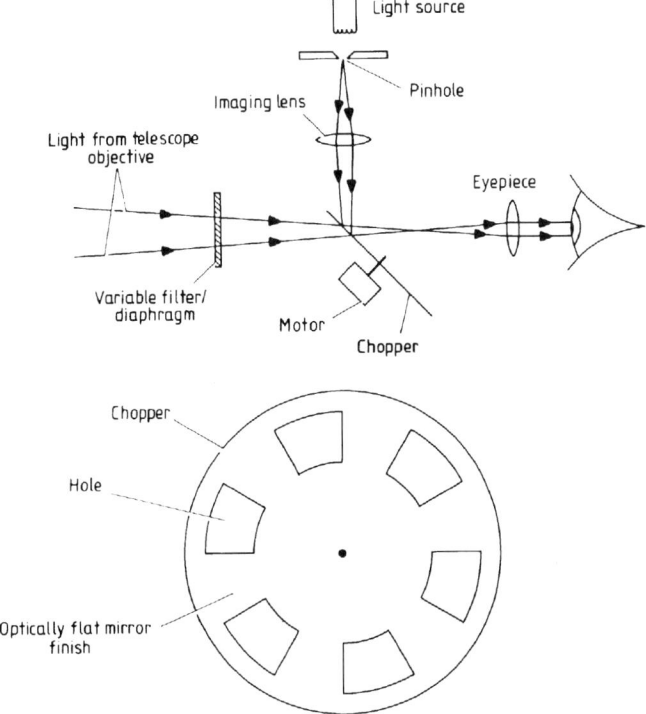

Figure 3.2.2. Schematic plan of a visual flicker photometer.

calibration curve. A more precise measurement than the actual diameter would be the distance between points in the image which are some specified density greater than that of the gross fog, or between the half-density points. Alternatively a combination of diameter and density may be measured. Automatic measuring machines scan the image (section 5.1), so that the total density may be found by integration. Another system is based upon a variable diaphragm. This is set over the image and adjusted until it just

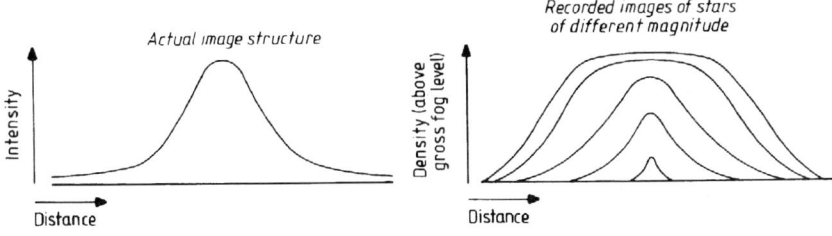

Figure 3.2.3. Schematic variation in image size and density for stars of differing magnitudes recorded on photographic emulsion.

Figure 3.2.4. Comparison of the standard UBV system and its photographic simulation.

contains that image. The transmitted intensity may then be measured and a calibration curve plotted from known standard stars as before.

The magnitudes obtained from a photograph are of course not the same as those of the standard photometric systems (section 3.1). Usually the magnitudes are photographic, or photovisual. However, with careful selection of the emulsion and the use of filters, a response that is close to that of some filter systems may be obtained (figure 3.2.4). Alternatively, a combination of emulsion types and filters may be used that gives a useful set of bands, but without trying to imitate any of the standard systems. The Palomar digital sky survey (DPOSS), for example, used blue, red and near infrared bands, obtained using the now discontinued IIIaJ, IIIaF and IVN plates combined with GG395, RG610 and RG9 filters respectively.

CCD and photoelectric photometers

The most precise photometry relies upon photoelectric devices of various types. These are primarily the CCD in the visual and near infrared regions, and infrared array detectors (section 1.1) at longer wavelengths. Photomultipliers continue to be used occasionally, as do p–i–n photodiodes in small photometers aimed at the amateur market. Photometry with a CCD differs little from ordinary imaging, except that several images of the same area need to be obtained through appropriate filters (section 3.1). Antiblooming CCDs should be avoided since their response is nonlinear for objects nearing saturation. The normal data reduction techniques such as dark signal subtraction and flat fielding (section 1.1) need to be applied to the images. The brightness of a star is obtained directly by adding together the intensities of the pixels covered by the star image and subtracting the intensities in a similar number of pixels covering a nearby area of sky background. Conversion to stellar magnitudes then requires the brightnesses of several standard stars also to be measured. With mosaics of several large area CCDs, there are likely to be several standard stars on any image.

However, if this is not the case, or with detectors covering smaller areas of the sky, one or more separate calibration images will be needed. These are as identical to the main image as possible—as close in time as possible, as close in altitude and azimuth as possible, identical exposures, filters and data processing etc. Small CCDs sold for the amateur market normally have software supplied by the manufacturer to enable stars' brightnesses to be found. This is accomplished by summing the intensities in a square of pixels that the user centres on to the stars' images one at a time. The size of the square is usually adjustable to take account of changes to the image sizes under different seeing conditions. Clearly the same size of square is needed to obtain the background reading. Images from larger CCDs are usually processed automatically using software such as IRAF (section 2.9) or specialized programs written for the individual CCD. Only when stars' images overlap may the observer need to intervene in this process. Integrated magnitudes for extended objects such as galaxies are obtained in a similar fashion, though more input from the observer may be required to ensure that the object is delineated correctly before the pixel intensities are added together. From a good site under good observing conditions, photometry to an accuracy of $\pm 0.001^m$ is now achievable with CCDs.

At near- and mid-infrared wavelengths the procedure is similar to that for CCDs. But in the long-wavelength infrared regions, array detectors are still small (section 1.1), and so only one or two objects are likely to be on each image, making separate calibration exposures essential. In many cases, the object being observed is much fainter than the background noise. For the MIR and FIR regions special observing techniques such as chopping rapidly between the source and the background, subtracting the background from the signal and integrating the result over a long period need to be used. Chopping may be via a rotating 'windmill' whose blades reflect the background (say) on to the detector, while the gaps between the blades allow the radiation from the source to fall on to the detector directly. Alternatively, some telescopes designed specifically for infrared work have secondary mirrors that can be oscillated to achieve this switching.

Photomultipliers (section 1.1) continue to be used when individual photons need to be detected as in the neutrino and cosmic ray Čerenkov detectors (sections 1.3 and 1.4) or when very rapid responses are required as in the observation of occultations (section 2.7). They may also be used on board spacecraft for ultraviolet measurements in the 10 nm to 300 nm region where CCDs are insensitive.

Observing techniques

Probably the single most important 'technique' for successful photometry lies in the selection of the observing site. Only the clearest and most consistent of skies are suitable for precision photometry. Haze, dust, clouds, excessive scintillation, light pollution etc. and the variations in these all render a site unsuitable for photometry. Furthermore, for infrared work, the amount of water vapour above the site should be as low as possible. For these reasons good photometric observing sites are rare. Generally they are at high altitudes and are located where the weather is particularly stable. Oceanic islands, and mountain ranges with a prevailing wind from an ocean, with the site above the inversion layer, are fairly typical of the best choices. Sites that are less than ideal can still be used for photometry, but their number of good nights will be reduced. Restricting the observations to near the zenith is likely to improve the results obtained at a mediocre observing site.

The second most vital part of photometry lies in the selection of the comparison star(s). This must be non-variable, close to the star of interest, of similar apparent magnitude and spectral class, and have its own magnitude known reliably on the photometric system that is in use. Among the brighter stars there is usually some difficulty in finding a suitable standard that is close enough. With fainter stars the likelihood that a suitable comparison star exists is higher, but the chances of its details being known are much lower. Thus ideal comparison stars are found only rarely. The best known variables already have lists of good comparison stars, which may generally be found from the literature. But for less well studied stars, and those being investigated for the first time, there is no such useful information to hand. It may even be necessary to undertake extensive prior investigations, such as checking back through archive records to find non-variable stars in the region, and then measuring these to obtain their magnitudes before studying the star itself. An additional star may need to be observed several times throughout an observing session to supply the data for the correction of atmospheric absorption (see below).

A single observatory can clearly only observe objects for a fraction of the time, even when there are no clouds. Rapidly changing objects, however, may need to be monitored continuously over 24 hours. A number of observatories distributed around the globe have therefore instituted various cooperative programmes so that such continuous photometric observations can be made. The Whole Earth Telescope (WET) for example is a consortium of (currently) 24 observatories studying variable stars (see also BISON and GONG, section 5.3).

Data reduction and analysis

The reduction of the data is performed in three stages—correction for the effects of the Earth's atmosphere, correction to a standard photometric system, and correction to heliocentric time.

The atmosphere absorbs and reddens the star's light, and its effects are expressed by Bouguer's law

$$m_{\lambda,0} = m_{\lambda,z} - a_\lambda \sec z \qquad (3.2.2)$$

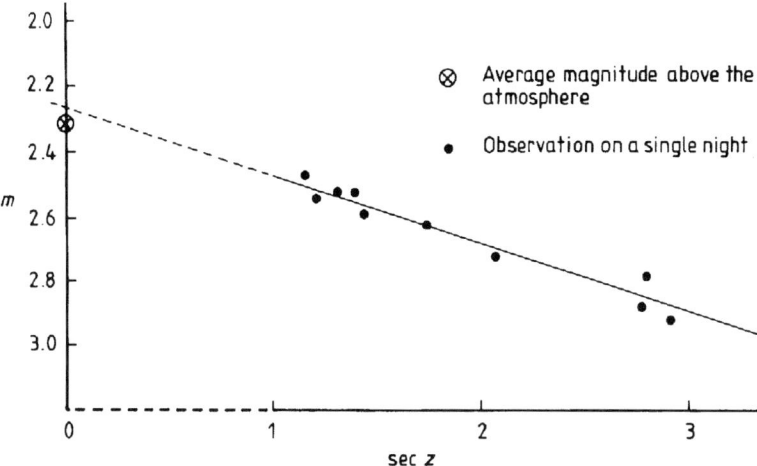

Figure 3.2.5. Schematic variation in magnitude of a standard star with zenith distance.

where $m_{\lambda,z}$ is the magnitude at wavelength λ and at zenith distance z. a_λ is a constant that depends on λ. The law is accurate for zenith distances up to about 60°, which is usually sufficient since photometry is rarely carried out on stars whose zenith distances are greater than 45°. Correction for atmospheric extinction may be simply carried out once the value of the extinction coefficient a_λ is known. Unfortunately a_λ varies from one observing site to another, with the time of year, from day to day, and even throughout the night. Thus its value must be found on every observing occasion. This is done by observing a standard star at several different zenith distances, and by plotting its observed brightness against $\sec z$. Now for zenith angles less than 60° or 70°, to a good approximation $\sec z$ is just a measure of the air mass along the line of sight, so that we may reduce the observations to above the atmosphere by extrapolating them back to an air mass of zero, ignoring the question of the meaning of a value of $\sec z$ of zero! (figure 3.2.5). For the same standard star, this brightness should always be the same on all nights, so that an additional point is available to add to the observations on any given night, which is the average of all the previous determinations of the above-atmosphere magnitude of the star. The extinction coefficient is simply obtained from the slope of the line in figure 3.2.5. The coefficient is strongly wavelength dependent (figure 3.2.6), and so it must be determined separately for every filter that is being used. Once the extinction coefficient has been found, the above-atmosphere magnitude of the star, m_λ, is given by

$$m_\lambda = m_{\lambda,z} - a_\lambda(1 + \sec z). \tag{3.2.3}$$

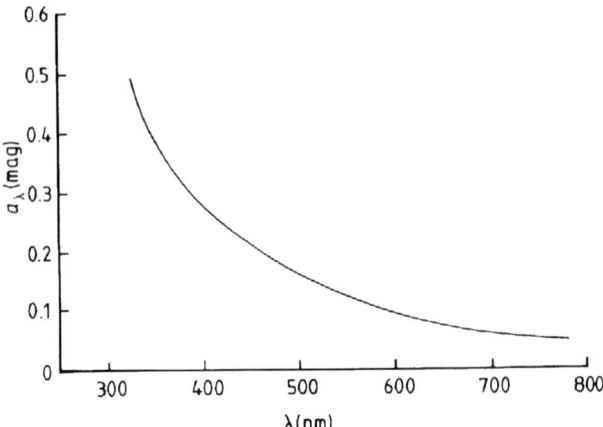

Figure 3.2.6. A typical dependence of the extinction coefficient with wavelength for a good observing site.

Thus the observations of the star of interest and its standards must all be multiplied by a factor k_λ,

$$k_\lambda = 10^{0.4 a_\lambda (1 + \sec z)} \qquad (3.2.4)$$

to correct them to their unabsorbed values. When the unknown star and its comparisons are very close together in the sky, the differential extinction will be negligible and this correction need not be applied. But the separation must be very small; for example at a zenith distance of $45°$, the star and its comparisons must be within $10'$ of arc of each other if the differential extinction is to be less than a thousandth of a magnitude. If E_λ and E'_λ are the original average signals for the star and its comparison, respectively, through the filter centred on λ then the corrected magnitude is given by

$$m_\lambda = m'_\lambda - 2.5 \log_{10} \left(\frac{E_\lambda 10^{0.4 a_\lambda (1 + \sec z)}}{E'_\lambda 10^{0.4 a_\lambda (1 + \sec z')}} \right) \qquad (3.2.5)$$

$$= m'_\lambda + a_\lambda (\sec z' - \sec z) - 2.5 \log_{10} \left(\frac{E_\lambda}{E'_\lambda} \right) \qquad (3.2.6)$$

where m_λ and m'_λ are the magnitudes of the unknown star and its comparison respectively, and z and z' are similarly their zenith distances. The zenith angle is given by

$$\cos z = \sin \phi \sin \delta + \cos \phi \cos \delta \cos(\text{LST} - \alpha) \qquad (3.2.7)$$

where ϕ is the latitude of the observatory, α and δ are the right ascension and declination of the star and LST is the local sidereal time of the observation.

If the photometer is working in one of the standard photometric systems, then the magnitudes obtained through equation (3.2.6) may be used directly, and colour indices, colour excesses etc. may be obtained as discussed in section 3.1. Often, however, the filters or the detector may not be of the standard type. Then the magnitudes must be corrected to a standard system. This can only be attempted if the difference is small, since absorption lines, ionization edges etc. will make any large correction exceedingly complex. The required correction is best determined empirically by observing a wide range of the standard stars of the photometric system. Suitable correction curves may then be plotted from the known magnitudes of these stars.

Finally, and especially for variable stars, the time of the observation should be expressed in heliocentric Julian days. The geocentric Julian date is tabulated in appendix 2, and if the time of the observation is expressed on this scale, then the heliocentric time is obtained by correcting for the travel time of the light to the Sun,

$$T_{\text{Sun}} = T + (5.757 \times 10^{-3})[\sin \delta_* \sin \delta_{\text{Sun}} - \cos \delta_* \cos \delta_{\text{Sun}} \cos(\alpha_{\text{Sun}} - \alpha_*)]$$

$$(3.2.8)$$

where T is the actual time of observation in geocentric Julian days, T_{Sun} is the time referred to the Sun, α_{Sun} and δ_{Sun} are the right ascension and declination of the Sun at the time of the observation, and α_* and δ_* are the right ascension and declination of the star. For very precise work the varying distance of the Earth from the Sun may need to be taken into account as well.

Further items in the reduction and analysis of the data such as the corrections for interstellar reddening, calculation of colour index, temperature etc. were covered in section 3.1.

Exercise

3.2.1   Show that if the differential extinction correction is to be less than Δm magnitudes, then the zenith distances of the star and its comparison must differ by less than Δz, where

$$\Delta z = \frac{\Delta m}{a_\lambda} \cot z \operatorname{cosec} z \quad \text{radians.}$$

Chapter 4

Spectroscopy

4.1   Spectroscopy

Introduction

Practical spectroscopes are usually based upon one or other of two quite separate optical principles—interference and differential refraction. The former produces instruments based upon diffraction gratings or interferometers, while the latter results in prism-based spectroscopes. There are also some hybrid designs. The details of the spectroscopes themselves are considered in section 4.2; here we discuss the basic optical principles that underlie their designs.

Diffraction gratings

The operating principle of diffraction gratings relies upon the effects of diffraction and interference of light waves. Logically, therefore, it might seem that they should be included within the section on interference-based spectroscopes below. Diffraction gratings, however, are in such widespread and common use in astronomy that they merit a section to themselves.

We have already seen in section 2.5 the structure of the image of a single source viewed through two apertures (figure 2.5.6). The angular distance of the first fringe from the central maximum is λ/d, where d is the separation of the apertures. Hence the position of the first fringe, and of course the positions of all the other fringes, is a function of wavelength. If such a pair of apertures were illuminated with white light, all the fringes apart from the central maximum would thus be short spectra with the longer wavelengths farthest from the central maximum. In an image such as that of figure 2.5.6, the spectra would be of little use since the fringes are so broad that they would overlap each other long before a useful dispersion could be obtained. However, if we add a third aperture in line with the first two, and again separated from the nearer of the original apertures by a distance d, then we find that the fringes remain stationary, but become narrower

Figure 4.1.1. Small portion of the image structure for a single point source viewed through two apertures (broken curve) and three apertures (full curve).

and more intense. Weak secondary maxima also appear between the main fringes (figure 4.1.1). The peak intensities are of course modulated by the pattern from a single slit when looked at on a larger scale, in the manner of figure 2.5.6. If further apertures are added in line with the first three and with the same separations, then the principal fringes continue to narrow and intensify, and further weak maxima appear between them (figure 4.1.2). The intensity of the pattern at some angle, θ, to the optical axis is given by (cf. equation (1.1.6))

$$I(\theta) = I(0) \left[\frac{\sin^2 \left(\frac{\pi D \sin \theta}{\lambda} \right)}{\left(\frac{\pi D \sin \theta}{\lambda} \right)^2} \right] \left[\frac{\sin^2 \left(\frac{N \pi d \sin \theta}{\lambda} \right)}{\sin^2 \left(\frac{\pi d \sin \theta}{\lambda} \right)} \right] \qquad (4.1.1)$$

Figure 4.1.2. Small portion of the image structure for a single point source viewed through 20 apertures.

where D is the width of the aperture and N the number of apertures. The term

$$\left[\frac{\sin^2 \left(\frac{\pi D \sin \theta}{\lambda} \right)}{\left(\frac{\pi D \sin \theta}{\lambda} \right)^2} \right] \tag{4.1.2}$$

represents the modulation of the image by the intensity structure for a single aperture, while the term

$$\left[\frac{\sin^2 \left(\frac{N \pi d \sin \theta}{\lambda} \right)}{\sin^2 \left(\frac{\pi d \sin \theta}{\lambda} \right)} \right] \tag{4.1.3}$$

represents the result of the interference between N apertures. We may write

$$\Delta = \left(\frac{\pi D \sin \theta}{\lambda} \right) \tag{4.1.4}$$

and

$$\delta = \left(\frac{\pi d \sin \theta}{\lambda} \right) \tag{4.1.5}$$

and equation (4.1.1) then becomes

$$I(\theta) = I(0) \frac{\sin^2 \Delta}{\Delta^2} \frac{\sin^2 (N\delta)}{\sin^2 \delta}. \tag{4.1.6}$$

Now consider the interference component as δ tends to $m\pi$, where m is an integer. Putting

$$P = \delta - m\pi \tag{4.1.7}$$

we have

$$\lim_{\delta \to m\pi} \left(\frac{\sin(N\delta)}{\sin \delta} \right) = \lim_{P \to 0} \left(\frac{\sin[N(P + m\pi)]}{\sin(P + m\pi)} \right) \tag{4.1.8}$$

$$= \lim_{P \to 0} \left(\frac{\sin(NP)\cos(Nm\pi) + \cos(NP)\sin(Nm\pi)}{\sin P \cos(m\pi) + \cos P \sin(m\pi)} \right) \tag{4.1.9}$$

$$= \lim_{P \to 0} \left(\pm \frac{\sin(NP)}{\sin P} \right) \tag{4.1.10}$$

$$= \pm N \lim_{P \to 0} \left(\frac{\sin(NP)}{NP} \frac{P}{\sin P} \right) \tag{4.1.11}$$

$$= \pm N. \tag{4.1.12}$$

Hence integer multiples of π give the values of δ for which we have a principal fringe maximum. The angular positions of the principal maxima are given by

$$\theta = \sin^{-1}\left(\frac{m\lambda}{d}\right) \qquad (4.1.13)$$

and m is usually called the order of the fringe. The zero intensities in the fringe pattern will be given by

$$N\delta = m'\pi \qquad (4.1.14)$$

where m' is an integer, but excluding the cases where $m' = mN$ that are the principal fringe maxima. Their positions are given by

$$\theta = \sin^{-1}\left(\frac{m'\lambda}{Nd}\right). \qquad (4.1.15)$$

The angular width of a principal maximum, W, between the first zeros on either side of it is thus given by

$$W = \frac{2\lambda}{Nd\cos\theta}. \qquad (4.1.16)$$

The width of a fringe is therefore inversely proportional to the number of apertures, while its peak intensity, from equations (4.1.6) and (4.1.12), is proportional to the square of the number of apertures. Thus for a bichromatic source observed through a number of apertures we obtain the type of image structure shown in figure 4.1.3. The angular separation of fringes of the same order for the two wavelengths, for small values of θ, can be seen from equation (4.1.13) to be proportional to both the wavelength and to the order of the fringe, while the fringe width is independent of the order (equation (4.1.16)). For a white light source, by a simple extension of figure 4.1.3, we may see that the image will consist of a series of spectra on either side of a white central image. The Rayleigh resolution within this

Figure 4.1.3. A portion of the image structure for a single bichromatic point source viewed through several apertures.

image is obtained from equation (4.1.16):

$$W' = \frac{\lambda}{Nd \cos \theta} \tag{4.1.17}$$

and is independent of the fringe order. The ability of a spectroscope to separate two wavelengths is called the spectral resolution and is denoted by W_λ, and it may now be found from equation (4.1.17):

$$W_\lambda = W' \frac{d\lambda}{d\theta} \tag{4.1.18}$$

but from equation (4.1.13)

$$\frac{d\lambda}{d\theta} = \frac{d}{m} \cos \theta \tag{4.1.19}$$

so that

$$W_\lambda = \frac{\lambda}{Nm}. \tag{4.1.20}$$

The spectral resolution thus improves directly with the fringe order because of the increasing dispersion of the spectra.

More commonly the resolution is expressed as the ratio of the operating wavelength to the spectral resolution, and denoted by R:

$$R = \frac{\lambda}{W_\lambda} = Nm. \tag{4.1.21}$$

The resolution for a series of apertures is thus just the product of the number of apertures and the order of the spectrum. It is independent of the width and spacing of the apertures.

From figure 4.1.3 we may see that at higher orders the spectra are overlapping. This occurs at all orders when white light is used. The difference in wavelength between two superimposed wavelengths from adjacent spectral orders is called the free spectral range, Σ. From equation (4.1.13) we may see that if λ_1 and λ_2 are two such superimposed wavelengths, then

$$\sin^{-1} \left(\frac{m\lambda_1}{d} \right) = \sin^{-1} \left(\frac{(m+1)\lambda_2}{d} \right). \tag{4.1.22}$$

That is, for small angles,

$$\Sigma = \lambda_1 - \lambda_2 \approx \frac{\lambda_2}{m}. \tag{4.1.23}$$

For small values of m, Σ is therefore large, and the unwanted wavelengths in a practical spectroscope may be rejected by the use of filters. Some spectroscopes, such as those based on Fabry–Pérot etalons and echelle gratings, however, operate at very high spectral orders and both of the overlapping wavelengths may be desired. Then it is necessary to use a cross disperser so that the final spectrum consists of a two-dimensional array of short sections of the spectrum (see the discussion later in this section).

A practical device for producing spectra by diffraction uses a large number of closely spaced, parallel, narrow slits or grooves, and is called a diffraction grating. Typical gratings for astronomical use have between 100 and 1000 grooves per millimetre, and 1000 to 50 000 grooves in total. They are used at orders ranging from one up to two hundred or so. Thus the resolutions range from 10^3 to 10^5. Although the earlier discussion was based upon the use of clear apertures, a small plane mirror can replace each aperture without altering the results. Thus diffraction gratings can be used either in transmission or reflection modes. Most astronomical spectro-scopes are in fact based upon reflection gratings. Often the grating is inclined to the incoming beam of light, but this changes the discussion only margin-ally. There is a constant term, $d \sin i$, added to the path differences, where i is the angle made by the incoming beam with the normal to the grating. The whole image (figure 4.1.3) is shifted an angular distance i along the image plane. Equation (4.1.13) then becomes

$$\theta = \sin^{-1}\left[\left(\frac{m\lambda}{d}\right) - \sin i\right] \qquad (4.1.24)$$

and in this form is often called the grating equation.

Volume Phase Holographic Gratings (VPHGs) are currently starting to be used within astronomical spectroscopes. These have a grating in the form of a layer of gelatine within which the refractive index changes (section 4.2), with the lines of the grating produced by regions of differing refractive indices. VPHGs operate through Bragg diffraction (section 1.3, figure 1.3.15 and equation (1.3.9)). Their efficiencies can thus be up to 95% in the first order. They can be used either as transmission or reflection gratings and replace conventional gratings in spectroscopes at the appropriate Bragg angle for the operating wavelength.

To form a part of a spectroscope, a grating must be combined with other elements. The basic layout is shown in figure 4.1.4, and practical designs are discussed in section 4.2. The grating is illuminated by parallel light that is usually obtained by placing a slit at the focus of a collimating lens, but some-times may be obtained simply by allowing the light from a very distant object to fall directly on to the grating. After reflection from the grating, the light is focused by the imaging lens to form the required spectrum, and this may then be recorded, observed through an eyepiece, projected on to a screen etc. as desired. The collimator and imaging lenses may be simple lenses as shown, in which case the spectrum will be tilted with respect to the optical axis because of chromatic aberration, or they may be achromats or mirrors (see the next section).

The angular dispersion of a grating is not normally used as a parameter of a spectroscopic system. Instead it is combined with the focal length of the

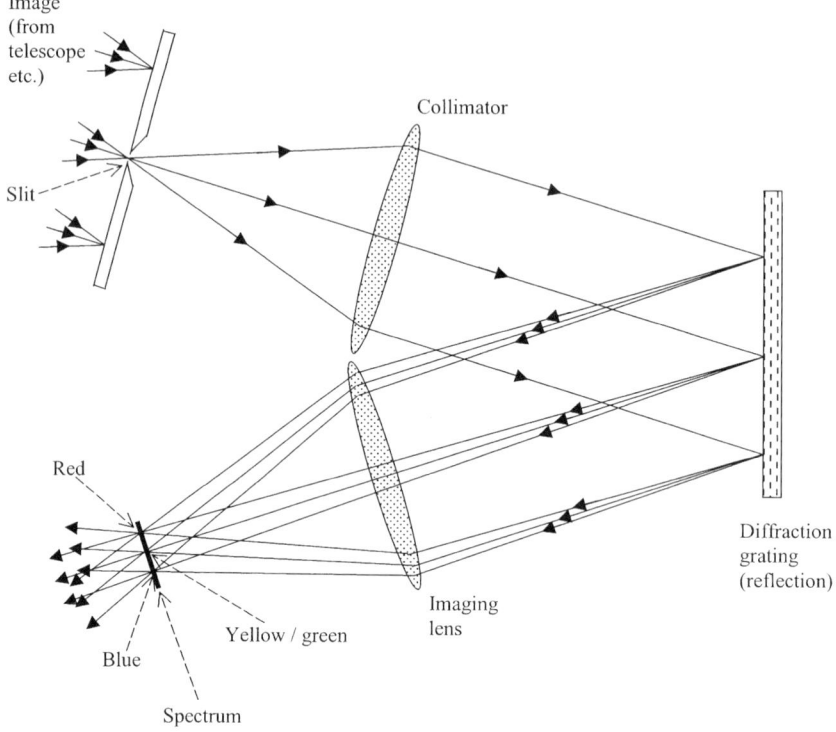

Figure 4.1.4. Basic optical arrangement of a reflection grating spectroscope.

imaging element to give either the linear dispersion or the reciprocal linear dispersion. If x is the linear distance along the spectrum from some reference point, then we have for an achromatic imaging element of focal length f_2,

$$\frac{\mathrm{d}x}{\mathrm{d}\lambda} = f_2 \frac{\mathrm{d}\theta}{\mathrm{d}\lambda} \qquad (4.1.25)$$

where θ is small and is measured in radians. From equation (4.1.24), the linear dispersion within each spectrum is thus given by

$$\frac{\mathrm{d}x}{\mathrm{d}\lambda} = \pm \frac{mf_2}{d \cos \theta}. \qquad (4.1.26)$$

Now since θ varies very little over an individual spectrum, we may write

$$\frac{\mathrm{d}x}{\mathrm{d}\lambda} \approx \text{constant}. \qquad (4.1.27)$$

The dispersion of a grating spectroscope is thus roughly constant compared with the strong wavelength dependence of a prism spectroscope (equation (4.1.69)). More commonly, the reciprocal linear dispersion, $\mathrm{d}\lambda/\mathrm{d}x$, is quoted and used. For practical astronomical spectrometers, this usually has values in the range

$$10^{-7} < \frac{d\lambda}{dx} < 5 \times 10^{-5}. \qquad (4.1.28)$$

The commonly used units are nanometres change of wavelength per millimetre along the spectrum so that the above range is from 0.1 to 50 nm mm^{-1}. The use of Å mm^{-1} is still fairly common practice among astronomers; the magnitude of the dispersion is then a factor of ten larger than the conventional measure.

The resolving power of a spectroscope is limited by the spectral resolution of the grating, the resolving power of its optics (section 1.1), and by the projected slit width. The spectrum is formed from an infinite number of monochromatic images of the entrance slit. It is easy to see that the width of one of these images, S, is given by

$$S = s\frac{f_2}{f_1} \qquad (4.1.29)$$

where s is the slit width, f_1 is the collimator's focal length and f_2 is the imaging element's focal length. In wavelength terms, the slit width, $S(d\lambda/dx)$, is sometimes called the spectral purity of the spectroscope. The entrance slit must have a physical width of s_{max} or less, if it is not to degrade the spectral resolution, where

$$s_{max} = \frac{\lambda f_1}{Nd\cos\theta} \qquad (4.1.30)$$

(cf. equation (4.1.77)).

If the optics of the spectroscope are well corrected then we may ignore their aberrations and consider only the diffraction limit of the system. When the grating is fully illuminated, the imaging element will intercept a rectangular beam of light. The height of the beam is just the height of the grating, and has no effect upon the spectral resolution. The width of the beam, D, is given by

$$D = L\cos\theta \qquad (4.1.31)$$

where L is the length of the grating and θ the angle of the exit beam to the normal to the plane of the grating. The diffraction limit is just that of a rectangular slit of width D. So, from figure 1.1.26, the linear Rayleigh limit of resolution, W'', is given by

$$W'' = \frac{f_2\lambda}{D} \qquad (4.1.32)$$

$$= \frac{f_2\lambda}{L\cos\theta}. \qquad (4.1.33)$$

If the beam is limited by some other element of the optical system, and/or is of circular cross section, then D must be evaluated as may be appropriate,

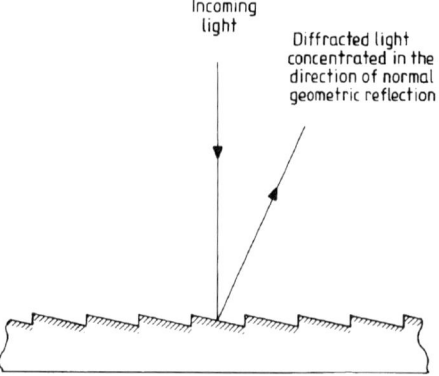

Figure 4.1.5. Enlarged section through a blazed reflection grating.

or the Rayleigh criterion for the resolution through a circular aperture (equation (1.1.56)) used in place of that for a rectangular aperture. Optimum resolution occurs when

$$S = W''$$ (4.1.34)

i.e.

$$s = \frac{f_1 \lambda}{D}$$ (4.1.35)

$$= \frac{f_1 \lambda}{L \cos \theta}.$$ (4.1.36)

The major disadvantage of a grating as a dispersing element is immediately obvious from figure 4.1.3: the light from the original source is spread over a large number of spectra. The grating's efficiency in terms of the fraction of light concentrated into the spectrum of interest is therefore very low. This disadvantage, however, may be largely overcome with reflection gratings through the use of the technique of blazing the grating. In this technique, the individual mirrors that comprise the grating are angled so that they concentrate the light into a single narrow solid angle (figure 4.1.5). For instruments based upon the use of gratings at low orders, the angle of the mirrors is arranged so that the light is concentrated into the spectrum to be used, and by this means up to 90% efficiency can be achieved. In terms of the interference patterns, the grating is designed so that central peak due to an individual aperture just extends over the width of the desired spectrum. The blaze angle then shifts that peak along the array of spectra until it coincides with the desired order.

For those spectroscopes that use gratings at high orders, the grating can still be blazed, but then the light is concentrated into short segments of many different orders of spectra. By a careful choice of parameters, these short segments can be arranged so that they overlap slightly at their ends, and so coverage of a much wider spectral region may be obtained by producing a montage of the segments. Transmission gratings can also be blazed although this is less common. Each of the grooves then has the cross section of a small prism, the apex angle of which defines the blaze angle. Blazed transmission gratings for use at infrared wavelengths can be produced by etching the surface of a block of silicon in a similar manner to the way in which integrated circuits are produced.

Another problem that is intrinsically less serious, but which is harder to counteract, is that of shadowing. If the incident and/or reflected light makes a large angle to the normal to the grating, then the steplike nature of the surface (figure 4.1.5) will cause a significant fraction of the light to be intercepted by the vertical portions of the grooves, and so lost to the final spectrum. There is little that can be done to eliminate this problem except either to accept the light loss, or to design the system so that large angles of incidence or reflection are not needed.

Curved reflection gratings are frequently produced. By making the curve that of an optical surface, the grating itself can be made to fulfil the function of the collimator and/or the imaging element of the spectroscope, thus reducing light losses and making for greater simplicity of design and reduced costs. The grooves should be ruled so that they appear parallel and equally spaced when viewed from infinity. The simplest optical principle employing a curved grating is that due to Rowland. The slit, grating and spectrum all lie on a single circle that is called the Rowland circle (figure 4.1.6). This has a diameter equal to the radius of curvature of the grating. The use of a curved grating at large angle to its optical axis introduces astigmatism, and spectral lines may also be curved due to the varying angles of incidence for rays from the centre and ends of the slit (cf. figure 4.1.14). Careful design, however, can reduce or eliminate these defects, and there are several practical designs for spectroscopes based upon the Rowland circle (section 4.2). Aspherical curved gratings are also possible and can be used to provide very highly corrected designs with few optical components.

Higher dispersion can be obtained by using immersed reflection gratings. The light in these interacts with the grating within a medium other than air (or a vacuum). The refractive index of the medium shortens the wavelength so that the groove spacing is effectively increased relative to the wavelength and so the dispersion is augmented (equation (4.1.19)). One approach to producing an immersion grating is simply to illuminate the grating from the back—i.e. through the transparent substrate. A second approach is to flood the grating with a thin layer of oil kept in place by a cover sheet.

A grating spectrum generally suffers from unwanted additional features superimposed upon the desired spectrum. Such features are usually much fainter than the main spectrum and are called ghosts. They arise from a

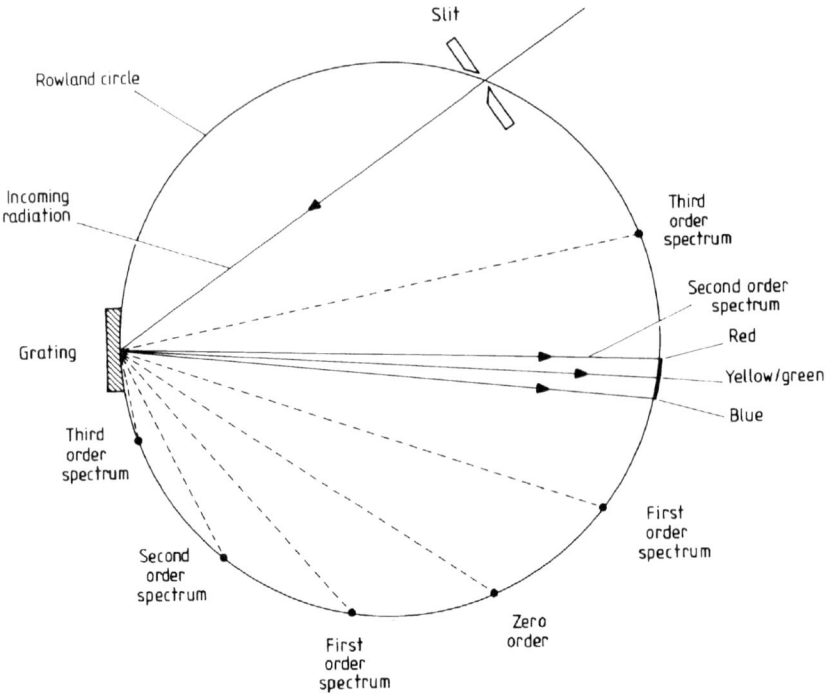

Figure 4.1.6. Schematic diagram of a spectroscope based upon a Rowland circle, using a curved grating blazed for the second order.

variety of causes. They may be due to overlapping spectra from higher or lower orders, or to the secondary maxima associated with each principal maximum (figure 4.1.2). The first of these is usually simple to eliminate by the use of filters since the overlapping ghosts are of different wavelengths from the overlapped main spectrum. The second source is usually unimportant since the secondary maxima are very faint when more than a few tens of apertures are used, though they still contribute to the wings of the PSF. Of more general importance are the ghosts that arise through errors in the grating. Such errors most commonly take the form of a periodic variation in the groove spacing. A variation with a single period gives rise to Rowland ghosts that appear as faint lines close to and on either side of strong spectrum lines. Their intensity is proportional to the square of the order of the spectrum. Thus echelle gratings (see below) must be of very high quality since they may use spectral orders of several hundred. If the error is multi-periodic, then Lyman ghosts of strong lines may appear. These are similar to the Rowland ghosts, except that they can be formed at large distances from the line that is producing them. Some compensation for these errors can be obtained through deconvolution of the PSF

Figure 4.1.7. Enlarged view of an echelle grating.

(section 2.1), but for critical work the only real solution is to use a grating without periodic errors, such as a holographically produced grating.

Wood's anomalies may also sometimes occur. These do not arise through grating faults, but are due to light that should go into spectral orders behind the grating (were that to be possible) reappearing within lower order spectra. The anomalies have a sudden onset and a slower decline towards longer wavelengths and are almost 100% plane polarized. They are rarely important in efficiently blazed gratings.

By increasing the angle of a blazed grating, we obtain an echelle grating (figure 4.1.7). This is illuminated more or less normally to the groove surfaces and therefore at a large angle to the normal to the grating. It is usually a very coarse grating—ten lines per millimetre is not uncommon—so that the separation of the apertures, d, is very large. The reciprocal linear dispersion

$$\frac{d\lambda}{dx} = \pm \frac{d \cos^3 \theta}{mf_2} \qquad (4.1.37)$$

is therefore also very large. Such gratings concentrate the light into many overlapping high-order spectra, and so from equation (4.1.21) the resolution is very high. A spectroscope that is based upon an echelle grating requires a second low dispersion grating or prism whose dispersion is perpendicular to that of the echelle and so is called a cross-disperser in order to separate out each of the orders (section 4.2).

A quantity known variously as throughput, etendu, or light gathering power, is useful as a measure of the efficiency of the optical system. It is the amount of energy passed by the system when its entrance aperture is illuminated by unit intensity per unit area per unit solid angle, and it is denoted by u:

$$u = \tau A \Omega \qquad (4.1.38)$$

where Ω is the solid angle accepted by the instrument, A is the area of its aperture and τ is the fractional transmission of its optics, incorporating losses due to scattering, absorption, imperfect reflection etc. For a spectroscope, Ω is the solid angle subtended by the entrance slit at the collimator or, for slitless spectroscopes, the solid angle accepted by the telescope–spectroscope combination. A is the area of the collimator or the effective area of the dispersing element, whichever is the smaller; τ will depend critically upon the design of the system, but as a reasonably general rule it may be taken to be the product of the transmissions for all the surfaces. These will usually be in the region of 0.8 to 0.9 for each surface, so that τ for the design illustrated in figure 4.1.4 will have a value of about 0.4. The product of resolution and throughput, P, is a useful figure for comparing the performances of different spectroscope systems:

$$P = Ru. \tag{4.1.39}$$

Normally it will be found that, other things being equal, P will be largest for Fabry–Pérot spectroscopes (see below), of intermediate values for grating-based spectroscopes, and lowest for prism-based spectroscopes.

Prisms

Pure prism-based spectroscopes will rarely be encountered today, except within instruments constructed some time ago. However, they are used in conjunction with gratings in some modern instruments. The combination is known as a 'grism' and the deviation of the light beam by the prism is used to counteract that of the grating, so that the light passes straight through the instrument. The spectroscope can then be used for direct imaging just by removing the grism and without having to move the camera. Prisms are also often used as cross-dispersers for high spectral order spectroscopes based upon echelle gratings or etalons, and may be used non-spectroscopically for folding light beams.

When monochromatic light passes through an interface between two transparent isotropic media at a fixed temperature, then we can apply the well-known Snell's law relating the angle of incidence, i, to the angle of refraction, r, at that interface

$$\mu_1 \sin i = \mu_2 \sin r \tag{4.1.40}$$

where μ_1 and μ_2 are constants that are characteristic of the two media. When $\mu_1 = 1$, which strictly only occurs for a vacuum, but which holds to a good approximation for most gases, including air, we have

$$\frac{\sin i}{\sin r} = \mu_2 \tag{4.1.41}$$

and μ_2 is known as the refractive index of the second medium. Now we have already seen that the refractive index varies with wavelength for many media (section 1.1, equation (1.1.32) etc.). The manner of this variation may, over a restricted wavelength interval, be approximated by the Hartmann dispersion formula, in which A, B and C are known as the Hartmann constants (see also the Cauchy formula, equation (4.2.18)):

$$\mu_\lambda = A + \frac{B}{\lambda - C}. \tag{4.1.42}$$

If the refractive index is known at three different wavelengths, then we can obtain three simultaneous equations for the constants, from equation (4.1.42), giving

$$C = \frac{\left[\left(\dfrac{\mu_1 - \mu_2}{\mu_2 - \mu_3}\right)\lambda_1(\lambda_2 - \lambda_3) - \lambda_3(\lambda_1 - \lambda_2)\right]}{\left[\left(\dfrac{\mu_1 - \mu_2}{\mu_2 - \mu_3}\right)(\lambda_2 - \lambda_3) - (\lambda_2 - \lambda_3)\right]} \tag{4.1.43}$$

$$B = \frac{\mu_1 - \mu_2}{\left(\dfrac{1}{\lambda_1 - C} - \dfrac{1}{\lambda_2 - C}\right)} \tag{4.1.44}$$

$$A = \mu_1 - \frac{B}{\lambda_1 - C}. \tag{4.1.45}$$

The values for the constants for the optical region for typical optical glasses are:

	A	B	C
Crown glass	1.477	3.2×10^{-8}	-2.1×10^{-7}
Dense flint glass	1.603	2.08×10^{-8}	1.43×10^{-7}

Thus white light refracted at an interface is spread out into a spectrum with the longer wavelengths refracted less than the shorter ones. This phenomenon was encountered in section 1.1 as chromatic aberration, and there we were concerned with eliminating or minimizing its effects; for spectroscopy by contrast we are interested in maximizing the dispersion.

Consider, then, a prism with light incident upon it as shown in figure 4.1.8. For light of wavelength λ, the deviation, θ, is given by

$$\theta = i_1 + r_2 - \alpha \tag{4.1.46}$$

so by using equation (4.1.42) and using the relations

$$\mu_\lambda = \frac{\sin i_1}{\sin r_1} = \frac{\sin r_2}{\sin i_2} \tag{4.1.47}$$

and

$$\alpha = r_1 + i_2 \tag{4.1.48}$$

we get

$$\theta = i_1 - \alpha + \sin^{-1}\left\{\left(A + \frac{B}{\lambda - C}\right)\sin\left[\alpha - \sin^{-1}\left(\frac{\sin i_1}{A + \dfrac{B}{\lambda - C}}\right)\right]\right\}. \tag{4.1.49}$$

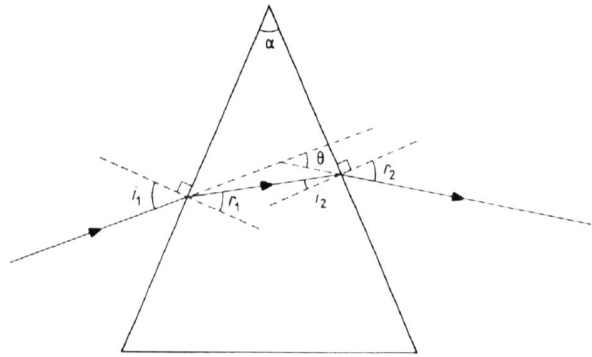

Figure 4.1.8. Optical path in a prism.

Now we wish to maximize $\partial\theta/\partial\lambda$, which we could study by differentiating equation (4.1.49), but which is easier to obtain from

$$\frac{\Delta\theta}{\Delta\lambda} = \frac{\theta_{\lambda_1} - \theta_{\lambda_2}}{\lambda_1 - \lambda_2} \tag{4.1.50}$$

so that

$$\frac{\Delta\theta}{\Delta\lambda} = \frac{\left\{ \sin^{-1}\left[\mu_{\lambda_1}\sin\left(\alpha - \sin^{-1}\left\{\frac{\sin i_1}{\mu_{\lambda_1}}\right\}\right)\right] - \sin^{-1}\left[\mu_{\lambda_2}\sin\left(\alpha - \sin^{-1}\left\{\frac{\sin i_1}{\mu_{\lambda_2}}\right\}\right)\right]\right\}}{\lambda_1 - \lambda_2}. \tag{4.1.51}$$

The effect of altering the angle of incidence or the prism angle is now most simply followed by an example. Consider a dense flint prism for which

$$\lambda_1 = 4.86 \times 10^{-7}\,\text{m} \qquad \mu_{\lambda_1} = 1.664 \tag{4.1.52}$$

$$\lambda_2 = 5.89 \times 10^{-7}\,\text{m} \qquad \mu_{\lambda_2} = 1.650 \tag{4.1.53}$$

and

$$\frac{\Delta\theta}{\Delta\lambda} = \frac{\left\{ \sin^{-1}\left[1.664\sin\left(\alpha - \sin^{-1}\left\{\frac{\sin i_1}{1.664}\right\}\right)\right] - \sin^{-1}\left[1.650\sin\left(\alpha - \sin^{-1}\left\{\frac{\sin i_1}{1.650}\right\}\right)\right]\right\}}{1.03 \times 10^{-7}} \tag{4.1.54}$$

(in $^\circ\,\text{m}^{-1}$). Figure 4.1.9 shows the variation of $\Delta\theta/\Delta\lambda$ with angle of incidence for a variety of apex angles as given by equation (4.1.54). From this figure it is reasonably convincing to see that the maximum dispersion of $1.02 \times 10^8\,^\circ\,\text{m}^{-1}$ occurs for an angle of incidence of $90°$ and an apex angle of $73.8776°$. This represents the condition of glancing incidence and exit

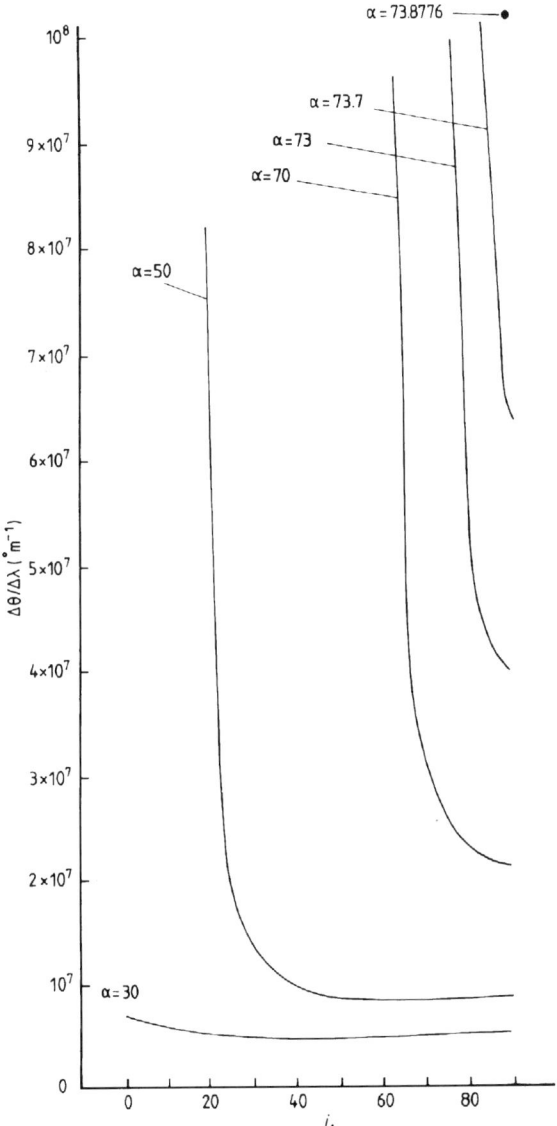

Figure 4.1.9. Variation of $\Delta\theta/\Delta\lambda$ with angle of incidence for a dense flint prism, for various apex angles, α.

from the prism (figure 4.1.10), and the ray passes symmetrically through the prism.

The symmetrical passage of the ray through the prism is of importance apart from being one of the requirements for maximum $\Delta\theta/\Delta\lambda$. It is only

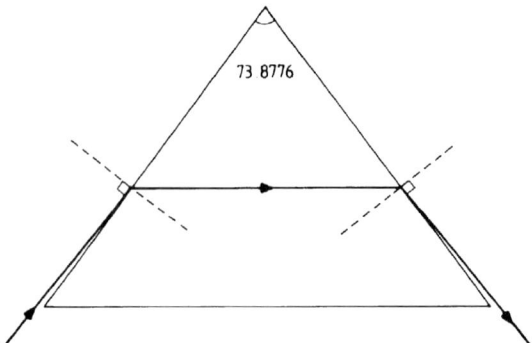

Figure 4.1.10. Optical path for maximum dispersion in a dense flint prism.

when this condition applies that the astigmatism introduced by the prism is minimized. The condition of symmetrical ray passage for any prism is more normally called the position of minimum deviation. Again an example quickly illustrates why this is so. For a dense flint prism with an apex angle of $30°$, and at a wavelength of 500 nm, figure 4.1.11 shows the variation of the deviation with angle of incidence. The minimum value of θ occurs for

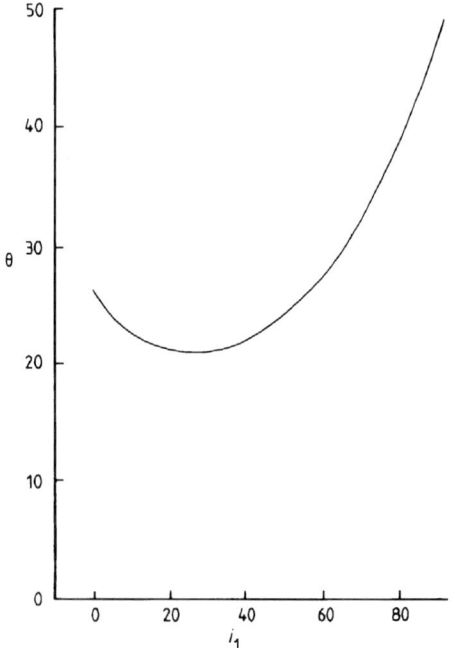

Figure 4.1.11. Deviation of dense flint prism with an apex angle of $30°$ at a wavelength of 500 nm.

$i_1 = 25.46°$, from which we rapidly find that $r_1 = 15°$, $i_2 = 15°$ and $r_2 = 25.46°$, and so the ray is passing through the prism symmetrically when its deviation is a minimum. More generally we may write θ in terms of r_1,

$$\theta = \sin^{-1}(\mu_\lambda \sin r_1) + \sin^{-1}[\mu_\lambda \sin(\alpha - r_1)] - \alpha \qquad (4.1.55)$$

so that

$$\frac{\partial \theta}{\partial r_1} = \frac{\mu_\lambda \cos r_1}{\sqrt{1 - \mu_\lambda^2 \sin^2 r_1}} - \frac{\mu_\lambda \cos(\alpha - r_1)}{\sqrt{1 - \mu_\lambda^2 \sin^2(\alpha - r_1)}}. \qquad (4.1.56)$$

Since $\partial \theta / \partial r_1 = 0$ for θ to be an extremum, we obtain at this point, after some manipulation,

$$[\cos(2r_1) - \cos(2\alpha - 2r_1)](1 - 2\mu_\lambda^2) = 0 \qquad (4.1.57)$$

giving

$$r_1 = \alpha/2 \qquad (4.1.58)$$

and so also

$$i_2 = \alpha/2 \qquad (4.1.59)$$

and the minimum deviation always occurs for a symmetrical passage of the ray through the prism.

In practice the maximum dispersion conditions of glancing incidence and exit are unusable because most of the light will be reflected at the two interactions and not refracted through the prism. Anti-reflection coatings (section 1.1) cannot be employed because these can only be optimized for a few wavelengths and so the spectral distribution would be disturbed. Thus the apex angle must be less than the optimum, but the minimum deviation condition must be retained in order to minimize astigmatism. A fairly common compromise, then, is to use a prism with an apex angle of 60°, which has advantages from the manufacturer's point of view in that it reduces the amount of waste material if the initial blank is cut from a billet of glass. For an apex angle of 60°, the dense flint prism considered earlier has a dispersion of $1.39 \times 10^7 \, °\,m^{-1}$ for the angle of incidence of 55.9° that is required to give minimum deviation ray passage. This is almost a factor of ten lower than the maximum possible value.

With white light, it is obviously impossible to obtain the minimum deviation condition for all the wavelengths, and the prism is usually adjusted so that this condition is preserved for the central wavelength of the region of interest. To see how the deviation varies with wavelength, we consider the case of a prism with a normal apex angle of 60°. At minimum deviation we then have

$$r_1 = 30°. \qquad (4.1.60)$$

Putting these values into equation (4.1.55) and using equation (4.1.42), we get

$$\theta = 2\sin^{-1}\left\{\frac{1}{2}\left[A + \frac{B}{\lambda - C}\right]\right\} - 60 \quad \text{degrees} \tag{4.1.61}$$

so that

$$\frac{d\theta}{d\lambda} = \frac{-180B}{\pi(\lambda - C)^2\left\{1 - \frac{1}{4}\left[A + \frac{B}{\lambda - C}\right]^2\right\}^{1/2}} \quad {}^{\circ}\,\text{m}^{-1}. \tag{4.1.62}$$

Now

$$A + \frac{B}{\lambda - C} = \mu_\lambda \approx 1.5 \tag{4.1.63}$$

and so

$$\left\{1 - \frac{1}{4}\left[A + \frac{B}{\lambda - C}\right]^2\right\}^{1/2} \approx A + \frac{B}{\lambda - C}. \tag{4.1.64}$$

Thus

$$\frac{d\theta}{d\lambda} \approx \frac{-180AB}{\pi(\lambda - C)^2} - \frac{180B^2}{(\lambda - C)^3}. \tag{4.1.65}$$

Now the first term on the right-hand side of equation (4.1.65) has a magnitude about 30 times larger than that of the second term for typical values of λ, A, B and C. Hence

$$\frac{d\theta}{d\lambda} \approx \frac{-180AB}{\pi(\lambda - C)^2} \quad {}^{\circ}\,\text{m}^{-1} \tag{4.1.66}$$

and hence

$$\frac{d\theta}{d\lambda} \propto (\lambda - C)^{-2} \tag{4.1.67}$$

and so the dispersion of a prism increases rapidly towards shorter wavelengths. For the example we have been considering involving a dense flint prism, the dispersion is nearly five times larger at 400 nm than at 700 nm.

To form a part of a spectrograph, a prism, like a diffraction grating, must be combined with other elements. The basic layout is shown in figure 4.1.12; practical designs are discussed in section 4.2.

The linear dispersion of a prism is obtained in a similar way to that for the grating. If x is the linear distance along the spectrum from some reference point, then we have for an achromatic imaging element of focal length f_2,

$$\frac{dx}{d\lambda} = f_2\frac{d\theta}{d\lambda} \tag{4.1.68}$$

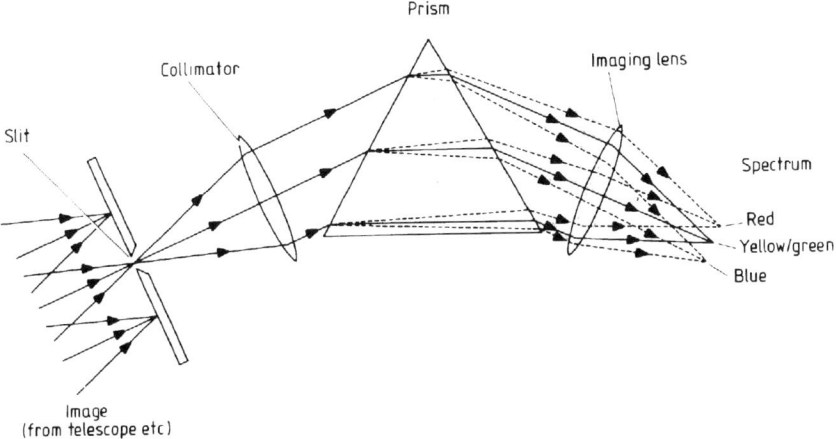

Figure 4.1.12. Basic optical arrangement of a prism spectroscope.

where θ is small and is measured in radians. Thus from equation (4.1.66)

$$\frac{dx}{d\lambda} = \frac{-180ABf_2}{\pi(\lambda - C)^2}.$$ (4.1.69)

The reciprocal linear dispersion, $d\lambda/dx$, is thus

$$\frac{d\lambda}{dx} = \frac{\pi(\lambda - C)^2}{-180ABf}.$$ (4.1.70)

In a similar way to the grating spectroscope, the resolving power of a prism spectroscope is limited by the spectral resolution of the prism, the resolving power of its optics (section 1.1), and by the projected slit width. The spectrum is formed from an infinite number of monochromatic images of the entrance slit. The width of one of these images, S, is again given by

$$S = s\frac{f_2}{f_1}$$ (4.1.71)

where s is the slit width, f_1 is the collimator's focal length and f_2 is the imaging element's focal length. If the optics of the spectroscope are well corrected then we may ignore their aberrations and consider only the diffraction limit of the system. When the prism is fully illuminated, the imaging element will intercept a rectangular beam of light. The height of the beam is just the height of the prism, and has no effect upon the spectral resolution. The width of the beam, D, is given by

$$D = L\left[1 - \mu_\lambda^2 \sin^2\left(\frac{\alpha}{2}\right)\right]^{1/2}$$ (4.1.72)

where L is the length of a prism face, and the diffraction limit is just that of a rectangular slit of width D, so from figure 1.1.26 the linear Rayleigh limit of resolution, W'', is given by

$$W'' = \frac{f_2 \lambda}{D} \tag{4.1.73}$$

$$= \frac{f_2 \lambda}{L\left[1 - \mu_\lambda^2 \sin^2\left(\dfrac{\alpha}{2}\right)\right]^{1/2}}. \tag{4.1.74}$$

If the beam is limited by some other element of the optical system, and/or is of circular cross section, then D must be evaluated as may be appropriate, or the Rayleigh criterion for the resolution through a circular aperture (equation 1.1.56) used in place of that for a rectangular aperture. Optimum resolution occurs when

$$S = W'' \tag{4.1.75}$$

i.e.

$$s = \frac{f_1 \lambda}{D} \tag{4.1.76}$$

$$= \frac{f_1 \lambda}{L\left[1 - \mu_\lambda^2 \sin^2\left(\dfrac{\alpha}{2}\right)\right]^{1/2}}. \tag{4.1.77}$$

The spectral resolution may now be found from equations (4.1.68) and (4.1.73)

$$W_\lambda = W'' \frac{d\lambda}{dx} \tag{4.1.78}$$

$$= \frac{\lambda}{d} \frac{d\lambda}{d\theta} \tag{4.1.79}$$

$$\approx \frac{\lambda(\lambda - C)^2}{ABL\left\{1 - \left[A + \dfrac{B}{\lambda - C}\right]^2 \sin^2\left(\dfrac{\alpha}{2}\right)\right\}^{1/2}} \tag{4.1.80}$$

and the resolution is

$$R = \frac{\lambda}{W_\lambda} \tag{4.1.81}$$

$$\approx \frac{ABL\left\{1 - \left[A + \dfrac{B}{\lambda - C}\right]^2 \sin^2\left(\dfrac{\alpha}{2}\right)\right\}^{1/2}}{(\lambda - C)^2}. \tag{4.1.82}$$

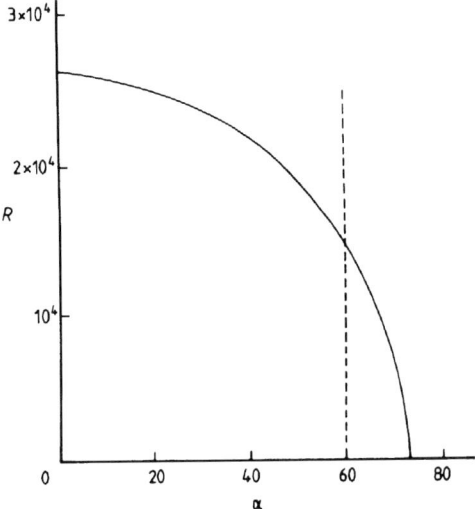

Figure 4.1.13. Resolution of a dense flint prism with a side length of 0.1 m at a wavelength of 500 nm, with changing apex angle α.

For a dense flint prism with an apex angle of 60° and a side length of 0.1 m, we then obtain in the visible

$$R \approx 1.5 \times 10^4 \tag{4.1.83}$$

and this is a fairly typical value for the resolution of a prism-based spectroscope. We may now see another reason why the maximum dispersion (figure 4.1.9) is not used in practice. Working back through the equations, we find that the term $L\{1 - [A + B/(\lambda - C)]^2 \sin^2(\alpha/2)\}^{1/2}$ involved in the numerator of the right-hand side of equation (4.1.82) is just the width of the emergent beam from the prism. Now for maximum dispersion, the beam emerges at 90° to the normal to the last surface of the prism. Thus, however large the prism may be, the emergent beam width is zero, and thus R is zero as well. The full variation of R with α is shown in figure 4.1.13. A 60° apex angle still preserves 60% of the maximum resolution and so is a reasonable compromise in terms of resolution as well as dispersion. The truly optimum apex angle for a given type of material will generally be close to but not exactly 60°. Its calculation will involve a complex trade-off between resolution, dispersion, and throughput of light, assessed in terms of the final amount of information available in the spectrum, and is not usually attempted unless absolutely necessary.

The resolution varies slightly across the width of the spectrum, unless cylindrical lenses or mirrors are used for the collimator, and these have severe disadvantages of their own. The variation arises because the light rays from the ends of the slit impinge on the first surface of the prism at

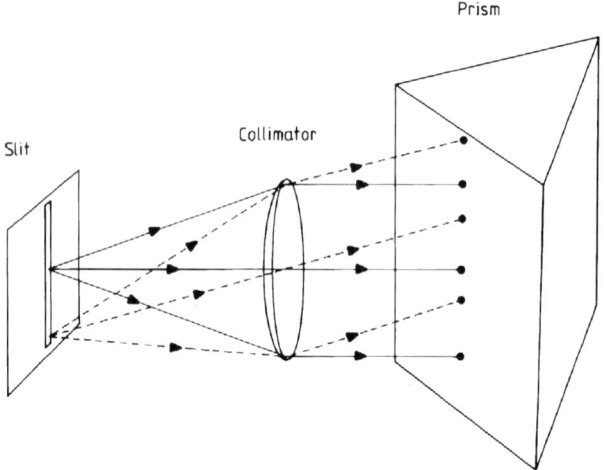

Figure 4.1.14. Light paths of rays from the centre and edge of the slit of a spectroscope.

an angle to the optical axis of the collimator (figure 4.1.14). The peripheral rays therefore encounter a prism whose effective apex angle is larger than that for the paraxial rays. The deviation is also increased for such rays, and so the ends of the spectrum lines are curved towards shorter wavelengths. Fortunately astronomical spectra are mostly so narrow that both these effects can be neglected, although the wider beams encountered in integral field and multi-object spectroscopes (section 4.2) may be affected in this way.

The material used to form prisms depends upon the spectral region that is to be studied. In the visual region, the normal types of optical glass may be used, but these mostly start to absorb in the near ultraviolet. Fused silica and crystalline quartz can be formed into prisms to extend the limit down to 200 nm. Crystalline quartz, however, is optically active (section 5.2), and therefore must be used in the form of a Cornu prism. This has the optical axis parallel to the base of the prism so that the ordinary and extraordinary rays coincide, and is made in two halves cemented together. The first half is formed from a right-handed crystal, and the other half from a left-handed crystal, the deviations of left- and right-hand circularly polarized beams are then similar. If required, calcium fluoride or lithium fluoride can extend the limit down to 140 nm or so, but astronomical spectroscopes working at such short wavelengths are normally based upon gratings. In the infrared, quartz can again be used for wavelengths out to about 3.5 μm. Rock salt can be used at even longer wavelengths, but it is extremely hygroscopic which makes it difficult to use. More commonly, Fourier spectroscopy (see below) is applied when high resolution spectroscopy is required in the far infrared.

Interferometers

We only consider here in any detail the two main types of spectroscopic interferometry that are of importance in astronomy: the Fabry–Pérot interferometer or etalon, and the Michelson interferometer or Fourier-transform spectrometer. Other systems exist but at present are of little importance for astronomy.

Fabry–Pérot interferometer

Two parallel, flat, partially reflecting surfaces are illuminated at an angle θ (figure 4.1.15). The light undergoes a series of transmissions and reflections as shown, and pairs of adjoining emergent rays differ in their path lengths by ΔP, where

$$\Delta P = 2t \cos \theta. \tag{4.1.84}$$

Constructive interference between the emerging rays will then occur at those wavelengths for which

$$\mu \, \Delta P = m\lambda \tag{4.1.85}$$

where m is an integer, i.e.

$$\lambda = \frac{2t\mu \cos \theta}{m}. \tag{4.1.86}$$

If such an interferometer is used in a spectroscope in place of the prism or grating (figure 4.1.16), then the image of a point source is still a point. However, the image is formed from only those wavelengths for which equation (4.1.86) holds. If the image is then fed into another spectroscope it will be broken up into a series of monochromatic images. If a slit is now

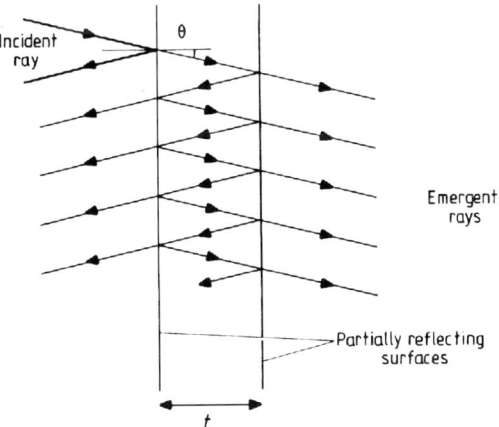

Figure 4.1.15. Optical paths in a Fabry–Pérot interferometer.

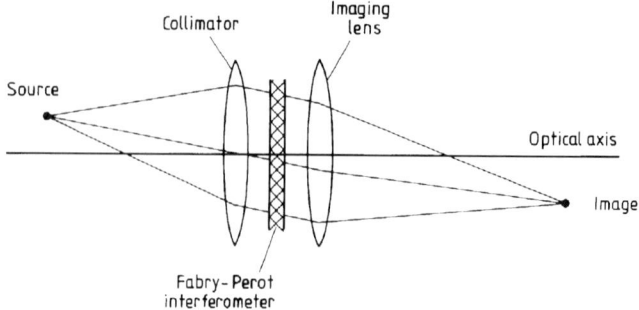

Figure 4.1.16. Optical paths in a Fabry–Pérot spectroscope.

used as the source, the rays from the different points along the slit will meet the etalon at differing angles, and the image will then consist of a series of superimposed short spectra. If the second spectroscope is then set so that its dispersion is perpendicular to that of the etalon (a cross-disperser), then the final image will be a rectangular array of short parallel spectra. The widths of these spectra depend upon the reflectivity of the surfaces. For a high reflectivity, we get many multiple reflections, while with a low reflectivity the intensity becomes negligible after only a few reflections. The monochromatic images of a point source are therefore not truly monochromatic but are spread over a small wavelength range in a similar, but not identical, manner to the intensity distributions for several collinear apertures (figures 4.1.1 and 4.1.2). The intensity distribution varies from that of the multiple apertures since the intensities of the emerging beams decrease as the number of reflections required for their production increases. Examples of the intensity distribution with wavelength are shown in figure 4.1.17. In practice reflectivities of about 90% are usually used.

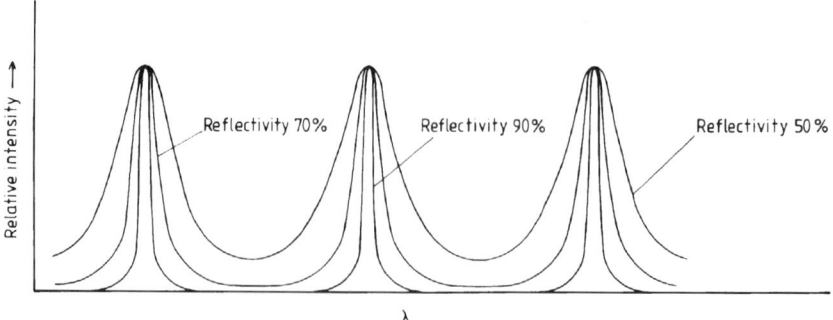

Figure 4.1.17. Intensity versus wavelength in the image of a white light point source in a Fabry–Pérot spectroscope, assuming negligible absorption.

In the absence of absorption the emergent intensity at a fringe peak is equal to the incident intensity at that wavelength. This often seems a puzzle to readers who, on inspection of figure 4.1.15, might expect there to be radiation emerging from the left as well as the right, so that at the fringe peak the total emergent intensity would appear to be greater than the incident intensity. If we examine the situation more closely, however, we find that when at a fringe peak for the light emerging on the right, there is zero intensity in the beam emerging on the left. If the incident beam has intensity I, and amplitude a ($I = a^2$), then the amplitudes of the successive beams on the left in figure 4.1.15 are

$$-aR^{1/2}, \ aR^{1/2}T, \ aR^{3/2}T, \ aR^{5/2}T, \ aR^{7/2}T, \ldots$$

where the first amplitude is negative because it results from an internal reflection. It has therefore an additional phase delay of $180°$ compared with the other reflected beams. T is the fractional intensity transmitted and R the fractional intensity reflected by the reflecting surfaces (note that $T + R = 1$ in the absence of absorption). Summing these terms (assumed to go on to infinity) gives zero amplitude and therefore zero intensity for the left-hand emergent beam. Similarly the beams emerging on the right have amplitudes

$$aT, \ aTR, \ aTR^2, \ aTR^3, \ aTR^4, \ldots. \tag{4.1.87}$$

Summing these terms to infinity gives, at a fringe peak, the amplitude on the right as a. This is the amplitude of the incident beam, and so at a fringe maximum the emergent intensity on the right equals the incident intensity (see also equation (4.1.95)).

The dispersion of an etalon may easily be found by differentiating equation (4.1.86)

$$\frac{d\lambda}{d\theta} = -\frac{2t\mu}{m} \sin\theta. \tag{4.1.88}$$

Since the material between the reflecting surfaces is usually air, and the etalon is used at small angles of inclination, we have

$$\mu \approx 1 \tag{4.1.89}$$

$$\sin\theta \approx \theta \tag{4.1.90}$$

and from equation (4.1.86)

$$\frac{2t}{m} \approx \lambda \tag{4.1.91}$$

so that

$$\frac{d\lambda}{d\theta} \approx \lambda\theta. \tag{4.1.92}$$

Thus the reciprocal linear dispersion for a typical system, with $\theta = 0.1°$, $f_2 = 1$ m, and used in the visible, is 0.001 nm mm^{-1}, which is a factor of 100 or so larger than that achievable with more common dispersing elements.

The resolution of an etalon is rather more of a problem to estimate. Our usual measure—the Rayleigh criterion—is inapplicable since the minimum intensities between the fringe maxima do not reach zero, except for a reflectivity of 100%. However, if we consider the image of two equally bright point sources viewed through a telescope at its Rayleigh limit (figure 2.5.3) then the central intensity is 81% of that of either of the peak intensities. We may therefore replace the Rayleigh criterion by the more general requirement that the central intensity of the envelope of the images of two equal sources falls to 81% of the peak intensities. Consider, therefore, an etalon illuminated by a monochromatic slit source perpendicular to the optical axis (figure 4.1.16). The image will be a strip, also perpendicular to the optical axis, and the intensity will vary along the strip accordingly as the emerging rays are in or out of phase with each other. The intensity variation is given by

$$I(\theta) = \frac{T^2 I_\lambda}{(1 - R)^2 + 4R \sin^2 \left(\dfrac{2\pi t \mu \cos \theta}{\lambda} \right)} \qquad (4.1.93)$$

where I_λ is the incident intensity at wavelength λ. The image structure will resemble that shown in figure 4.1.18. If the source is now replaced with a bichromatic one, then the image structure will be of the type shown in figure 4.1.19. Consider just one of these fringes; its angular distance, θ_{max}, from the optical axis is, from equation (4.1.86),

$$\theta_{max} = \cos^{-1} \left(\frac{m\lambda}{2t\mu} \right) \qquad (4.1.94)$$

Figure 4.1.18. Image structure in a Fabry–Pérot spectroscope viewing a monochromatic slit source, with $T = 0.1$, $R = 0.9$, $t = 0.1$ m, $\mu = 1$ and $\lambda = 550$ nm.

Figure 4.1.19. Image structure in a Fabry–Pérot spectroscope viewing a bichromatic slit source, with $T = 0.1$, $R = 0.9$, $t = 0.1$ m, $\mu = 1$ and $\lambda = 550$ nm (full curve) and $\lambda = 550.001$ nm (broken curve).

and so the peak intensity from equation (4.1.93) is

$$I(\theta_{max}) = \frac{T^2 I_\lambda}{(1 - R)^2} \qquad (4.1.95)$$

$$= I_\lambda \quad \text{(when there is no absorption).} \qquad (4.1.96)$$

Let the angular half width of a fringe at half intensity be $\Delta\theta$, then a separation of twice the half-half width of the fringes gives a central intensity of 83% of either of the peak intensities. So that if α is the resolution by the extended Rayleigh criterion, we may write

$$\alpha \approx 2\Delta\theta = \frac{\lambda(1 - R)}{2\pi\mu t \sqrt{R}\theta_{max}\cos\theta_{max}}. \qquad (4.1.97)$$

Hence from equation (4.1.92) we obtain the spectral resolution

$$W_\lambda = \alpha \frac{d\lambda}{d\theta} \qquad (4.1.98)$$

$$= \frac{\lambda^2(1 - R)}{2\pi\mu t \sqrt{R}\cos\theta_{max}} \qquad (4.1.99)$$

and so the resolution of the system (previously given the symbol R) is

$$\frac{\lambda}{\Delta\lambda} = \frac{2\pi\mu t \sqrt{R}\cos\theta_{max}}{\lambda(1 - R)} \qquad (4.1.100)$$

or, since θ_{max} is small and μ is usually close to unity,

$$\frac{\lambda}{\Delta\lambda} \approx \frac{2\pi t \sqrt{R}}{\lambda(1 - R)}. \qquad (4.1.101)$$

Thus for typical values of $t = 0.1$ m, $R = 0.9$ and for visible wavelengths we have

$$\frac{\lambda}{\Delta\lambda} \approx 10^7 \qquad (4.1.102)$$

which is almost two orders of magnitude higher than typical values for prisms and gratings, and is comparable with the resolution for a large echelle grating while the device is much less bulky. An alternative measure of the resolution that may be encountered is the finesse. This is the reciprocal of the half-width of a fringe measured in units of the separation of the fringes from two adjacent orders. It is given by

$$\text{finesse} = \frac{\pi\sqrt{R}}{1 - R} = \frac{\lambda}{2t} \times \text{resolution}. \qquad (4.1.103)$$

For a value of R of 0.9, the finesse is therefore about 30.

The free spectral range of an etalon is small since it is operating at very high spectral orders. From equation (4.1.94) we have

$$\Sigma = \lambda_1 - \lambda_2 = \frac{\lambda_2}{m} \qquad (4.1.104)$$

where λ_1 and λ_2 are superimposed wavelengths from adjacent orders (cf. equation (4.1.23)). Thus the device must be used with a cross-disperser as already mentioned and/or the free spectral range increased. The latter may be achieved by combining two or more different etalons, then only the maxima that coincide will be transmitted through the whole system, and the intermediate maxima will be suppressed.

Practical etalons are made from two plates of glass or quartz whose surfaces are flat to 1 or 2% of their operating wavelength, and held accurately parallel to each other by low thermal expansion spacers, with a spacing in the region of 10 to 200 mm. The inner faces are mirrors, which are usually produced by a metallic or dielectric coating. The outer faces are inclined by a very small angle to the inner faces so that the plates are the basal segments of very low angle prisms, thus any multiple reflections other than the desired ones are well displaced from the required image. The limit to the resolution of the instrument is generally imposed by departures of the two reflecting surfaces from absolute flatness. This limits the main use of the instrument to the visible and infrared regions. The absorption in metallic coatings also limits the short-wave use, so that 200 nm represents the shortest practicable wavelength even for laboratory usage. Etalons are commonly used as scanning instruments. By changing the air pressure by a few per cent, the refractive index of the material between the plates is changed and so the wavelength of a fringe at a given place within the image is altered (equation (4.1.93)). The astronomical applications of Fabry–Pérot spectroscopes are comparatively few for direct observations. However, the instruments are used extensively in determining oscillator strengths and transition probabilities upon which much of the more conventional astronomical spectroscopy is based.

Another important application of etalons, and one that does have some direct applications for astronomy, is in the production of narrow-band filters. These are usually known as interference filters and are etalons in which the separation of the two reflecting surfaces is very small. For materials with refractive indices near 1.5, and for near normal incidence, we see from equation (4.1.86) that if t is 167 nm (say) then the maxima will occur at wavelengths of 500, 250, 167 nm and so on, accordingly as m is 1, 2, 3 etc. From equation (4.1.97), the widths of these transmitted regions will be 8.4, 2.1, 0.9 nm etc. for 90% reflectivity of the surfaces. Thus a filter centred upon 500 nm with a bandwidth of 8.4 nm can be made by combining such an etalon with a simple dye filter to eliminate the shorter-wavelength transmission regions. Other wavelengths and bandwidths can easily be chosen by changing t, μ and R. Such a filter would be constructed by evaporating a partially reflective layer on to a sheet of glass. A second layer of an appropriate dielectric material such as magnesium fluoride or cryolite is then evaporated on top of this to the desired thickness, followed by a second partially reflecting layer. A second sheet of glass is then added for protection. The reflecting layers may be silver or aluminium, or they may be formed from a double layer of two materials with very different refractive indices in order to improve the overall filter transmission. In the far infrared, pairs of inductive meshes can be used in a similar way for infrared filters. The band-passes of interference filters can be made squarer by using several superimposed Fabry–Pérot layers. Recently, tunable filters have been developed that are especially suited to observing the emission lines of gaseous nebulae. The reflecting surfaces are mounted on stacks of piezo-electric crystals so that their separations can be altered. The TAURUS[52] tunable filter used at the Anglo–Australian observatory, for example, can scan between 370 and 960 nm with selectable band-passes between 6 and 0.6 nm (see also the Lyot birefringent filter, section 5.3).

Michelson interferometer

This Michelson interferometer should not be confused with the Michelson stellar interferometer that was discussed in section 2.5. The instrument discussed here is similar to the device used by Michelson and Morley to try and detect the Earth's motion through the aether. Its optical principles are shown in figure 4.1.20. The light from the source is split into two beams by the beam splitter, and then recombined as shown. For a particular position of the movable mirror and with a monochromatic source, there will be a

[52] Taylor Atherton Uariable resolution Radial Uelocity System.

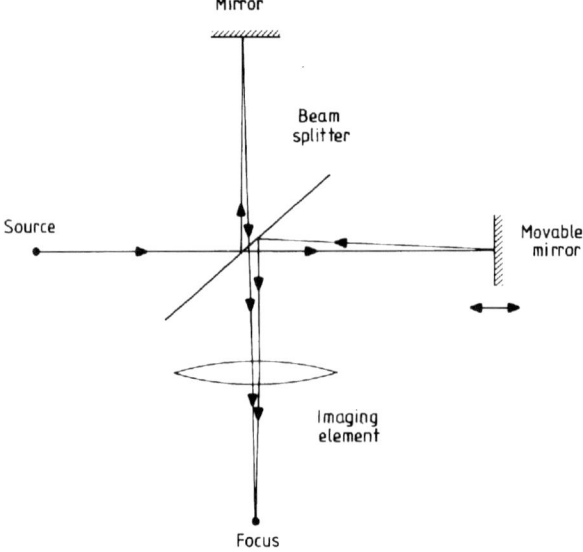

Figure 4.1.20. Optical pathways in a Michelson interferometer.

path difference, ΔP, between the two beams at their focus. The intensity at the focus is then

$$I_{\Delta P} = I_{\mathrm{m}} \left[1 + \cos\left(\frac{2\pi\Delta P}{\lambda} \right) \right] \tag{4.1.105}$$

where I_{m} is a maximum intensity. If the mirror is moved, then the path difference will change, and the final intensity will pass through a series of maxima and minima (figure 4.1.21). If the source is bichromatic, then two

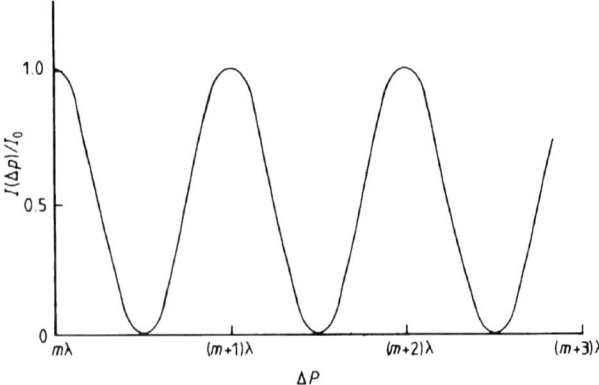

Figure 4.1.21. Variation of fringe intensity with mirror position in a Michelson interferometer.

Figure 4.1.22. Output of a Michelson interferometer observing a bichromatic source.

such variations will be superimposed with slightly differing periods and the final output will then have a beat frequency (figure 4.1.22). The difference between the two outputs (figures 4.1.20 and 4.1.22) gives the essential principle of the Michelson interferometer when it is used as a spectroscope. Neither output in any way resembles an ordinary spectrum, yet it would be simple to recognize the first as due to a monochromatic source, and the second as due to a bichromatic source. Furthermore the spacing of the fringes could be related to the original wavelength(s) through equation (4.1.105). More generally, of course, sources emit a broad band of wavelengths, and the final output will vary in a complex manner. To find the spectrum of an unknown source from such an output therefore requires a rather different approach than this simple visual inspection.

Let us consider, therefore, a Michelson interferometer in which the path difference is ΔP observing a source whose intensity at wavelength λ is I_λ. The intensity in the final image due to the light of a particular wavelength, $I'_{\Delta P}(\lambda)$ is then

$$I'_{\Delta P}(\lambda) = KI(\lambda)\left[1 + \cos\left(\frac{2\pi\Delta P}{\lambda}\right)\right] \qquad (4.1.106)$$

where K is a constant that takes account of the losses at the various reflections, transmissions etc. Thus the total intensity in the image for a given path difference is just

$$I'_{\Delta P} = \int_0^\infty I'_{\Delta P}(\lambda)\,\mathrm{d}\lambda \qquad (4.1.107)$$

$$= \int_0^\infty KI(\lambda)\,\mathrm{d}\lambda + \int_0^\infty KI(\lambda)\cos\left(\frac{2\pi\Delta P}{\lambda}\right)\mathrm{d}\lambda. \qquad (4.1.108)$$

Now the first term on the right-hand side of equation (4.1.108) is independent of the path difference, and is simply the mean intensity of the image. We may therefore disregard it and concentrate instead on the deviations from this average level, $I(\Delta P)$. Thus

$$I(\Delta P) = K\int_0^\infty I(\lambda)\cos\left(\frac{2\pi\Delta P}{\lambda}\right)\mathrm{d}\lambda \qquad (4.1.109)$$

or in frequency terms

$$I(\Delta P) = K^{\mathrm{I}} \int_0^\infty I(\nu) \cos\left(\frac{2\pi\Delta P\nu}{c}\right) \mathrm{d}\nu. \tag{4.1.110}$$

Now the Fourier transform, $F(u)$, of a function, $f(t)$ (see also section 2.1), is defined by

$$\mathscr{F}(f(t)) = F(u) = \int_{-\infty}^{\infty} f(t)\, \mathrm{e}^{-2\pi \mathrm{i} u t}\, \mathrm{d}t \tag{4.1.111}$$

$$= \int_{-\infty}^{\infty} f(t)\cos(2\pi u t)\, \mathrm{d}t - \mathrm{i}\int_{-\infty}^{\infty} f(t)\sin(2\pi u t)\, \mathrm{d}t. \tag{4.1.112}$$

Thus we see that the output of the Michelson interferometer is akin to the real part of the Fourier transform of the spectral intensity function of the source. Furthermore by defining

$$I(-\nu) = I(\nu) \tag{4.1.113}$$

we have

$$I(\Delta P) = \frac{1}{2} K^{\mathrm{I}} \int_{-\infty}^{\infty} I(\nu)\cos\left(\frac{2\pi\Delta P\nu}{c}\right) \mathrm{d}\nu \tag{4.1.114}$$

$$= K^{\mathrm{II}} \operatorname{Re}\left\{ \int_{-\infty}^{\infty} I(\nu)\exp\left[-\mathrm{i}\left(\frac{2\pi\Delta P}{c}\right)\nu\right]\mathrm{d}\nu \right\} \tag{4.1.115}$$

where K^{II} is an amalgam of all the constants. Now by inverting the transformation

$$\mathscr{F}^{-1}(F(u)) = f(t) = \int_{-\infty}^{\infty} F(u)\, \mathrm{e}^{2\pi \mathrm{i} u t}\, \mathrm{d}u \tag{4.1.116}$$

and taking the real part of the inversion, we may obtain the function that we require—the spectral energy distribution, or as it is more commonly known, the spectrum. Thus

$$I(\nu) = K^{\mathrm{III}} \operatorname{Re}\left\{ \int_{-\infty}^{\infty} I\left(\frac{2\pi\Delta P}{c}\right)\exp\left[\mathrm{i}\left(\frac{2\pi\Delta P}{c}\right)\nu\right]\mathrm{d}\left(\frac{2\pi\Delta P}{c}\right) \right\} \tag{4.1.117}$$

or

$$I(\nu) = K^{\mathrm{IV}} \int_{-\infty}^{\infty} I\left(\frac{2\pi\Delta P}{c}\right)\cos\left(\frac{2\pi\Delta P\nu}{c}\right)\mathrm{d}(\Delta P) \tag{4.1.118}$$

where K^{III} and K^{IV} are again amalgamated constants. Finally by defining

$$I\left(\frac{-2\pi\Delta P}{c}\right) = I\left(\frac{2\pi\Delta P}{c}\right) \tag{4.1.119}$$

we have

$$I(\nu) = 2K^{\mathrm{IV}} \int_0^\infty I\left(\frac{2\pi\Delta P}{c}\right) \cos\left(\frac{2\pi\Delta P\nu}{c}\right) \mathrm{d}(\Delta P) \qquad (4.1.120)$$

and so the spectrum is obtainable from the observed output of the inter-ferometer as the movable mirror scans through various path differences. We may now see why a Michelson interferometer when used as a scanning spectroscope is often called a Fourier transform spectroscope. The inversion of the Fourier transform is carried out on computers using the fast Fourier transform algorithm.

In practice, of course, it is not possible to scan over path differences from zero to infinity, and also measurements are usually made at discrete intervals rather than continuously, requiring the use of the discrete Fourier transform equations (section 2.1). These limitations are reflected in a reduction in the resolving power of the instrument. To obtain an expression for the resolving power, we may consider the Michelson interferometer as equivalent to a two-aperture interferometer (figure 4.1.1) since its image is the result of two interfering beams of light. We may therefore write equation (4.1.20) for the resolution of two wavelengths by the Rayleigh criterion as

$$W_\lambda = \frac{\lambda^2}{2\Delta P}. \qquad (4.1.121)$$

However, if the movable mirror in the Michelson interferometer moves a dis-tance x, then ΔP ranges from 0 to $2x$, and we must take the average value of ΔP rather than the extreme value for substitution into equation (4.1.121). Thus we obtain the spectral resolution of a Michelson interferometer as

$$W_\lambda = \frac{\lambda^2}{2x} \qquad (4.1.122)$$

so that the system's resolution is

$$\frac{\lambda}{W_\lambda} = \frac{2x}{\lambda}. \qquad (4.1.123)$$

Since x can be as much as $2\,\mathrm{m}$, we obtain a resolution of up to 4×10^6 for such an instrument used in the visible region.

The sampling intervals must be sufficiently frequent to preserve the resolution, but not more frequent than this, or time and effort will be wasted. If the final spectrum extends from λ_1 to λ_2, then the number of useful intervals, n, into which it may be divided, is given by

$$n = \frac{\lambda_1 - \lambda_2}{W_\lambda} \qquad (4.1.124)$$

so that if λ_1 and λ_2 are not too different then

$$n \approx \frac{8x(\lambda_1 - \lambda_2)}{(\lambda_1 + \lambda_2)^2}. \qquad (4.1.125)$$

However, the inverse Fourier transform gives both $I(\nu)$ and $I(-\nu)$, so that the total number of separate intervals in the final inverse transform is $2n$. Hence we must have at least $2n$ samples in the original transformation, and therefore the spectroscope's output must be sampled $2n$ times. Thus the interval between successive positions of the movable mirror, Δx, at which the image intensity is measured, is given by

$$\Delta x = \frac{(\lambda_1 + \lambda_2)^2}{16(\lambda_1 - \lambda_2)}. \qquad (4.1.126)$$

A spectrum between 500 and 550 nm therefore requires step lengths of 1 µm, while between 2000 and 2050 nm a spectrum would require step lengths of 20 µm. This relaxation in the physical constraints required on the accuracy of the movable mirror for longer wavelength spectra, combined with the availability of other methods of obtaining visible spectra, has led to the major applications of Fourier transform spectroscopy to date being in the infrared and far infrared. Even there, though, the increasing size of infrared arrays is now leading to the use of more conventional diffraction grating spectroscopes.

The basic PSF of the Fourier transform spectroscope is of the form $(\sin \Delta\lambda)/\Delta\lambda$ (figure 4.1.23), where $\Delta\lambda$ is the distance from the central wavelength, λ, of a monochromatic source. This is not a particularly convenient form for the profile, and the secondary maxima may be large enough to be significant especially where the spectral energy undergoes abrupt changes, such as at ionization edges or over molecular bands. The effect of the PSF may be reduced at some cost to the theoretical resolution by a technique known as apodization (see also sections 1.1, 2.1, 2.5 and 5.1). The transform is weighted by some function, ω, known as the apodization function. Many functions can be used, but the commonest is probably a triangular weighting function, i.e.

$$\omega(\Delta P) = 1 - \frac{\Delta P}{2x}. \qquad (4.1.127)$$

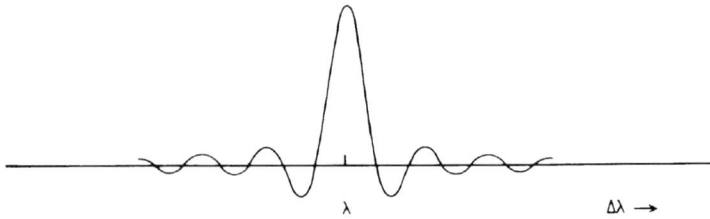

Figure 4.1.23. Basic instrumental profile (PSF) of a Fourier transform spectroscope.

The resolving power is halved, but the PSF becomes that for a single rectangular aperture (figure 1.1.26, equation (1.1.6)), and has much reduced secondary maxima.

A major advantage of the Michelson interferometer over the etalon when the latter is used as a scanning instrument lies in its comparative rapidity of use when the instrument is detector-noise limited. Not only is the total amount of light gathered by the Michelson interferometer higher (the Jacquinot advantage), but even for equivalent image intensities the total time required to obtain a spectrum is much reduced. This arises because all wavelengths are contributing to every reading in the Michelson interferometer, whereas a reading from the etalon gives information for just a single wavelength. The gain of the Michelson interferometer is called the multiplex or Fellget advantage, and is similar to the gain of the Hadamard masking technique over simple scanning (section 2.4). If t is the integration time required to record a single spectral element, then the etalon requires a total observing time of nt. The Michelson interferometer, however, requires a time of only $t/\sqrt{n/2}$ to record each sample since it has contributions from n spectral elements. It must obtain $2n$ samples so the total observing time for the same spectrum is therefore $2nt/\sqrt{n/2}$, and it has an advantage over the etalon of a factor of $\sqrt{n/8}$ in observing time.

Michelson interferometers have another advantage in that they require no entrance slit in order to preserve their spectral resolution. This is of considerable significance for large telescopes where the stellar image may have a physical size of a millimetre or more due to atmospheric turbulence, while the spectroscope's slit may be only a few tenths of a millimetre wide. Thus either much of the light is wasted or complex image dissectors (section 4.2) must be used.

Fibre optics

Fibre optic cables are now widely used to connect spectroscopes to telescopes, enabling the spectroscope to be mounted separately from the telescope. This reduces the problems of flexure that occur within telescope-mounted spectroscopes as gravitational loads change with the different telescope positions. It also enables the spectroscope to be kept in a temperature-controlled room and in some cases cooled and inside a vacuum chamber. Fibre optics can also be used to reformat stellar images so that all the light enters the spectroscope and to enable extended objects or multiple objects to be observed efficiently. Specific examples are discussed in the next section. Here we are concerned with the optics of fibre optic cables.

Fibre optic cables usually consist of a thin (10–500 μm) filament of glass encased in a cladding of another glass with a lower refractive index. One or more (sometimes thousands) of these strands make up the cable as a whole.

Light entering the core is transmitted through the core by multiple internal reflections off the interface between the two glasses, provided that its angle of incidence exceeds the critical angle for total internal reflection. The critical angle θ_c is given by

$$\theta_c = \sin^{-1} \sqrt{\mu_{core}^2 - \mu_{cladding}^2} \qquad (4.1.128)$$

where μ_{core} and $\mu_{cladding}$ are the refractive indices of the core and cladding respectively. Fibre optic cables are usually characterized by their numerical aperture (NA) and this is simply equal to $\sin(\theta_c)$. The minimum focal ratio, f_{min}, that will be transmitted by the core is then

$$f_{min} = \frac{\sqrt{1 - NA}}{2NA}. \qquad (4.1.129)$$

Commercially produced fibre optics have numerical apertures ranging from about 0.15 to 0.35, giving minimum focal ratios ranging from f3 to f0.6.

Silica glass is used as the core of the fibre for the spectral region 400–2000 nm and is sufficiently transparent that cables can be tens of metres long without significant absorption. Farther into the infrared, specialist glasses such as zirconium fluoride need to be used. Imperfections in the walls of the fibres and internal stresses lead to focal ratio degradation. This is a decrease in the focal ratio that can lead to light loss if the angle of incidence exceeds the critical angle. It may also cause problems in matching the output from the cable to the focal ratio of the instrument being used. Focal ratio degradation affects long focal ratio light beams worst, so short (faster) focal ratios are better for transmitting light into the fibres.

Cables comprising many individual strands will not transmit all the light that they receive, because of the area occupied by the cladding. This is typically 40% of the cross sectional area of the cable. Many astronomical applications therefore use single-strand cables with a core diameter sufficient to accept a complete stellar image. Multi-strand cables can be coherent or non-coherent. In the former the individual strands have the same relative positions at the input end of the cable as at the output. In non-coherent cables the relationship between strand positions at the input and output ends of the cable is random. For some special applications, such as re-formatting stellar images to match the shape of the spectroscope's entrance slit, the input and output faces of the fibre optic cable are of different shapes. Such cables need to be coherent in the sense that the positions of the strands at the output face are related in a simple and logical manner to those at the input face.

Exercises

4.1.1 Calculate the heliocentric radial velocity for a star in which the $H\alpha$ line (laboratory wavelength 656.2808 nm) is observed to have a wavelength of 656.1457 nm. At the time of the observation, the solar celestial longitude was 100°, and the star's right ascension and declination were $13^h\,30^m$, and $+45°$ respectively. The obliquity of the ecliptic is $23°\,27'$.

4.1.2 A prism spectroscope is required with a reciprocal linear dispersion of $1\,\text{nm}\,\text{mm}^{-1}$ or better at a wavelength of 400 nm. The focal length of the imaging element is limited to 1 m through the need for the spectroscope to fit on to the telescope. Calculate the minimum number of 60° crown glass prisms required to achieve this, and the actually resulting reciprocal linear dispersion at 400 nm.

Note: Prisms can be arranged in a train so that the emergent beam from one forms the incident beam of the next in line. This increases the dispersion in proportion to the number of prisms, but does not affect the spectral resolution, which remains at the value appropriate for a single prism.

4.1.3 Show that the reciprocal linear dispersion of a Fabry–Pérot etalon is given by

$$\frac{d\lambda}{dx} = \frac{\lambda l}{2f_2^2}$$

where l is the length of the slit, and the slit is symmetrical about the optical axis.

4.2 Spectroscopes

Basic design considerations

The specification of a spectroscope usually begins from just three parameters. One is the focal ratio of the telescope upon which the spectroscope is to operate, the second is the required spectral resolution, and the third is the required spectral range. Thus in terms of the notation used in section 4.1, we have f', W_λ and λ specified (where f' is the effective focal ratio of the telescope at the entrance aperture of the spectroscope), and we require the values of f_1, f_2, s, R, $d\lambda/d\theta$, L and D. We may immediately write down the resolution required of the dispersion element

$$R = \frac{\lambda}{W_\lambda}. \tag{4.2.1}$$

Now for a 60° prism from equation (4.1.82) we find that at 500 nm

$$R = 6 \times 10^4 L \qquad \text{(crown glass)} \tag{4.2.2}$$

$$R = 15 \times 10^4 L \qquad \text{(flint glass)} \tag{4.2.3}$$

where L is the length of a side of a prism, in metres. So we may write

$$R \approx 10^5 L \tag{4.2.4}$$

when the dispersing element is a prism. For a grating, the resolution depends upon the number of lines and the order of the spectrum (equation (4.1.21)). A reasonably typical astronomical grating (ignoring echelle gratings etc.) will operate in its third order and have some 500 lines mm^{-1}. Thus we may write the resolution of a grating as

$$R \approx 1.5 \times 10^6 L \qquad (4.2.5)$$

where L is the width of the ruled area of the grating, in metres. The size of the dispersing element will thus be given approximately by equations (4.2.4) and (4.2.5). It may be calculated more accurately in second and subsequent iterations through this design process, using equations (4.1.82) and (4.1.21).

The diameter of the exit beam from the dispersing element, assuming that it is fully illuminated, can then be found from

$$D = L \cos \phi \qquad (4.2.6)$$

where ϕ is the angular deviation of the exit beam from the perpendicular to the exit face of the dispersing element. For a prism ϕ is typically 60°, while for a grating used in the third order with equal angles of incidence and reflection, it is 25°. Thus we get

$$D = 0.5L \qquad \text{(prism)} \qquad (4.2.7)$$

$$D = 0.9L \qquad \text{(grating).} \qquad (4.2.8)$$

The dispersion can now be determined by setting the angular resolution of the imaging element equal to the angle between two just resolved wavelengths from the dispersing element

$$\frac{\lambda}{D} = W_\lambda \frac{\mathrm{d}\theta}{\mathrm{d}\lambda} \qquad (4.2.9)$$

giving

$$\frac{\mathrm{d}\theta}{\mathrm{d}\lambda} = \frac{R}{D}. \qquad (4.2.10)$$

Since the exit beam is rectangular in cross section, we must use the resolution for a rectangular aperture of the size and shape of the beam (equation (1.1.7)) for the resolution of the imaging element, and not its actual resolution, assuming that the beam is wholly intercepted by the imaging element, and that its optical quality is sufficient not to degrade the resolution below the diffraction limit.

The final parameters now follow easily. The physical separation of two just resolved wavelengths on the CCD or other imaging detector must be greater than or equal to the separation of two pixels. CCD pixels are typically around 15 to 20 μm in size, so the focal length in metres of the imaging element must be at least

$$f_2 = \frac{2 \times 10^{-5}}{W_\lambda} \frac{d\theta}{d\lambda}. \tag{4.2.11}$$

The diameter of the imaging element, D_2, must be sufficient to contain the whole of the exit beam. Thus for a square cross section exit beam

$$D_2 = \sqrt{2}D. \tag{4.2.12}$$

The diameter of the collimator, D_1, must be similar to that of the imaging element in general if the dispersing element is to be fully illuminated. Thus again

$$D_1 = \sqrt{2}D. \tag{4.2.13}$$

Now in order for the collimator to be fully illuminated in its turn, its focal ratio must equal the effective focal ratio of the telescope. Hence the focal length of the collimator, f_1, is given by

$$f_1 = \sqrt{2}Df'. \tag{4.2.14}$$

Finally from equation (4.1.76) we have the slit width

$$s = \frac{f_1\lambda}{D} \tag{4.2.15}$$

and a first approximation has been obtained to the design of the spectroscope.

These various stages in the process of specifying a spectroscope are summarized for convenience in table 4.2.1.

The low light levels involved in astronomy usually require the focal ratio of the imaging elements to be small, so that it is fast in imaging terms. Satisfying this requirement usually means a compromise in some other part of the design, so that an optimally designed system is rarely achievable in practice. The slit may also need to be wider than specified in order to use a reasonable fraction of the star's light.

The limiting magnitude of a telescope–spectroscope combination is the magnitude of the faintest star for which a useful spectrum may be obtained. This is a very imprecise quantity, for it depends upon the type of spectrum and the purpose for which it is required, as well as the properties of the instrument. For example, if strong emission lines in a spectrum are the features of interest, then fainter stars may be studied than if weak absorption lines are desired. Similarly spectra of sufficient quality to determine radial velocities may be obtained for fainter stars than if line profiles are wanted. A guide, however, to the limiting magnitude may be gained through the use of Bowen's formula

$$m = 12 + 2.5 \log_{10}\left(\frac{sD_1 T_\mathrm{D} gqt(d\lambda/d\theta)}{f_1 f_2 \alpha H}\right) \tag{4.2.16}$$

Table 4.2.1. Stages in the specification of a spectroscope.

Stage	Equation	Notes
1	f', W_λ, λ	Initial requirements
2	$R = \lambda / W_\lambda$	
3a	$L = 10^{-5} R$ metres	Prism—approximate value for 500 nm wavelength. Use equation (4.1.82) for more accurate determinations.
3b	$L = 6.7 \times 10^{-7} R$ metres	Grating—used in the third order, 500 grooves mm^{-1}, wavelength 500 nm. Use equation (4.1.21) for more accurate determinations.
4a	$D = 0.5L$ metres	Prism—reasonable approximation.
4b	$D = 0.9L$ metres	Grating—details as for 3b, equal angles of incidence and reflection. More precise determinations must take account of the exact angles and the blaze angle.
5	$\dfrac{d\theta}{d\lambda} = \dfrac{R}{D}$ rad m^{-1}	
6	$f_2 = \dfrac{2 \times 10^{-5}}{W_\lambda} \dfrac{d\theta}{d\lambda}$ metres	For a typical CCD.
7	$D_2 = \sqrt{2}D$ metres	Exit beam of square cross-section.
8	$D_1 = D_2$	Only necessarily true if the incident and exit angles are equal. This is often not the case for a grating.
9	$f_1 = \sqrt{2}Df'$	
10	$s = \dfrac{f_1 \lambda}{D}$	
11	Return to stage 1	Repeated iterations through the design process are usually required in order to obtain precise and optimized values for the parameters.

where m is the faintest **B** magnitude that will give a usable spectrum in t seconds of exposure, T_D is the telescope objective's diameter and g is the optical efficiency of the system, i.e. the ratio of the usable light at the focus of the spectroscope to that incident upon the telescope. Typically it has a value of 0.2. Note, however, that g does not include the effect of the curtailment of the image by the slit. q is the quantum efficiency of the detector; typical values are 0.4 to 0.8 for CCDs (section 1.1). α is the angular size of the stellar image at the telescope's focus, typically 5×10^{-6} to 2×10^{-5} radians. H is the height of the spectrum. This formula gives quite good approximations for spectroscopes in which the star's image is larger than the slit and it is trailed along the length of the slit to broaden the spectrum. This is probably the commonest mode of use for astronomical spectroscopes. Other situations such as an untrailed image, or an image smaller than the slit, require the formula to be modified. Thus when the slit is wide enough for the whole stellar image to pass through it, the exposure varies inversely with the square of the telescope's diameter, while for extended sources it varies inversely with the square of the telescope's focal ratio (cf. equations (1.1.68) and (1.1.71) etc).

The slit is quite an important part of the spectroscope since in most astronomical spectroscopes it fulfils two functions. First, it acts as the entrance aperture of the spectroscope. For this purpose, its sides must be accurately parallel to each other and perpendicular to the direction of the dispersion. It is also usual for the slit width to be adjustable or slits of different widths provided so that alternative detectors may be used and/or changing observing conditions catered for. Although we have seen how to calculate the optimum slit width, it is usually better in practice to find the best slit width empirically. The slit width is optimized by taking a series of images of a sharp emission line in the comparison spectrum through slits of different widths. As the slit width decreases, the image width should also decrease at first, but should eventually become constant. The changeover point occurs when some other part of the spectroscope system starts to limit the resolution, and the slit width at changeover is the required optimum value. As well as allowing the desired light through, the slit must reject unwanted radiation. The jaws of the slit are therefore usually of a knife-edge construction with the chamfering on the inside of the slit so that light is not scattered or reflected into the spectroscope from the edges of the jaws. On some instruments a secondary purpose of the slit is to assist in the guiding of the telescope on the object. When the stellar image is larger than the slit width it will overlap on to the slit jaws. By polishing the front of the jaws to an optically flat mirror finish, these overlaps can be observed via an auxiliary detector, and the telescope driven and guided to keep the image bisected by the slit.

However, guiding in this manner is wasteful of the expensively collected light from the star of interest. Most modern instruments therefore use another star within the field of view to guide on. If the stellar image is then larger than the slit it can be reformatted so that all its light enters the spectroscope. There are several ways of reformatting the image. Early approaches such as that due to Bowen are still in use. The Bowen image slicer consists of a stack of overlapped mirrors (figure 4.2.1) that section the image and then rearrange the sections end to end to form a linear image suitable for matching to a spectroscope slit. The Bowen–Walraven image slicer uses multiple internal reflections. A prism with a chamfered side is used and has a thin plate attached to it. Because of the chamfered side, the plate only touches the prism along one edge. A light beam entering the plate is repeatedly internally reflected wherever the plate is not in contact with the prism, but is transmitted into the prism along the contact edge (figure 4.2.2). The simplest concept is a bundle of optical fibres whose cross section matches the slit at one end and the star's image at the other. Thus apart from the reflection and absorption losses within the fibres, all the star's light is

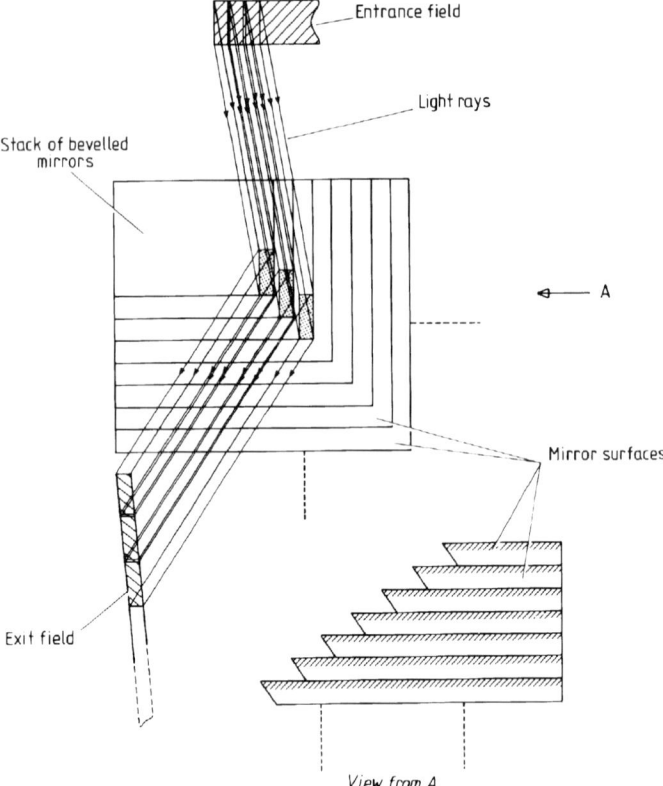

Figure 4.2.1. Bowen image slicer.

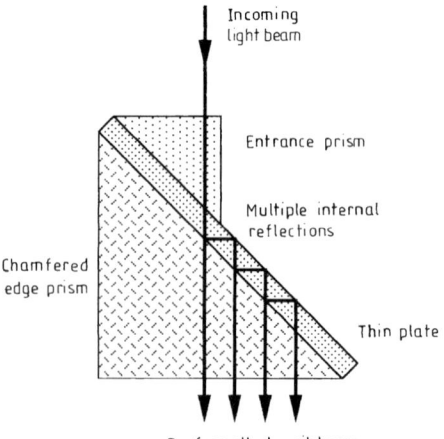

Figure 4.2.2. The Bowen–Walraven image slicer.

conducted into the spectroscope. Disadvantages of fibre optics are mainly the degradation of the focal ratio of the beam due to imperfections in the walls of the fibre so that not all the light is intercepted by the collimator, and the multilayered structure of the normal commercially available units leads to other light losses since only the central core of fibre transmits the light. Thus specially designed fibre optic cables are usually required and these are made 'in house' at the observatory needing them. They are usually much thicker than normal fibre optics and are formed from plastic or fused quartz. On telescopes with adaptive optics, the size of the star's image is much reduced and this allows not only the slit width to be smaller but also a more compact design to be used for the whole spectroscope.

For extended sources it is common practice to use long slits. Provided that the image does not move with respect to the slit (and no spectrum widener is used), then each point in that portion of the image falling on to the slit has its individual spectrum recorded at the appropriate height within the final spectrum. Several slits can be used as the entrance aperture of the spectroscope provided that the spectra do not overlap. Then, all the information that is derivable from a single-slit spectrogram is available, but the whole source can be covered in a fraction of the time. The ultimate development of this procedure, known as integral field spectroscopy or three-dimensional spectroscopy, is to obtain a spectrum for every resolved point within an extended source, and this is discussed further below.

For several purposes the slit may be dispensed with and some specific designs are considered later in this section. Apart from the Fourier transform spectroscope (section 4.1), they fall into two main categories. In the first, the projected image size on the spectrum is smaller than some other constraint on the system's resolution. The slit and the collimator may be discarded and parallel light from the source allowed to impinge directly on to the dispersing element. In the second type of slitless spectroscope, the source is producing a nebular type of spectrum (i.e. a spectrum consisting almost entirely of emission lines with little or no continuum). If the slit alone is then eliminated from the telescope–spectroscope combination, the whole of the image of the source passes into the spectroscope. The spectrum then consists of a series of monochromatic images of the source in the light of each of the emission lines. Slitless spectroscopes are difficult to calibrate so that radial velocities can be found from their spectra, but they may be very much more optically efficient than a slit spectroscope. In the latter, perhaps 1 to 10% of the incident light is eventually used in the image, but some types of slitless spectroscope can use as much as 75% of the light. Furthermore some designs, such as the objective prism (see below), can image as many as 10^5 stellar spectra in one exposure. A system that is closely related to the objective prism places the disperser shortly before the focal point of the telescope. Although the light is no longer in a parallel beam, the additional aberrations that are produced may be tolerable if the focal ratio is long. Using a zero-deviation grism in combination with correcting optics enables a relatively wide field to be covered, without needing the large sizes required for objective prisms. With suitable blazing for the grating part of the grism the zero-order images provide wavelength reference points for the spectra. Grisms used on the wide field imager of ESO's 2.2 m telescope enable spectra to be obtained in this way over an unvignetted area 19′ of arc across.

Spectroscopes, as we have seen, contain many optical elements that may be separated by large distances and arranged at large angles to each other. In order for the spectroscope to perform as expected, the relative positions of these various components must be stable to within very tight limits. The two major problems in achieving such stability arise through flexure and thermal expansion. Flexure primarily affects the smaller spectroscopes that are attached to telescopes at Cassegrain foci and so move around with the telescope. Their changing attitudes as the telescope moves cause the stresses within them to alter, so that if in correct adjustment for one position, they will be out of adjustment in other positions. The light beam from the telescope is often folded so that it is perpendicular to the telescope's optical axis. The spectroscope is then laid out in a plane that is parallel to the back of the main mirror. Hence the spectroscope components can be rigidly mounted on a stout metal plate which in turn is bolted flat to the back of the telescope. Such a design can be made very rigid, and the flexure reduced to acceptable levels. In some modern spectroscopes, active supports are used to compensate for flexure along the lines of those used for telescope mirrors (section 1.1).

Temperature changes affect all spectroscopes, but the relatively short light paths in the small instruments that are attached directly to telescopes mean that generally the effects are unimportant. If this is not the case, then a simple heating jacket and thermostat will usually suffice to eliminate the problem. However, this can then introduce problems of its own by causing convection and turbulence close to the light paths in the telescope. Thus ideally the thermostat should merely stabilize the temperature of the spectroscope a degree or two above the ambient temperature. On nights with rapidly changing ambient temperatures this may then require frequent resetting of the thermostat with consequent readjustment of the spectroscope settings.

The large fixed spectroscopes that operate at Coudé and Nasmyth foci are obviously unaffected by changing flexure and usually there is little difficulty other than that of cost in making them as rigid as desired. They experience much greater problems, however, from thermal expansion. The size of the spectrographs may be very large (see exercise 4.2.1 for example) with optical path lengths measured in tens of metres. Thus the temperature control must be correspondingly strict. A major problem is that the thermal inertia of the system may be so large that it may be impossible to stabilize the spectroscope at ambient temperature before the night has ended. Thus many such spectroscopes are in temperature-controlled sealed rooms. The light then has to be piped in through an optically flat window.

Any spectroscope except the Michelson interferometer can be used as a monochromator, that is, a device to observe the object in a very restricted range of wavelengths. Most scanning spectroscopes are in effect mono-chromators whose waveband may be varied. The most important use of the devices in astronomy, however, is in the spectrohelioscope. This builds up a picture of the Sun in the light of a single wavelength, and this is usually chosen to be coincident with a strong absorption line. Further details are given in section 5.3. A related instrument for visual use is called a prominence spectroscope. This has the spectroscope offset from the telescope's optical axis so that the (quite wide) entrance slit covers the solar limb. A small direct-vision prism or transmission grating then produces a spectrum and a second slit isolates an image in Hα light, allowing promi-nences and other solar features to be discerned.

Spectroscopy is undertaken throughout the entire spectrum. In the infrared and ultraviolet regions, techniques, designs etc. are almost identical to those for visual work except that different materials may need to be used. Some indication of these has already been given in section 4.1. Generally diffraction gratings and reflection optics are preferred since there is then no worry over absorption in the optical components. The technique of Fourier transform spectroscopy, as previously mentioned, has so far had its main applications in the infrared region. At very short ultraviolet and at x-ray wavelengths, glancing optics (section 1.3) and diffraction gratings can be used to produce spectra using the same designs as for visual spectro-scopes. The appearance and layout, however, will look very different even though the optical principles are the same, because of the off-axis optical elements. Radio spectroscopes have been described in section 1.2.

Prism-based spectroscopes

The basic layout of a prism-based spectroscope is shown in figure 4.1.12. Many instruments are constructed to this design with only slight modi-fications, the most important of which is the use of several prisms. If several identical prisms are used with the light passing through each along minimum deviation paths, then the total dispersion is just that of one of the prisms multiplied by the number of prisms. The resolution is unchanged and remains that for a single prism. Thus such an arrangement is of use when the resolution of the system is limited by some element of the spectroscope other than the prism. A rather more compact system than that shown in figure 4.1.12 can be made by replacing the 60° prism by one with a 30° apex angle that is aluminized on one surface (figure 4.2.3). The light therefore passes twice through the prism making its effect the equivalent of a single 60° prism. However, the minimum deviation path is no longer possible, so that

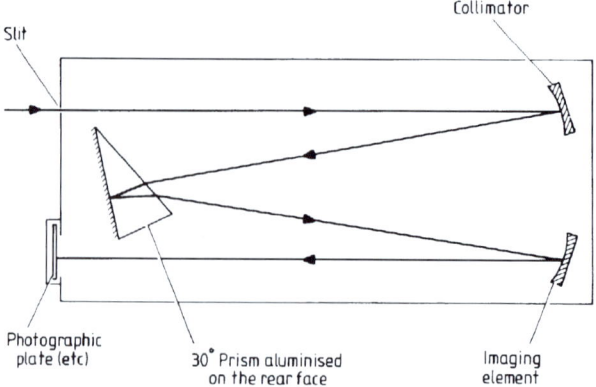

Figure 4.2.3. Compact design for the basic prism spectroscope.

some astigmatism is introduced into the image, but by careful design this can be kept lower than the resolution of the system as a whole.

Another similar arrangement, widely used for long-focus spectroscopes in laboratory and solar work, is called the Littrow spectroscope, or auto-collimating spectroscope. A single lens, or occasionally a mirror, acts as both the collimator and imaging element (figure 4.2.4) thus saving on both the cost and size of the system.

The deviation of the optical axis caused by the prism can be a disadvantage for some purposes. Direct-vision spectroscopes overcome this problem and have zero deviation for some selected wavelength. There are several designs but most consist of combinations of prisms made in two different types of glass with the deviations arranged so that they cancel out while some remnant of the dispersion remains. This is the inverse of the achromatic lens discussed in section 1.1, and therefore usually uses crown and flint glasses for its prisms. The condition for zero deviation, assuming that the

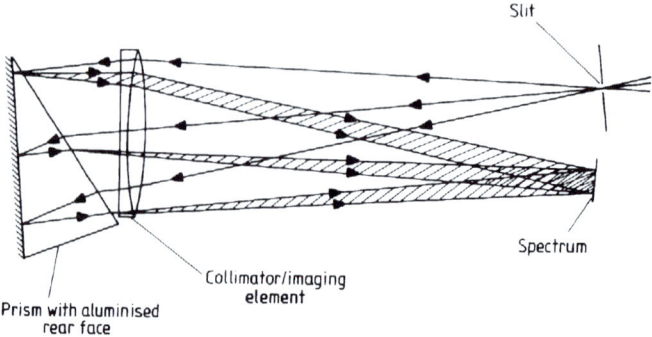

Figure 4.2.4. Light paths in a Littrow spectroscope.

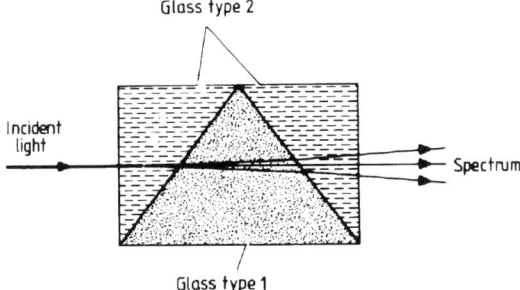

Figure 4.2.5. Treanor's direct-vision prism.

light passes through the prisms at minimum deviation, is

$$\sin^{-1}\left[\mu_1 \sin\left(\frac{\alpha_1}{2}\right)\right] - \frac{\alpha_1}{2} = \sin^{-1}\left[\mu_2 \sin\left(\frac{\alpha_2}{2}\right)\right] - \frac{\alpha_2}{2} \qquad (4.2.17)$$

where α_1 is the apex angle of prism number 1, α_2 is the apex angle of prism number 2, μ_1 is the refractive index of prism number 1 and μ_2 is the refractive index of prism number 2. In practical designs the prisms are cemented together so that the light does not pass through all of them at minimum deviation. Nonetheless, equation (4.2.17) still gives the conditions for direct vision to a good degree of approximation. Direct vision spectroscopes can also be based upon grisms, where the deviations of the prism and grating counteract each other. Applications of most direct-vision spectroscopes are non-astronomical since they are best suited to visual work. There is one ingenious astronomical application, however, due to Treanor, which enables many approximate stellar radial velocities to be found rapidly. The method is based upon a direct-vision prism that is formed from one prism with an apex angle of α, and two prisms in a different glass with apex angles of $\alpha/2$. The combination forms a block of glass whose incident and exit faces are perpendicular to the light paths (figure 4.2.5), so that there is no displacement of the undeviated ray. The two glasses are chosen so that their refractive indices are identical for the desired undeviated wavelength, but their refractive index gradients against wavelength differ. Table 4.2.2 shows one possible example that gives an undeviated ray at 530 nm that combines crown glass and borosilicate glass.

Table 4.2.2.

λ (nm)	μ (crown glass)	μ (borosilicate glass)
400	1.530	1.533
530	1.522	1.522
600	1.519	1.517

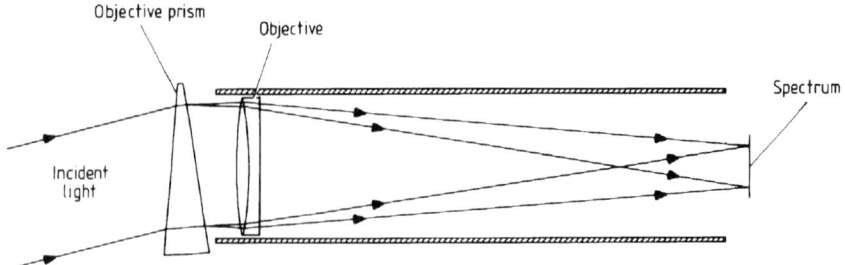

Figure 4.2.6. An objective prism spectroscope.

The prism is placed without a slit in a collimated beam of light from a telescope. After passing through the prism, the spectrum is then focused to produce an image as usual. Since no slit is used all the stars in the field of view of the telescope are imaged as short spectra. If the collimated beam is slightly larger than the prism, then the light not intercepted by the prism will be focused to a stellar image, and this image will be superimposed upon the spectrum at a position corresponding to the undeviated wavelength. Thus many stellar spectra may be imaged simultaneously, each with a reference point from which to measure the wavelengths of lines in the spectrum so enabling the radial velocities to be found. The dispersion of the system is low, but the efficiency is high compared with a slit spectroscope.

The simplest spectroscope of all is the objective prism. This is simply a thin prism that is large enough to cover completely the telescope's objective and is positioned in use immediately before the objective. The starlight is already parallel so that a collimator is unnecessary, while the scintillation disc of the star replaces the slit. The telescope acts as the imaging element (figure 4.2.6). Like Treanor's prism this has the enormous advantage that a spectrum is obtained for every star in the normal field of view. Thus if the telescope is a Schmidt camera, up to 10^5 spectra may be obtainable in a single exposure. The system has three main disadvantages: the dispersion is low, the observed star field is at an angle to the telescope axis, and there is no reference point for wavelength measurements. Additionally, a simple prism introduces distortion into the image. The first of these problems is usually unimportant because the objective prism spectroscope is mostly used for survey work. The low dispersion is sufficient to enable interesting objects to be identified that may then be studied in detail with other spectroscopes. The second problem can be overcome quite easily, at a cost of lowering the dispersion, by using two prisms of different glasses to form a low power direct-vision prism.

The third problem is more serious, and much effort has been expended in efforts to provide reference points in the spectra. While results comparable with those from slit spectroscopes cannot be matched, some useful systems have been devised. The simplest method of all is to use the absorption lines arising from terrestrial atmospheric gases. Unfortunately, these are strongest in the red and infrared regions where the low dispersion of a prism renders the spectra of little real use. Absorption lines can be introduced artificially through the use of filters with narrow absorption bands. One substance that has had some success in this way is neodymium chloride. This has a narrow absorption feature near 427 nm; however, it is not ideal since the precise position of the feature is temperature dependent, and it is asymmetrically shaped. Bright lines can be introduced into the spectrum through the use of a narrow band interference filter (section 5.3). If the band pass dye filters are omitted from such a filter then it is transparent to a series of very narrow wavebands over the whole spectrum. Placing it in the light beam from an objective prism so that a small fraction of the light beam is intercepted leads to bright lines in the spectrum at these same wavebands. This system also has the disadvantage of the line positions being temperature dependent, and they vary with the angle of inclination of the filter to the light beam. Furthermore the physical structure of the filter is much larger than its clear aperture so that a much larger fraction of the light beam is obstructed than is strictly necessary. A sharp cut-off point can be induced at the long or short wavelength end of the spectrum through the use of an appropriate filter. A quite different approach is to use two low angle grisms oriented so that their first-order spectra are side by side and with the directions of the dispersions reversed. The separation of the same spectrum line in the two spectra can then simply be related to radial velocity provided that some stars with known radial velocities are available in the field of view to enable the image to be calibrated. In an improved variant of this system due to Fehrenbach, two direct-vision prisms are used that give better efficiency, and with careful design enable the field distortion caused by the use of a simple prism to be reduced. The same principle for the determination of radial velocities can still be applied even if only a single objective prism is available. Then two separate exposures are made on the same plate with the prism rotated through 180° between them in order to produce the two spectra. Yet another approach uses a direct image of the same field, and by assuming the stars to have zero radial velocities, predicts the positions of the Ca(II) H and K lines within the spectra on the objective prism image. The difference between the predicted and actual positions of the lines then gives the actual radial velocity.

With the larger Schmidt cameras, a single objective prism becomes very large and heavy and can absorb a significant fraction of the light. A mosaic of identical co-aligned smaller prisms can then be used in place of the single larger one. The adjustment of such an array can pose problems, but with care these can be overcome, and several such systems have been used successfully.

Grating spectroscopes

Most of the gratings used in astronomical spectroscopes are of the reflection type. This is because the light can be concentrated into the desired order by blazing quite easily whereas, for transmission gratings, blazing is much more difficult and costly. Transmission gratings are, however, often used in grisms, and these are finding increasing use in integral field spectroscopes etc.

Plane gratings are most commonly used in astronomical spectroscopes and are almost invariably incorporated into one or other of two designs discussed in the previous section, with the grating replacing the prism. These are the compact basic spectroscope (figure 4.2.3), sometimes called a Czerny–Turner system when it is based upon a grating, and the Littrow spectroscope (figure 4.2.4) that is called an Ebert spectroscope when based upon grating and reflection optics.

Most of the designs of spectroscopes that use curved gratings are based upon the Rowland circle (figure 4.1.6). The Paschen–Runge mounting in fact is identical to that shown in figure 4.1.6. It is a common design for laboratory spectroscopes since wide spectral ranges can be accommodated, but its size and awkward shape make it less useful for astronomical purposes. A more compact design based upon the Rowland circle is called the Eagle spectroscope (figure 4.2.7). However, the vertical displacement of the slit and the spectrum (see the side view in figure 4.2.7) introduces some astigmatism. The Wadsworth design abandons the Rowland circle but still produces a stigmatic image through its use of a collimator (figure 4.2.8). The focal surface, however, becomes paraboloidal, and some spherical aberration and coma are introduced. Furthermore, the dispersion for a given grating, if it is mounted into a Wadsworth system, is only half what it could be if the same grating were mounted into an Eagle system, since the spectrum is produced at the prime focus of the grating and not at its radius of curvature. With designs such as the Wadsworth and its variants, the imaging element is likely to be a Schmidt camera system (section 1.1) in order to obtain high quality images with a fast system. Some recent spectroscopes, though, have used dioptric cameras to avoid the light loss involved with the plate holder in a Schmidt system. Exotic optical materials such as calcium fluoride often need to be used in these designs to achieve the required imaging speed, flat field and elimination of other aberrations. Gratings can also be used as discussed earlier in various specialized applications such as nebular and prominence spectroscopes.

Spectroscopes are usually optimized for one spectral region. If spectra are needed that extend over a wider range than is covered by the spectroscope, then it may need adjusting to operate in another region, or a different instrument entirely may be required. To overcome this problem to some extent, several spectroscopes have two channels optimized for different wavelength ranges. The incoming light from the telescope is split into the channels

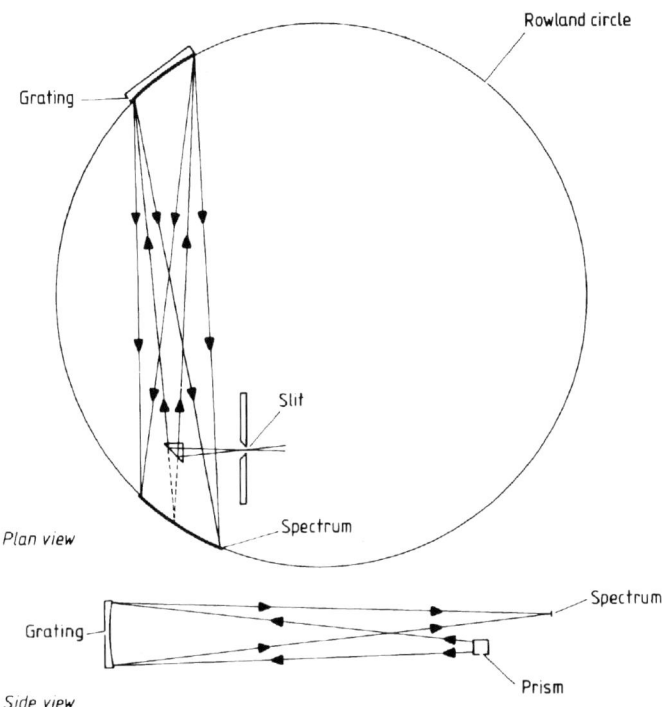

Figure 4.2.7. Optical arrangement of an Eagle spectroscope.

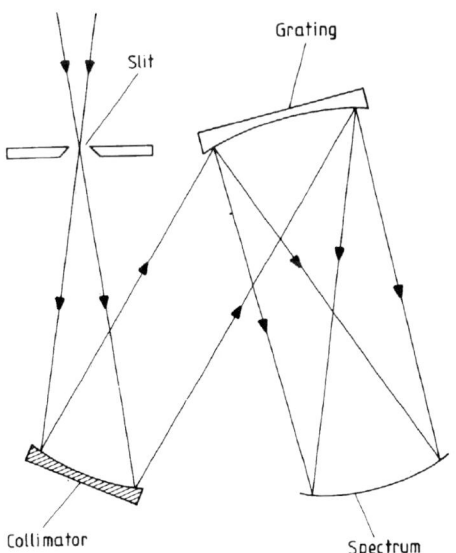

Figure 4.2.8. The Wadsworth spectroscope mounting.

by dichroic mirrors, so that (with some designs) the spectra can be obtained simultaneously. In other designs, the spectra are obtained in quick succession with little down time needed to adjust the spectroscope. ESO's UVES (Ultraviolet/Visual Echelle Spectroscope), for example, has a blue channel covering 300 to 500 nm and a red channel covering 420 to 1100 nm.

The design of a spectroscope is generally limited by the available size and quality of the grating, and these factors are of course governed by cost. The cost of a grating in turn is dependent upon its method of production. The best gratings are originals, which are produced in the following manner. A glass or other low expansion substrate is over-coated with a thin layer of aluminium. The grooves are then scored into the surface of the aluminium by lightly drawing a diamond across it. The diamond's point is precisely machined and shaped so that the required blaze is imparted to the rulings. An extremely high quality machine is required for controlling the diamond's movement, since not only must the grooves be straight and parallel, but also their spacings must be uniform if Rowland and Lyman ghosts are to be avoided. The position of the diamond is therefore controlled by a precision screw thread, and is monitored interferometrically. We have seen that the resolution of a grating is dependent upon the number of grooves, while the dispersion is a function of the groove spacing (section 4.1). Thus ideally a grating should be as large as possible, and the grooves should be as close together as possible (at least until their separation is less than their operating wavelength). Unfortunately both of these parameters are limited by the wear on the diamond point. Thus it is possible to have large coarse gratings and small fine gratings, but not large fine gratings. The upper limits on size are about 0.5 m square, and on groove spacing, about 1500 lines mm^{-1}. A typical grating for an astronomical spectroscope might be 0.1 m across and have 500 lines mm^{-1}.

Echelle gratings are used for many recently built spectroscopes. The rectangular format of the group of spectral segments after the cross-disperser matches well to the shape of large CCD arrays, so that the latter may be used efficiently. The UVES for ESO's VLT, for example, operates in the blue and the red and near infrared regions with two 0.2×0.8 m echelle gratings. The gratings have 41 and 31 lines mm^{-1} and spectral resolutions of 80 000 and 115 000 respectively.

Often a replica grating will be adequate, especially for lower resolution instruments. Since many replicas can be produced from a single original their cost is a small fraction of that of an original grating. Some loss of quality occurs during replication, but this is acceptable for many purposes. A replica improves on an original in one way, however, and that is in its reflection efficiency. This is better than that of the original because the most highly burnished portions of the grooves are at the bottoms of the grooves on the original, but are transferred to the tops of the grooves on the replicas. The original has of course to be the inverse of the finally desired grating. Covering the original with a thin coat of liquid plastic or epoxy resin, which is stripped off after it has set, produces the replicas. The replica is then mounted on to a substrate, appropriately curved if necessary, and then aluminized.

More recently, large high-quality gratings have been produced holographically. An intense monochromatic laser beam is collimated and used to illuminate a photoresist-covered surface. The reflections from the back of the blank interfere with the incoming radiation, and the photoresist is exposed along the nodal planes of the interference field. Etching away the surface then leaves a grating that can be aluminized and used directly, or it can have replicas formed from it as above. The wavelength of the illuminating radiation and its inclination can be altered to give almost any required groove spacing and blaze angle.

A related technique to the last produces volume phase holographic gratings (VPHGs—see also Section 4.1). A layer of gelatine around $10 \, \mu m$ thick and containing a small proportion of ammonia or potassium dichromate is coated on to a glass substrate. It is then illuminated with an interference pattern from two laser beams. Instead of being etched, however, the layer is then treated in water and alcohol baths so that the refractive index within the layer varies according to the exposure it has received. The layer thus forms a grating with the lines of the grating produced by regions of differing refractive indices. Since the gelatine is hygroscopic, it must be protected after production of the grating by a cover sheet. VPHGs can have up to 95% efficiency in their blaze region, and currently can be produced up to 300 mm in diameter and with from 100 to 6000 lines mm^{-1}. They can be used both as transmission and reflection, and they seem likely to find increasing use within astronomical spectroscopes. It is also possible to produce two different gratings within a single element. Such gratings can be tilted with respect to each other so that the spectra are separated. Different wavelength regions can then be observed using a single spectroscope.

Integral field spectroscopy

Where the spectra of several individual parts of an extended object such as a gaseous nebula or galaxy are needed, then, as already discussed, a long slit may be used so that a linear segment of the object is covered by a single exposure. Where spectra of every resolution element of an object are required then repeated adjacent exposures using a long slit spectroscope are one, but a time-consuming, approach. Several techniques that come under the heading of integral field spectroscopy or three-dimensional spectroscopy have thus been developed recently to obtain spectra for every resolved point within an extended source more efficiently.

Most simply several long slits can be used as the entrance aperture of the spectroscope provided that the spectra do not overlap. Then, all the information that is derivable from a single-slit spectrogram is available, but the source can be covered in a fraction of the time. Another approach is to use a scanning narrow band filter and obtain numerous images at slightly differing wavelengths. One example of this is the TAURUS tunable filter discussed in section 4.1. Alternatively imaging detectors, such as STJs (section 1.1), that are also intrinsically sensitive to wavelength can be used to obtain the whole data set in a single exposure. A colour photograph or image is, of course, essentially a low resolution three-dimensional spectrogram. A possible extension to existing colour imaging techniques which may have a high enough spectral resolution to be useful, is through the use of dye-doped polymers. These are thin films of polymers such as polyvinylbutyral, containing a dye such as chlorin that has a very narrow absorption band. Changing conditions within the substrate cause the wavelength of the absorption band to scan through a few nm, and potentially provide high efficiency direct imaging combined with spectral resolutions of perhaps 10^5 or more.

Most integral field spectroscopy, however, relies on three other approaches. The first is to use an image slicer, though this will need to cover a larger area than that of the stellar image slicers discussed previously. A stack of long plane mirrors is used, whose widths match that of the slit and that are slightly twisted with respect to each other around their short axes. The stack is placed at the focal pane of the telescope and the segments of the sliced image then re-imaged into a line along the length of the slit. The second approach is to use a large fibre optic bundle with a square or circular distribution of the fibres at the input and a linear arrangement at the output. The arrangement of the fibres at the output must correspond in a simple and logical fashion to their arrangement at the input so that the position within the final spectrum of the individual spectrum from a point in the image can be found. As with the stellar image slicers, the cladding of the fibres would mean that significant amounts of light would be lost if the fibre optics were to be placed directly at the telescope's focus. Instead an array of small lenses is used to feed the image into the fibres. The array of square or hexagonal lenses is placed some distance behind the telescope focus and each lens images the telescope pupil on to the end of a fibre. Providing that these sub-images are contained within the transmitting portions of the fibres, no light is lost due to the cladding. SPIRAL[53] on the AAT, for example, uses an array of 512 square lenses to provide spectroscopy over an $11'' \times 22''$ area of the sky. The third approach also uses an array of small lenses but dispenses with the spectroscope slit. The lenses produce a grid of images that is fed directly into the spectroscope and results in a grid of spectra. By orienting the lens array at a small angle to the dispersion, the spectra can be arranged to lie side by side and not to overlap. OASIS[54] that was originally used on the 3.6 m Canada–France–Hawaii Telescope (CFHT)

---

[53] Segmented Pupil Image Reformatting Array Lens.
[54] Optically Adaptive System for Imaging Spectroscopy.

and is now on the 4.2 m William Herschel Telescope (WHT) uses an 1100 hexagonal array of lenses in this fashion.

Multi-object spectroscopy

Some versions of integral field spectroscopes, especially the earlier designs, use multiple slits to admit several parts of the image of the extended object into the spectroscope simultaneously. Such an instrument can easily be extended to obtain spectra of several different objects simultaneously by arranging for the slits to be positioned over the images of those objects within the telescope's image plane. Practical devices that used this approach employ a mask with narrow linear apertures cut at the appropriate points to act as the entrance slits. However, a new mask is needed for every field to be observed, or even for the same field if different objects are selected, and cutting the masks is generally a precision job requiring several hours of work for a skilled technician. This approach has therefore rather fallen out of favour, though the Gemini telescopes' Multi Object Spectroscopes (GMOSs) continue to use masks. Their masks are generated automatically using a laser cutter, with the slit positions determined from a direct image of the required area of the sky obtained by GMOS operating in an imaging mode. Nonetheless most multi-object spectroscopy is now undertaken using fibre optics to transfer the light from the telescope to the spectroscope,

The individual fibre optic strands are made large enough to be able to contain the whole of the seeing disc of a stellar image. Each strand then has one of its ends positioned in the image plane of the telescope so that it intercepts one of the required images. The other ends are aligned along the length of the spectroscope slit, so that hundreds of spectra may be obtained in a single exposure. Initially the fibre optics were positioned by being plugged into holes drilled through a metal plate. But this has the same drawbacks as cutting masks for integral field spectroscopes and has been superseded. There are now two main methods of positioning the fibres, both of which are computer controlled and allow four or five hundred fibres to be re-positioned in a matter of minutes. The first approach is to attach the input ends of the fibre optic strands to small magnetic buttons that cling to a steel back plate. The fibres are re-configured one at a time by a robot arm that picks up a button and moves it to its new position as required. A positional accuracy for the buttons of 10 to 20 µm is needed. This approach is used with the two-degree field (2dF) for the AAT that uses correcting lenses to provide an unvignetted 2° field of view, over which 400 fibres can be positioned to feed a spectroscope. The OzPoz[55] system for ESO's VLT can

---

[55] The device is based upon the positioner developed for the 2dF on the AAT—hence Oz for Australia and Poz for positioner.

position up to 560 fibres in the same way, although at the moment only a maximum of 132 are utilized. It uses two backing plates so that one can be in use while the other is being reset. Swapping the plates takes about 5 minutes, thus minimizing the dead time between observations. The second approach mounts the fibres on the ends of computer-controlled arms that can be moved radially and/or from side to side. The arms may be moved by small motors, or as in the Subaru telescope's Echidna system by an electromagnetic system. The latter has 400 arms (or spines to fit in with the device's name), each of which can be positioned within a 7 mm circle to within ±10 μm. Typically about 90% of 400 target objects can be reached in any one configuration. Re-positioning the spines takes only 10 minutes.

Schmidt cameras can also be used with advantage for multi-object spectroscopy because their wide fields of view provide more objects for observation, and their short focal ratios are well matched to transmission through optical fibres. Thus for example the six-degree field (6dF) project developed from FLAIR (Fibre-Linked Array-Image Reformatter) on the UK Schmidt camera can obtain up to 150 spectra simultaneously over a 40 square-degree field of view with automatic positioning of the fibre optics using the magnetic button system.

Once the light is travelling through the fibre optic cables, it can be led anywhere. There is thus no requirement for the spectroscope to be mounted on the telescope, although this is the case in a few systems. More frequently the fibre optics take the light to a fixed spectroscope that can be several tens of metres away. This has the advantage that the spectroscope is not subject to changing gravitational loads and so does not flex, and it can be in a temperature-controlled room. Subaru's Echidna feeds a spectroscope 50 m away from the telescope, for example. Fibre optic links to fixed spectroscopes can also be used for single-object instruments. ESO's HARPS[56] spectroscope on the 3.6 m telescope is intended for extra-solar planet finding and so needs to determine stellar velocities to within ±1 m/s. Thus a very stable instrument is required and HARPS is not only mounted 38 m away from the telescope but is enclosed in a temperature-controlled vacuum chamber.

Techniques of spectroscopy

There are several problems and techniques which are peculiar to astronomical spectroscopy and that are essential knowledge for the intending astrophysicist.

One of the greatest problems, and one which has been mentioned several times already, is that the image of the star may be broadened by atmospheric

---

[56] High Accuracy Radial velocity Planet Searcher.

turbulence until its size is several times the slit width. Only a small percentage of the already pitifully small amount of light from a star therefore enters the spectroscope. The design of the spectroscope, and in particular the use of a large focal length of the collimator in comparison with that of the imaging element, can enable wider slits to be used. Even then, the slit is generally still too small to accept the whole stellar image, and the size of the spectroscope may have to become very large if reasonable resolutions and dispersions are to be obtained. The alternative approach to the use of a large collimator focal length lies in the use of an image slicer and/or adaptive optics, as discussed earlier.

Another problem in astronomical spectroscopy concerns the width of the spectrum. If the stellar image is held motionless at one point of the slit during the exposure, then the final spectrum may only be a few microns high. Not only is such a narrow spectrum very difficult to measure, but also individual spectral features are recorded by only a few pixels and so the noise level is high. Both of these difficulties can be overcome, though at the expense of longer exposure times, by widening the spectrum artificially. There are several ways of doing this, such as introducing a cylindrical lens to make the images astigmatic, or by trailing the telescope during the exposure so that the image moves along the slit. Both these methods have drawbacks, however, and an improved and generally preferred method is based upon the rocking of a plane piece of glass. The glass block's sides are accurately plane-parallel, and it is placed a short distance in front of the slit. If the block is at an angle to the optical axis of the telescope, then the image is displaced from its normal position but otherwise unaffected (figure 4.2.9). It is easy to arrange for this displacement to be along the line of the slit. Rocking the block backwards and forwards about an axis perpendicular to both the

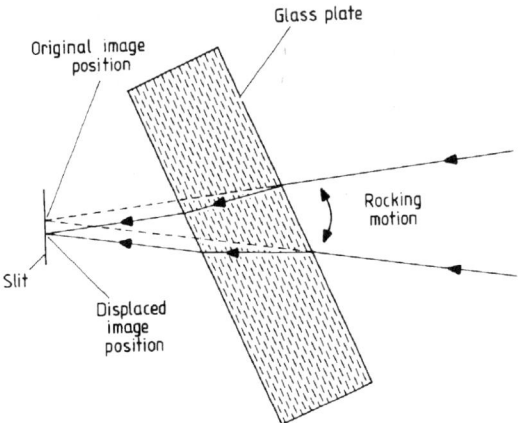

Figure 4.2.9. Image displacement by a plane-parallel glass plate.

line of the slit and the telescope's optical axis then causes the star's image to track up and down the slit. Provided that steep angles of incidence on to the block are avoided, there is little loss of light in the system. The length of the track of the star on the slit can be adjusted by varying the limits of the rocking motion, or by using blocks of differing thicknesses. Thus the final spectrum can be widened as desired, and in a uniform manner, provided that the exposure is much longer than the length of the period of the block's motion. Astronomical spectra are typically widened to about 0.1 to 1 mm.

Spectrographs operating in the near-infrared (1–5 μm) are of conventional designs, but need to be cooled to reduce the background noise level. Typically instruments such as UIST (UKIRT Imager SpecTrometer) and ESO's ISAAC (Infrared Spectrometer And Array Camera) have their main components cooled to 70 K, while the detectors are held at 4 K or less and the whole instrument is enclosed in a vacuum chamber. At mid-infrared wavelengths, the entire spectroscope may need to be cooled to less than 4 K.

Atmospheric dispersion is another difficulty that needs to be considered. Refraction in the Earth's atmosphere changes the observed position of a star from its true position (equation (5.1.17)). But the refractive index varies with wavelength. For example, at standard temperature and pressure, we have Cauchy's formula (cf. the Hartmann formula, equation (4.1.42))

$$\mu = 1.000287566 + \frac{1.3412 \times 10^{-18}}{\lambda^2} + \frac{3.777 \times 10^{-32}}{\lambda^4} \qquad (4.2.18)$$

so that the angle of refraction changes from one wavelength to another, and the star's image is drawn out into a very short vertical spectrum. In a normal basic telescope system (i.e. without the extra mirrors etc. required, for example, by the Coudé system), the long wavelength end of the spectrum will be uppermost. To avoid spurious results, particularly when undertaking spectrophotometry, the atmospheric dispersion of the image must be arranged to lie along the slit, otherwise certain parts of the spectrum may be preferentially selected from the image. Alternatively, an atmospheric dispersion corrector may be used. This is a low but variable dispersion direct vision spectroscope that is placed before the entrance slit of the main spectroscope and whose dispersion is equal and opposite to that of the atmosphere. The problem is usually only significant at large zenith distances, so that it is normal practice to limit spectroscopic observations to zenith angles of less than 45°. The increasing atmospheric absorption, and the tendency for telescope tracking to deteriorate at large zenith angles, also contributes to the wisdom of this practice.

In many cases it will be necessary to try to remove the degradation introduced into the observed spectrum because the spectroscope is not perfect. The many techniques and their ramifications for this process of deconvolution are discussed in section 2.1.

Figure 4.2.10. The wavelength comparison spectrum.

In order to determine radial velocities, it is necessary to be able to measure the actual wavelengths of the lines in the spectrum, and to compare these with their laboratory wavelengths. The difference, $\Delta\lambda$, then provides the radial velocity via the Doppler shift formula

$$v = \frac{\Delta\lambda}{\lambda} c = \frac{\lambda_{\text{observed}} - \lambda_{\text{laboratory}}}{\lambda_{\text{laboratory}}} c \qquad (4.2.19)$$

where c is the velocity of light. The observed wavelengths of spectrum lines are most normally determined by comparison with the positions of emission lines in an artificial spectrum. This comparison spectrum is usually that of an iron or copper arc or comes from a low-pressure gas emission lamp such as sodium or neon. The light from the comparison source is fed into the spectroscope and appears as two spectra on either side of the main spectrum (figure 4.2.10). The emission lines in the comparison spectra are at their rest wavelengths and are known very precisely. The observed wavelengths of the stellar (or other object) spectrum lines are found by comparison with the positions of the artificial emission lines.

For rapid high-precision determinations of radial velocity the cross-correlation spectroscope, originally devised by Roger Griffin, is often used. This places a mask over the image of the spectrum, and re-images the radiation passing through the mask on to a point source detector. The mask is a negative version (either a photograph or artificially generated) of the spectrum. The mask is moved across the spectrum, and when the two coincide exactly there is a very sharp drop in the output from the detector. The position of the mask at the correlation point can then be used to determine the radial velocity with a precision, in the best cases, of a few metres per second. More conventional spectroscopes can provide velocities to a precision of a few metres per second for bright stars through the use of an absorption cell containing HF or iodine. This is placed in the light path, and superimposes artificial absorption lines with very precisely known wavelengths on to the stellar spectrum. With this system there is no error due to differing light paths as there may be with a conventional comparison spectrum.

The guiding of a telescope–spectroscope combination has already been mentioned. For bright objects the reflected image from slit jaws polished to a mirror finish can be viewed through an optical inspection system. Guiding becomes more difficult for faint objects or when an image dissector is in use. Normally an accurate offset guiding system on some nearby brighter star must be used.

Future developments

The major foreseeable developments in spectroscopy seem likely to lie in the direction of improving the efficiency of existing systems rather than in any radically new systems or methods. The lack of efficiency of a typical spectro-scope arises mainly from the loss of light within the overall system. To improve on this requires gratings of greater efficiency, reduced scattering and surface reflection etc. from the optical components, and so on. Since these factors are already quite good, and it is mostly the total number of components in the telescope–spectroscope combination that reduces the efficiency, improvements are thus likely to be slow and gradual.

Adaptive optics is already used on one or two spectroscopic telescopes to reduce the seeing disc size of stellar images, and so allow more compact spectroscopes to be used. This approach is likely to become much more wide-spread in the near future. Direct energy detectors such as STJs are likely to be used more extensively, although at present they have relatively low spectral resolutions, and require extremely low operating temperatures. The use of integral field and multi-object spectroscopes is likely to become more common, with wider fields of view and more objects being studied for individual instruments. The extension of high-resolution spectroscopy to longer infrared wavelengths is likely to be developed, even though this may involve cooling the fibre optic connections and large parts of the telescope as well as most of the spectroscope. Plus, of course, the continuing increase in the power and speed of computers will make real-time processing of the data from spectroscopes much more commonplace.

Exercises

4.2.1 Design a spectroscope (i.e. calculate its parameters) for use at a Coudé focus where the focal ratio is f25. A resolution of 10^{-3} nm is required over the spectral range 500 to 750 nm. A grating with 500 lines mm^{-1} is available, and the final recording of the spectrum is to be by photography.

4.2.2 Calculate the limiting B magnitude for the system designed in problem (4.2.1) when it is used on a 4 m telescope. The final spectrum is widened to 0.2 mm, and the longest practicable exposure is 8 hours.

4.2.3 Calculate the apex angle of the dense flint prism required to form a direct-vision spectroscope in combination with two 40° crown glass prisms, if the undeviated wavelength is to be 550 nm. (See section 1.1 for some of the data.)

Chapter 5

Other techniques

5.1 Astrometry

Introduction

Astrometry is probably the most ancient branch of astronomy, dating back to at least several centuries BC and possibly to a couple of millennia BC. Indeed until the late 18th century, astrometry was the whole of astronomy. Yet although it is such an ancient sector of astronomy, it is still alive and well today and employing the most modern techniques, plus some that William Herschel would recognize. Astrometry is the science of measuring the positions in the sky of stars, planets, comets, asteroids, and recently, spacecraft. From these positional measurements come determinations of distance via parallax, motions in space via proper motion, orbits and hence sizes and masses within binary systems, and a reference framework that is used by the whole of the rest of astronomy and astrophysics, as well as by space scientists to direct and navigate their spacecraft. Astrometry also leads to the production of catalogues of the positions, and sometimes the natures, of objects that are then used for other astronomical purposes.

From the invention of the telescope until the 1970s, absolute positional accuracies of about $0.1''$ ($=100$ mas)[57] were the best that astrometry could deliver. That has now improved to better than 1 mas, and space missions planned for the next couple of decades should improve that by a factor of $\times 100$ at least. Positional accuracies of a few µas potentially allow proper motions of galaxies to be determined, though the diffuse nature of galaxy images may render this difficult. However, such accuracies will enable the direct measurement of stellar distances throughout the whole of the Milky Way galaxy, and out to the Magellanic clouds and the Andromeda galaxy (M31).

---

[57] The units of milli-arc seconds (mas) and micro-arc seconds (µas) are widely used in astrometry and are generally more convenient. $1'' = 1000$ mas and 1 mas $= 1000$ µas. To put these units in perspective, a resolution of 1 µas would mean that the headlines in a newspaper held by an astronaut walking on the Moon could be read from the Earth.

Astrometry may be absolute (sometimes called fundamental) when the position of a star is determined without knowing the positions of other stars, or relative when the star's position is found with respect to the positions of its neighbours. Relative astrometry may be used to give the absolute positions of objects in the sky, provided that some of the reference stars have their absolute positions known. It may also be used for determinations of parallax, proper motion and binary star orbital motion, without needing to convert to absolute positions. An important modern application of relative astrometry is to enable the optical fibres of multi-object spectroscopes (section 4.2) to be positioned correctly in the focal plane of the telescope in order to intercept the light from the objects of interest. The absolute reference frame is called the ICRS (International Celestial Reference System) and is now defined using 212 extra-galactic compact radio sources with their positions determined by radio interferometry (see below). Until 1998, the ICRS was based upon optical astrometric measurements, and when space-based interferometric systems (see below) produce their results, the definition may well revert to being based upon optical measurements. The practical realization of the ICRS for optical work is to be found in the Hipparcos catalogue (see below).

Background

Coordinate systems

The measurement of a star's position in the sky must be with respect to some coordinate system. There are several such systems in use by astronomers, but the commonest is that of right ascension and declination. This system, along with most of the others, is based upon the concept of the celestial sphere, i.e. a hypothetical sphere, centred upon the Earth and enclosing all objects observed by astronomers. The space position of an object is related to a position on the celestial sphere by a radial projection from the centre of the Earth. Henceforth in this section we talk about the position of an object as its position on the celestial sphere, and we ignore the differing radial distances that may be involved. We also extend the polar axis, the equatorial and orbital planes of the Earth until these too meet the celestial sphere (figure 5.1.1). These intersections are called the celestial north pole, the celestial equator etc. Usually there is no ambiguity if the 'celestial' qualification is omitted, so that they are normally referred to as the north pole, the equator and so on. The intersections (or nodes) of the ecliptic and the equator are the vernal and autumnal equinoxes. The former is also known as the first point of Aries from its position in the sky some two thousand years ago. The ecliptic is also the apparent path of the Sun across the sky during a year, and the vernal equinox is defined as the node at which the Sun passes from the southern to the northern hemisphere.

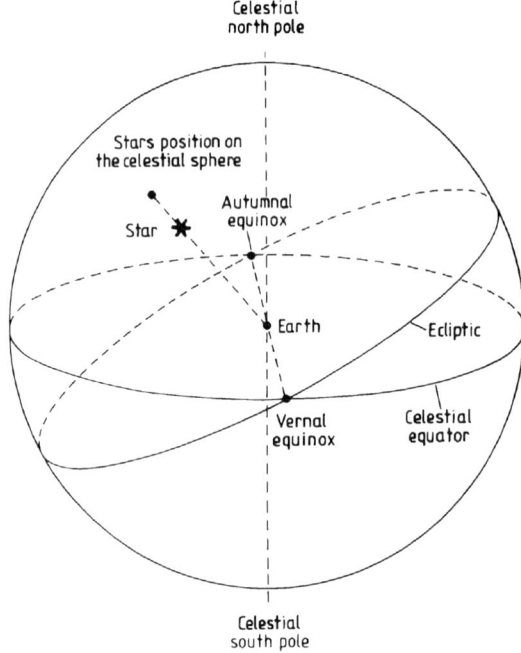

Figure 5.1.1. The celestial sphere.

This passage occurs within a day of 21 March each year. The position of a star or other object is thus given with respect to these reference points and planes.

The declination of an object, δ, is its angular distance north or south of the equator. The right ascension, α, is its angular distance around from the meridian (or great circle) that passes through the vernal equinox and the poles, measured in the same direction as the solar motion (figure 5.1.2). By convention, declination is measured from $-90°$ to $+90°$ in degrees, minutes and seconds of arc, and is positive to the north of the equator and negative to the south. Right ascension is measured from $0°$ to $360°$, in units of hours, minutes and seconds of time where

$$1 \text{ hour} = 15° \tag{5.1.1}$$

$$1 \text{ minute} = 15' \tag{5.1.2}$$

$$1 \text{ second} = 15'' \tag{5.1.3}$$

The Earth's axis moves in space with a period of about 25 750 years, a phenomenon known as precession. Hence the celestial equator and poles also move. The positions of the stars therefore slowly change with time. Catalogues of stars customarily give the date, or epoch, for which the stellar positions that they list are valid. To obtain the position at some

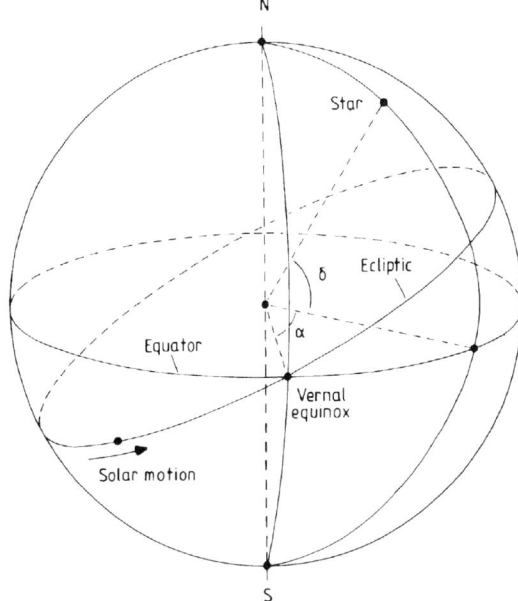

Figure 5.1.2. Right ascension and declination.

other date, the effects of precession must be added to the catalogue positions

$$\delta_T = \delta_E + (\theta \sin \varepsilon \cos \alpha_E)T \qquad (5.1.4)$$

$$\alpha_T = \alpha_E + [\theta(\cos \varepsilon + \sin \varepsilon \sin \alpha_E \tan \delta_E)]T \qquad (5.1.5)$$

where α_T and δ_T are the right ascension and declination of the object at an interval T years after the epoch E, α_E, δ_E are the coordinates at the epoch and θ is the precession constant

$$\theta = 50.40'' \text{ yr}^{-1}. \qquad (5.1.6)$$

ε is the angle between the equator and the ecliptic—more commonly known as the obliquity of the ecliptic

$$\varepsilon = 23° 27' 8''. \qquad (5.1.7)$$

Frequently used epochs are the beginnings of the years 1900, 1950, 2000 etc., with 1975 and 2025 also being encountered. Other effects upon the position, such as nutation, proper motion etc. may also need to be taken into account in determining an up-to-date position for an object.

Two alternative coordinate systems that are in use are, first, celestial latitude (β) and longitude (λ), that are respectively the angular distances up or down from the ecliptic and around the ecliptic from the vernal equinox. Second, galactic latitude (b) and longitude (l), that are respectively

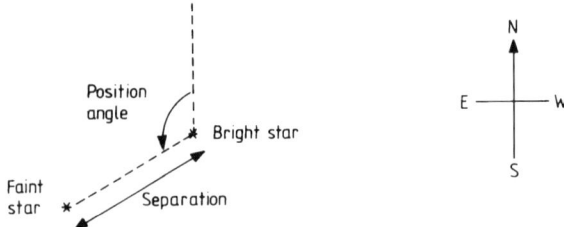

Figure 5.1.3. Position angle and separation of a visual double star (as seen on the sky directly).

the angular distances above or below the galactic plane, and around the plane of the galaxy measured from the direction to the centre of the galaxy.[58]

Position angle and separation

The separation of a visual double star is just the angular distance between its components. The position angle is the angle from the north, measured in the sense north → east → south → west → north, from $0°$ to $360°$ (figure 5.1.3) of the fainter star with respect to the brighter star. The separation and position angle are related to the coordinates of the star by

$$\text{separation} = \{[(\alpha_F - \alpha_B)\cos\delta_B]^2 + (\delta_F - \delta_B)^2\}^{1/2} \qquad (5.1.8)$$

$$\text{position angle} = \tan^{-1}\left(\frac{\delta_F - \delta_B}{(\alpha_F - \alpha_B)\cos\delta}\right) \qquad (5.1.9)$$

where α_B and δ_B are the right ascension and declination of the brighter star and α_F and δ_F are the right ascension and declination of the fainter star with the right ascensions of both stars being converted to normal angular measures.

Transit telescopes

Transit telescopes (figure 5.1.4), which are also sometimes called meridian circles, provide the basic absolute measurements of stellar positions for the calibration of other astrometric methods. They developed from the mural quadrants that were used before the invention of the telescope. These were

[58] The system is based upon a position for the galactic centre of RA_{2000} 17h 45m 36s, Dec_{2000} $28° 56' 18''$. It is now known that this is in error by about $4'$, but the incorrect position continues to be used. Prior to about 1960, the intersection of the equator and the galactic plane was used as the zero point and this is about $30°$ away from the galactic centre. Coordinates based upon this old system are sometimes indicated by a superscript 'I', and those using the current system by a superscript 'II', i.e. b^I or b^{II} and l^I or l^{II}.

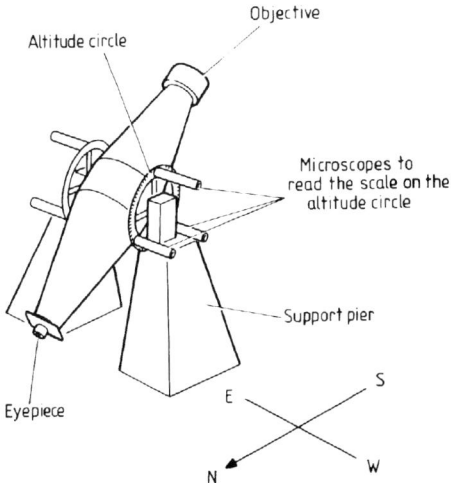

Figure 5.1.4. A transit telescope.

just sighting bars, pivoted at one end, restricted to moving in just one plane and with a divided arc to measure the altitude at the other end. Nonetheless, they provided surprisingly good measurements when in capable hands. The one used by Tycho Brahe (1546–1601), for example, was mounted on a north–south aligned wall and had a 2 m radius. He was able to measure stellar positions using it to a precision of about $\pm 30''$.

The principle of the transit telescope is simple, but great care is required in practice if it is to produce reliable results. The instrument is almost always a refractor, because of the greater stability of the optics, and it is pivoted only about a horizontal east–west axis. The telescope is thereby constrained to look only at points on the prime meridian. A star is observed as it crosses the centre of the field of view (i.e. when it transits the prime meridian), and the precise time of the passage is noted. The declination of the star is obtainable from the altitude of the telescope, while the right ascension is given by the local sidereal time at the observatory at the instant of transit,

$$\delta = A + \phi - 90° \qquad (5.1.10)$$

$$\alpha = \text{LST} \qquad (5.1.11)$$

where A is the altitude of the telescope, ϕ is the latitude of the observatory and LST is the local sidereal time at the instant of transit. In order to achieve an accuracy of $0.1''$ of arc in the absolute position of the star, a very large number of precautions and corrections are needed. A few of the more important ones are: temperature control; use of multiple or driven cross wires in the micrometer eyepiece; reversal of the telescope on its bearings; corrections for flexure, non-parallel objective and eyepiece focal planes, rotational axis not precisely horizontal, rotational axis not precisely east–west, errors in the setting circles, incorrect position of the micrometer cross wire, personal setting errors, and so on. Then of course all the normal corrections for refraction, aberration etc. have also to be added. Modern versions of the instrument, such as the Carlsberg Meridian Telescope (CMT), use CCDs for detecting transits. The CMT is a refractor with a 17.8 cm diameter objective and a 2.66 m focal length sited on La Palma in the Canary Islands. It is operated remotely via the internet, and measures between 100 000 and 200 000 star positions in a single night. The CCD detector on the CMT uses charge transfer to move the accumulating electrons from pixel to pixel at the same rate as the images drift over the detector (also known as time-delayed integration (TDI), drift scanning and image tracking—see section 1.1 (liquid mirror telescopes) and section 2.7). This enables the instrument to detect stars down to 17^m, and also to track several stars simultaneously. The main programme of the CMT is to link the positions of the bright stars measured by Hipparcos with those of fainter stars.

Photographic zenith tube and the impersonal astrolabe

Two more modern instruments that until recently have performed the same function as the transit telescope are the photographic zenith tube (PZT) and the astrolabe. Both of these use a bath of mercury in order to determine the zenith position very precisely. The PZT obtained photographs of stars that transit close to the zenith and provided an accurate determination of time and the latitude of an observatory, but it was restricted in its observations of stars to those that culminated within a few tens of minutes of arc of the zenith. The PZT has been superseded by very long baseline radio interferometers (below and section 2.5) and measurements from spacecraft that are able to provide positional measurements for far more stars and with significantly higher accuracies.

The astrolabe observes near an altitude of $60°$ and so can give precise positions for a wider range of objects than the PZT. In its most developed form (figure 5.1.5) it is known as the Danjon or impersonal astrolabe since its measurements are independent of focusing errors. Two separate beams of light are fed into the objective by the $60°$ prism, one of the beams having been reflected from a bath of mercury. A Wollaston prism (section 5.2) is used to produce two focused beams parallel to the optic axis with the other two non-parallel emergent beams being blocked off. Two images of the star are then visible in the eyepiece. The Wollaston prism is moved along the optical axis to compensate for the sidereal motion of the star. The measurement of the star's position is accomplished by moving the $60°$ prism along the optical axis until the two images merge into one. The position of the prism at a given time for this coincidence to occur can then be

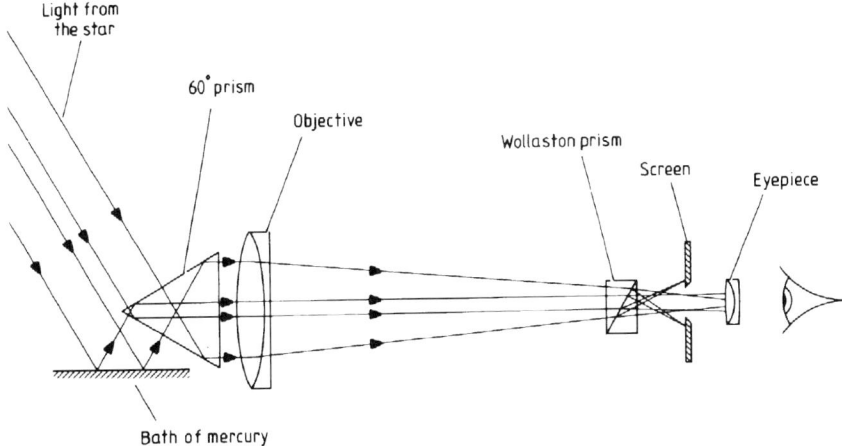

Figure 5.1.5. Optical paths in the impersonal astrolabe.

converted into the position of the star in the sky. The astrolabe has also largely fallen out of use, but one or two examples are still in operation such as the Mark III astrolabe at the Beijing National Observatory.

Micrometers

While the addition of one or more cross wires to an eyepiece can convert it into a micrometer when it is used in conjunction with a clock, it is the bifilar micrometer (sometimes called the filar micrometer) that is more normally brought to mind by the term micrometer. Although for most purposes the bifilar micrometer has been superseded by CCDs and photography, it is still a good method for measuring close visual double stars and is often used by amateur double-star observers. It also finds application as a means of off-setting from a bright object to a faint or invisible one. Designs for the device vary in their details, but the most usual model comprises a fixed cross wire at the focus of an eyepiece, with a third thread parallel to one of the fixed ones. The third thread is displaced slightly from the others, although still within the eyepiece's field of sharp focus, and it may be moved perpendicularly to its length by a precision screw thread. The screw has a calibrated scale so that the position of the third thread can be determined to within 1 μm. The whole assembly is mounted on a rotating turret whose angular position may also be measured (figure 5.1.6).

To measure a double star with such a micrometer, the following procedure should be followed.

(a) Align one of the fixed cross wires in an east–west direction. This is accomplished by turning off the telescope drive and rotating the turret

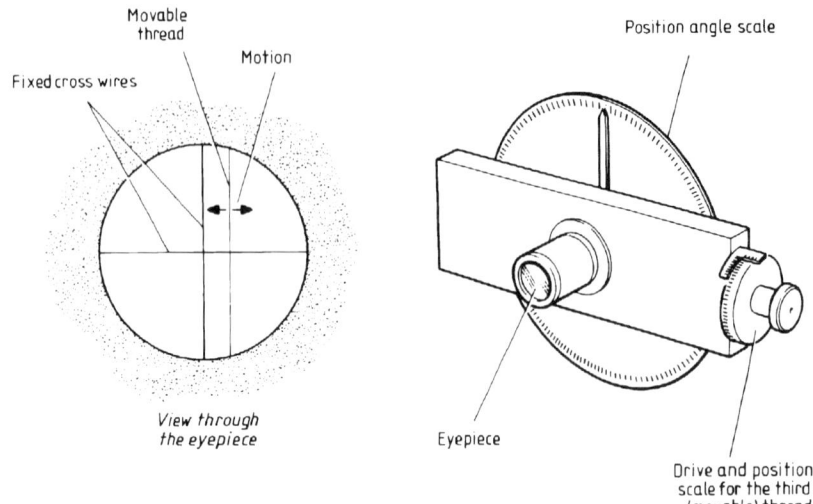

Figure 5.1.6. Bifilar micrometer.

until the wire is parallel to the drifting motion of the star. Note the reading of the position angle scale.
(b) Centre the cross wires on the brighter star and rotate the turret until the cross wire which was aligned east–west now passes through both stars. Again note the reading of the position angle scale.
(c) Calculate the position angle (figure 5.1.3). The precise relationship between the two readings just obtained and the position angle will depend upon the optical path of the telescope. In cases of doubt a little experimentation with the telescope's motions will indicate the orientation of the field of view.
(d) Calibrate the micrometer screw reading. This is achieved by aligning the movable cross wire in a north–south direction (cf. step (a)) and moving it until it is superimposed upon the fixed wire and noting the reading. It is then moved to, say, half way to the edge of the field of view, and the screw reading again noted. With the telescope drive turned off, the sidereal drift time, t, for a star to cross between the two parallel wires is measured. The angular distance between the parallel wires is then given by

$$\theta = 15t \cos \delta \qquad (5.1.12)$$

where θ is in seconds of arc and is the angular separation of the two parallel wires, t is in sidereal seconds of time and δ is the declination of the star. Alternatively if t is measured in solar seconds of time, we have

$$\theta = 15.04t \cos \delta \qquad (5.1.13)$$

Dividing this angle by the difference between the two screw scale readings then gives the angular distance in seconds of arc that corresponds to one unit of the screw scale.

(e) The separation is then simply measured by setting the cross wires on to one star, aligning the movable wire so that it is perpendicular to the line joining the stars, and moving it until it bisects the other star's image. The calibration of the screw scale then allows the readings to be converted directly into the star's angular separation.

If the micrometer is used to determine the absolute position of one star from the known position of another, then the same procedure is used, and the unknown star's position, providing that the separation is small, is given by

$$\alpha = \alpha_0 + r \sin \theta \sec \delta_0 \qquad (5.1.14)$$

$$\delta = \delta_0 + r \cos \theta \qquad (5.1.15)$$

where α and δ are the required right ascension and declination, α_0 and δ_0 are the right ascension and declination of the comparison star, θ is the position angle and r is the separation. The practical use of the micrometer requires only a few additional steps. The most important of these is to make multiple observations at all stages of the procedure, either by the use of repeated observations or by using several parallel co-moving wires or both. An improved method of measuring the separation is to proceed as indicated through to stage (e). The second screw reading, however, is then obtained by resetting the cross wires on to the second star and moving the movable wire until it bisects the image of the first star. The difference in the readings is then twice the separation. The advantage of this procedure is that if there is any systematic difference between the setting error of the cross wire and that of the movable wire, it will tend to cancel out. The cross wires are frequently formed from spiders' webs or from the individual fibres in dental floss, and so must be illuminated in order to be visible. This may either be by direct illumination, when the wires are visible as bright lines against a dark field of view, or by illumination of the field of view, when they are visible as silhouettes. Other features that may be added include adjustments to centre the cross wires on the rotational axis of the turret, the ability to move or change the eyepiece, devices to reduce the contrast between stars of differing magnitudes, filters and filter holders etc.

Another type of micrometer that has the advantage of being far less affected by scintillation is the double-image micrometer. Two images of each star are produced, and the device is adjusted so that one image of one star is superimposed upon one image of the other. The settings of the device can be calibrated, and then the separation and position angle follow directly. Commonly used methods of splitting the images are an objective cut into two along a diameter whose two halves may be tilted with respect to each other, and a Nicol prism (section 5.2). The advantage of systems of this type derives from the instantaneous determination of both separation and position angle, compared with the delay of several seconds or even minutes for obtaining the various readings from a bifilar micrometer, and from the reduced effect of scintillation since both images are likely to be affected by it in a similar manner. Very few examples of double-image micrometers now remain in use, but an amateur programme using one to track binary star systems is still in progress at the Observatori Astronomic del Garraf in Spain.

Astrographs and other telescopes

Almost any telescope can be used to obtain images from which the positions of objects in the sky can be obtained, but some telescopes are optimized for the work. The astrograph has a wide (a degree or more) field of view so that many faint stars' positions can be determined from those of a few brighter reference stars. Astrographs designed around the beginning of the 20th century were refractors with highly corrected multi-element objectives. Their apertures were in the region of 10 to 25 cm, and focal lengths between 2 and 4 m. Measurements using these instruments formed the basis of the ICRS. More recently, Schmidt cameras and conventional reflectors with correcting lenses have also been used.

Smaller fields of view are adequate for parallax determination. Long-focus refractors are favoured for this work because of their greater stability when compared with reflectors, because of their closed tubes, because their optics may be left undisturbed for many years and because flexure is generally less deleterious in them. A few specially designed reflectors, especially the US Naval Observatory's 1.55 m Strand reflector, are used successfully for astrometry, however. Against the advantages of the refractor their disadvantages must be set. These are primarily their residual chromatic aberration and coma, temperature effects, and slow changes in the optics such as the rotation of the elements of the objective in their mounting cells. Refractors in use for astrometry usually have focal ratios of f15 to f20, and range in diameter from about 0.3 to 1.0 m. Their objectives are achromatic (section 1.1), and a combination of the CCD's or photographic emulsion's spectral sensitivity (sections 1.1 and 2.2) and filters is usually used so that the bandwidth is limited to a few tens of nanometres either side of their optimally corrected wavelength. Coma may need to be limited by stopping down the aperture, by introducing correcting lenses near the focal plane, or by compensating the measurements later, in the analysis stages, if the magnitudes and spectral types of the stars are well known. Observations are usually obtained within a few degrees of the meridian in order to minimize the effects of atmospheric refraction and dispersion and to reduce changes in flexure. The position of the focus may change with temperature and needs to be checked frequently throughout the night. Other corrections for such things as plate tilt collimation errors, astigmatism, distortion, emulsion movement etc. need to be known and their effects allowed for when reducing the raw data.

Interferometers

As we have seen (section 2.5), interferometers have the capability of providing much higher resolutions than single-dish-type instruments since their baselines can be extended easily and cheaply. High resolution can be translated into high positional accuracy provided that the mechanical construction of the interferometer is adequate. Interferometers also provide absolute positions for objects. The disadvantage of interferometers for astrometry is that only one object can be observed at a time, and each observation requires several hours. Both radio and optical interferometers are used for the purpose.

Optical interferometers such as COAST (section 1.1) and the US Naval Observatory's NPOI (Navy Prototype Optical Interferometer) are essentially Michelson stellar interferometers (section 2.5) incorporating a delay line into one or both light beams. The delay line is simply an arrangement of several mirrors that reflects the light around an additional route. By moving one or more of the mirrors in the delay line, the delay can be varied. The observational procedure is to point both telescopes (or the flat mirrors feeding the telescopes) at the same star and adjust the delay until the fringe visibility (section 2.5) is maximum. The physical length of the delay line then gives the path difference between the light beams to the two telescopes. Combining this with the baseline length gives the star's altitude. Several such observations over a night will then enable the star's position in the sky to be calculated. Currently, absolute positions may be obtained in this manner to accuracies of a few mas, with the relative positions of close stars being determined to perhaps an order of magnitude better precision. Relative measurements for widely separated objects may be obtained from the variation of the delay over several hours. The delay varies sinusoidally over a 24 hour period, and the time interval between the maxima for two stars equals the difference in their right ascensions. As with conventional interferometers the use of more than two telescopes enables several delays to be measured simultaneously, speeding up the process and enabling instrumental effects to be corrected. The optical interferometer on Mount Wilson recently completed for CHARA (Center for High Angular Resolution Astronomy) for example uses six 1 m telescopes in a two-dimensional array and has a maximum baseline of 350 m.

Radio interferometers are operated in a similar manner for astrometry, and VLBI systems (section 2.5) provide the highest accuracies at the time of writing. For some of the radio sources this accuracy may reach 100 μas. The measurements by VLBI of some 212 compact radio sources have thus been used to define the ICRS since 1 January 1998.

Speckle interferometry (section 2.6) provides relative astrometry for close binary stars with accuracies of 10 mas or so.

Space-based systems

By operating instruments in space and so removing the effects of the atmosphere and gravitational loading, astrometry is expected to reach accuracies of a few μas in the next decade or so. Space astrometry missions, actual and planned, divide into two types: scanning (or survey) and point and stare. Scanning means that exposures are short, and so only the brighter stars can be observed and with relatively low accuracy, and the accuracy also depends upon the star's brightness. However, large numbers of star positions may be measured quickly. Point and stare missions, as the name suggests, look at individual stars (or a few stars very close together) for long periods of time. Such missions can observe relatively few star positions, but do so with high accuracy and to very faint magnitudes. Hipparcos and GAIA (see below) are examples of scanning systems, while the Hubble space telescope and SIM are point and stare.

The first astrometric spacecraft, called Hipparcos (High-Precision Parallax Collecting Satellite), was launched by ESA in 1990. Unfortunately the apogee boost motor failed and instead of going into a geostationary orbit, it eventually ended up in a $10\frac{1}{2}$ hour orbit with a perigee of 6800 km and apogee of 42 000 km. However, by using more ground stations and extending the mission duration to 3 years, the original aim of measuring the positions of 120 000 stars to an accuracy of 2 mas was achieved. The Hipparcos telescope was fed by two flat mirrors at an angle of 29° to each other enabling two areas of the sky 58° apart to be observed simultaneously. It determined the relative phase difference between two stars viewed via the two mirrors, in order to obtain their angular separation. A precision grid at the focal plane modulated the light from the stars, and the modulated light was then detected from behind the grid to determine the transit times (see also the MAP instrument below). The satellite rotated once every 2 hours and measured all the stars down to 9.0^{m} about 120 times, observing each star for about 4 seconds on each occasion. Each measurement was from a slightly different angle and the final positional catalogue was obtained by the processing of some 10^{12} bits of such information. The data from Hipparcos are to be found in the Hipparcos, Tycho and Tycho-2[59] catalogues. These may be accessed at http://archive.ast.cam.ac.uk/hipp/. The Hipparcos catalogue contains the positions, distances, proper motions and magnitudes of some 118 000 stars to an accuracy of ±0.7 mas. The Tycho catalogue contains the positions and magnitudes for about one million

[59] A separate instrument, the star mapper, observed many more stars, but for only 0.25 s, and its data produced the Tycho and Tycho-2 catalogues.

stars and has positional accuracies ranging from ± 7 mas to ± 25 mas. The Tycho-2 catalogue utilizes ground-based data as well as that from Hipparcos and has positional accuracies ranging from ± 10 mas to ± 70 mas, depending upon the star's brightness. It contains positions, magnitudes and proper motions for some two and a half million stars.

The Hubble space telescope can use the WFPC2 (Second Wide Field and Planetary Camera) and FGS (Fine Guidance Sensor) instruments for astrometry. The planetary camera provides measurements to an accuracy of ± 1 mas for objects down to magnitude 26^m, but only has a $34''$ field of view. Its images are also under sampled so that dithering (section 1.1, CCDs) has to be used to reach this level of accuracy. The telescope has three fine guidance sensors but only needs two for guidance purposes. The third is therefore available for astrometry. The FGSs are interferometers based upon Köster prisms (section 2.5), and proved positional accuracies of ± 5 mas.

There are currently two significant spacecraft[60] being planned for astrometry—GAIA[61] and SIM.[62] SIM is a NASA project, due for launch in 2005/2006. It will orbit the Sun and be nearly 10^7 km away from the Earth towards the end of its planned 5 year mission. It will have a 10 m baseline Michelson optical interferometer and is expected to observe 40 000 stars' positions to a limiting visual magnitude of 20^m and to an accuracy of ± 4 μas. The principal aims for SIM are to detect nearby low mass extra-solar planets and to measure proper motions for stars in nearby galaxies. The GAIA mission is funded by ESA and may be launched as early as 2010. It will be positioned at the Sun–Earth inner Lagrange point some 1.5×10^6 km from the Earth, to avoid eclipses and occultations and so provide a stable thermal environment. It will operate in a similar fashion to Hipparcos but using two separate telescopes set at a fixed angle and with elongated apertures 1.7 m $\times$ 0.7 m. The long axis of the aperture in each case will be aligned along the scanning direction in order to provide the highest resolution. The read-out will be by CCDs using time-delayed integration. The aim is to determine positions for a billion stars to ± 10 μas accuracy at visual magnitude 15^m. GAIA will also determine radial velocities for stars down to 17^m using a separate spectroscope telescope. The objective for the GAIA mission is to establish a precise three-dimensional map of the galaxy and provide a massive data base for other investigations of the Milky Way.

---

[60] A third, FAME (Full-sky Astrometric Mapping Explorer) has recently been cancelled, although efforts are being made to reinstate it. It was to have measured 40 million stars' positions to between 50 μas and 500 μas accuracy using a similar operating principle to Hipparcos.

[61] Global Astrometric Interferometer for Astrophysics.

[62] Space Interferometry Mission.

Detectors

CCDs (section 1.1) are now widely used as detectors for astrometry. They have the enormous advantage that the stellar or other images can be identified with individual pixels whose positions are automatically defined by the structure of the detector. The measurement of the relative positions of objects within the image is therefore vastly simplified. Software to fit a suitable point spread function (section 2.1) to the image enables the position of its centroid to be defined to a fraction of the size of the individual pixels.

The disadvantage of electronic images is their small scale. Most CCD chips are only a few centimetres in size (though mosaics are larger) compared with tens of centimetres for astrographic photographic plates. They therefore cover only a small region of the sky and it can become difficult to find suitable reference stars for the measurements. However, the greater dynamic range of CCDs compared with photographic emulsion and the use of anti-blooming techniques (section 1.1) enable the positions of many more stars to be usefully measured. With transit telescopes (see above) time-delayed integration may be used to reach fainter magnitudes. TDI is also proposed for the GAIA spacecraft detection system (see above). Otherwise the operation of a CCD for positional work is conventional (section 1.1). In other respects, the processing and reduction of electronic images for astrometric purposes is the same as that for a photographic image.

Grid modulation is used by the Multichannel Astrometric Photometer (MAP) at the Allegheny observatory. Light from up to 12 stars is fed into separate detectors by fibre optics as a grating with 4 lines mm^{-1} is passed across the field. Timing the disappearances of the stars behind the bars of the grating provides their relative positions to an accuracy of 3 mas. A similar technique was used by the Hipparcos spacecraft (see above), although in this case it was the stars that moved across a fixed grid.

Photography is still used for astrometry, especially when large areas need to be covered, though it is becoming less common. However, much material in the archives is in the form of photographs and this is still essential to long-term programmes measuring visual binary star orbits or proper motion etc. Thus the reduction of data from photographs will continue to be needed for astrometry for some considerable time to come.

Photography or CCD imaging may usefully be used to observe double stars when the separation is greater than about 3″ of arc. Many exposures are made with a slight shift of the telescope between each exposure. Averaging the measurements can give the separation to better than a hundredth of a second of arc and the position angle to within a few minutes of arc. For double stars with large magnitude differences between their components, an objective grating can be used (see below), or the shape of the aperture can be changed so that the fainter star lies between two diffraction spikes and has an improved noise level. The latter technique is a variation of the technique of apodization mentioned in sections 1.1, 1.2, 2.5, 4.1 and 5.3.

Measurement and reduction

Transit telescopes and interferometers give absolute positions directly as discussed above. In most other cases the position of an unknown star is obtained by comparison with reference stars that also appear on the image and whose positions are known. If absolute positions are to be determined, then the reference stars' positions must have been found via a transit instrument, interferometer or astrolabe etc. For relative positional work, such as that involved in the determination of parallax and proper motion, any very distant star may be used as a comparison star.

It is advantageous to have the star of interest and its reference stars of similar brightnesses. This may be effected by the use of variable density filters, by the use of a rotating chopper in front of the detector, or by the use of an objective grating (section 4.2). The latter device is arranged so that pairs of images, which in fact are the first-order or higher-order spectra of the star, appear on either side of it. By adjusting the spacing of the grating the brightness of these secondary images for the brighter stars may be arranged to be comparable with the brightness of the primary images of the fainter stars. The position of the brighter star is then taken as the average of the two positions of its secondary images. The apparent magnitude, m, of such a secondary image resulting from the nth-order spectrum is given by

$$m = m_0 + 5\log_{10}(N\eta\cosec\eta) \qquad (5.1.16)$$

where m_0 is the apparent magnitude of the star, η is given by

$$\eta = \pi Nd/D \qquad (5.1.17)$$

where D is the separation of the grating bars, d is the width of the gap between the grating bars and N is the total number of slits across the objective.

Once an image of a star field has been obtained, the positions of the stars' images must be measured to a high degree of accuracy and the measurements converted into right ascension and declination or separation and position angle. On photographs the measuring is undertaken in a manner similar to that used to measure spectra, except that two-dimensional positions are required. The measuring procedure is a tedious process and, as in the case of spectroscopic measurement, a number of automatic measuring machines have been developed over the past decade. In some ways the problem of automating the measurement of these plates is easier than in the spectroscopic case since almost all the images are the same shape and are isolated from each other, hence completely automatic measuring machines have been in use for several years. The machines are usually computer controlled and also use the computer to undertake some or all of the data reduction. Their designs vary quite widely, but most have the photograph mounted on a table that moves slowly in one axis, while it is scanned rapidly in the other. The scanning spot can be from a laser or from a television system or a variety of other sources. Alternatively a wider light beam may be used, and a CCD used to pick up the features on the plate. The position of an image is determined by fitting a one- or two-dimensional Gaussian profile to the output. The UK's Automatic Plate Measuring Machine (APM), for example, uses a laser to produce the scanning spot and has a linear accuracy of better than $0.5\,\mu m$ and a photometric accuracy better than $0.001D$ (0.2%). As well as measuring position, some of the machines also measure brightness and can sort the images into types of object (stars, galaxies, asteroid trails, etc). The Royal Observatory Edinburgh's SuperCOSMOS[63] machine can analyse a $350\,mm$ square Schmidt plate containing $2\,Gb$ of data in just 2 hours. It is based upon a moving granite table on to which the plate is mounted and a 2048 pixel linear CCD giving $10\,\mu m$ linear resolution and about $50\,mas$ accuracy in stars' positions. The US Naval Observatory's Precision Measuring Machine (PMM) uses two CCDs to scan a plate and processes the data in real time. Other examples of automatic plate measuring machines include MAMA (Machine Automatique à Mesurer pour l'Astronomie) at the Observatoire de Paris and HAM1 at Hamburger Sternwarte.

CCDs give positions directly (see above) once the physical structure of the device has been calibrated. Usually the position will be found by fitting a suitable point spread function (section 2.1) for the instrument used to obtain the overall image to the stellar images when they spread across several pixels. The stars' positions may then be determined to sub-pixel accuracy.

However the raw data may have been obtained, the process of converting the measurements into the required information is known as reduction; for accurate astrometry it is a lengthy process. A number of corrections have already been mentioned, and in addition to those, there are the errors of the measuring machine itself to be corrected, and the distortion introduced by the image to be taken into account. The latter is caused by the projection of the curved sky on to the flat photographic plate or CCD (known as the tangential plane), or in the case of Schmidt cameras its projection on to the curved focal plane of the camera, followed by the flattening of the plate after it is taken out of the plate holder. For the simpler case of the astrograph, the projection is shown in figure 5.1.7, and the relationship between the coordinates on the flat image and in the curved sky is

---

[63] CoOrdinates, Sizes, Magnitudes, Orientations, and Shapes.

Figure 5.1.7. Projection on to a photographic plate or flat CCD etc.

given by

$$\rho = \tan^{-1}\left(\frac{y}{F}\right) \tag{5.1.18}$$

$$\tau = \tan^{-1}\left(\frac{x}{F}\right) \tag{5.1.19}$$

where x and y are the coordinates on the image, τ and ρ are the equivalent coordinates on the image of the celestial sphere and F is the focal length of the telescope.

One instrument, which, while it is not strictly of itself a part of astrometry, is very frequently used to identify the objects that are to be measured, is called the blink comparator. In this machine the worker looks alternately at two aligned images of the same star field obtained some time apart from each other, the frequency of the interchange being about 2 Hz. Stars, asteroids, comets etc. which have moved in the interval between the photographs then call attention to themselves by appearing to jump backwards and forwards while the remainder of the stars are stationary. The device can also pick out variable stars for further study by photometry (chapter 3) since although these remain stationary, they appear to blink on and off (hence the name of the instrument). Early blink comparators were basically binocular microscopes that had a mechanical arrangement for swapping the view of two photographic plates. Nowadays, the images are usually viewed on a computer screen and software aligns the images and provides the alternating views. Software can also be used to identify the moving or changing objects directly. But this is still one area where the human eye–brain functions as efficiently as the computer.

Sky surveys and catalogues

The end result of most astrometry is a catalogue of positions and other properties of (usually) a large number of objects in the sky. The Hipparcos, Tycho and Tycho-2 (see above) catalogues are just the latest examples of the process (see also section 5.5). Other recent astrometric catalogues include the second issue of the US Naval Observatory's Twin Astrographic Catalog (TAC 2.0) based upon photographic plates obtained with the twin astrograph and containing over 700 000 stellar positions to between ±50 mas and ±120 mas accuracy, the USNO's A2.0 catalogue containing 526 million entries, and UCAC (USNO CCD Astrograph Catalog) that aims to measure the positions of some 40 million stars in the 10^m to 14^m range to ±20 mas accuracy. Older astrometric catalogues include the Fundamental Katalog series that culminated in FK5 in 1998 containing 1500 stars with positional accuracies of better than ±100 mas, the Astrographic Catalogue (AC) series, the Astronomische Gesellschaft Katalog (AGK) series, plus arguably the Argelander, Flamsteed, Brahe and even the original Hipparchus catalogues since these gave state-of-the-art stellar positions for their day.

There have been many other sky surveys and catalogues and more are being produced all the time that are non-astrometric. That is to say, their purpose is other than that of providing accurate positions, and the position, if determined at all for the catalogue, is well below the current levels of astrometric accuracy. Indeed many such catalogues just use the already-known astrometric positions. Non-astrometric catalogues are also produced for all regions of the spectrum, not just from optical and radio sources. There are tens of thousands of non-astrometric catalogues produced for almost as many different reasons, and ranging in content from a few tens to half a billion objects. Examples of such catalogues include the Hubble space telescope second Guide Star Catalogue (GSC2) containing 500 million objects, the Two-Micron All Sky Survey (2MASS), DENIS (DEep Near Infrared Survey of the southern sky) in the infrared, and the Sloan digital sky survey that aims to obtain a million red shifts of galaxies and quasars.

Exercises

5.1.1 If the axis of a transit telescope is accurately horizontal, but is displaced north or south of the east–west line by E seconds of arc, then show that for stars on the equator

$$\Delta\alpha^2 + \Delta\delta^2 \approx E^2 \sin^2 \phi$$

where $\Delta\alpha$ and $\Delta\delta$ are the errors in the determinations of right ascension and declination in seconds of arc and ϕ is the latitude of the observatory.

5.1.2 Calculate the widths of the slits required in an objective grating for use on a 1 m telescope if the third-order image of Sirius A is to be of the same brightness as the primary image of Sirius B. The total number of slits in the grating is 99. The apparent magnitude of Sirius A is -1.5, while that of Sirius B is $+8.5$.

5.2 Polarimetry

Background

Although the discovery of polarized light from astronomical sources goes back to the beginning of the 19th century, when Arago detected its presence in moonlight, the extensive development of its study is a relatively recent phenomenon. This is largely due to the technical difficulties that are involved, and to the lack of any expectation of polarized light from stars by astronomers. Many phenomena, however, can contribute to the polarization of radiation, and so, conversely, its observation can potentially provide information upon an equally wide range of basic causes.

Stokes' parameters

Polarization of radiation is simply the non-random angular distribution of the electric vectors of the photons in a beam of radiation. Customarily two types are distinguished—linear and circular polarizations. In the former the electric vectors are all parallel and their direction is constant, while in the latter the angle of the electric vector rotates with time at the frequency of the radiation. These are not really different types of phenomenon, however, and all types of radiation may be considered to be different aspects of partially elliptically polarized radiation. This has two components, one of which is unpolarized, the other being elliptically polarized. Elliptically polarized light is similar to circularly polarized light in that the electric vector rotates at the frequency of the radiation, but in addition the magnitude varies at twice that frequency, so that plotted on a polar diagram the electric vector would trace out an ellipse (see figure 5.2.1 for example). The properties of partially elliptically polarized light are completely described by four parameters that are called the Stokes' parameters. These fix the intensity of the unpolarized light, the degree of ellipticity, the direction of the major axis of the ellipse, and the sense (left- or right-handed rotation) of the elliptically polarized light. If the radiation is imagined to be propagating along the z axis of a three-dimensional rectangular coordinate system, then the elliptically polarized component at a point along the z axis may have its electric vector resolved into components along the x and y axes (figure 5.2.1), these being given by

$$E_x(t) = e_1 \cos(2\pi\nu t) \tag{5.2.1}$$

$$E_y(t) = e_2 \cos(2\pi\nu t + \delta) \tag{5.2.2}$$

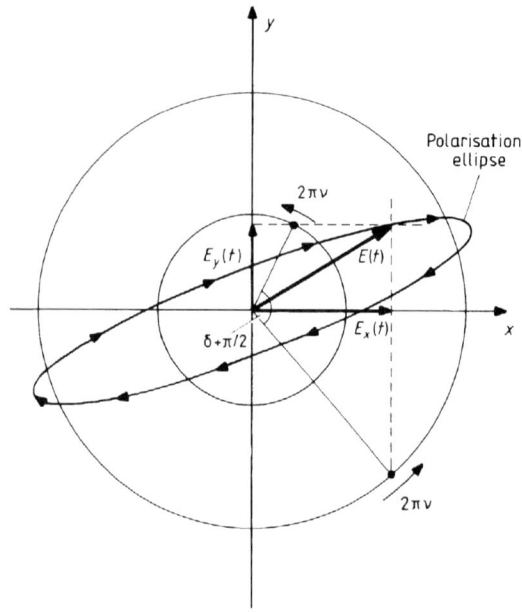

Figure 5.2.1. The x and y components of the elliptically polarized component of partially elliptically polarized light.

where ν is the frequency of the radiation, δ is the phase difference between the x and y components and e_1 and e_2 are the amplitudes of the x and y components. It is then a tedious but straightforward matter to show that

$$a = \left(\frac{(e_1^2 + e_2^2)}{1 + \tan^2[\frac{1}{2}\sin^{-1}\{[2e_1e_2/(e_1^2 + e_2^2)]\sin\delta\}]} \right)^{1/2} \tag{5.2.3}$$

$$b = a\tan\left[\frac{1}{2}\sin^{-1}\left\{ \left[\frac{2e_1e_2}{e_1^2 + e_2^2} \right] \sin\delta \right\} \right] \tag{5.2.4}$$

$$a^2 + b^2 = e_1^2 + e_2^2 \tag{5.2.5}$$

$$\psi = \frac{1}{2}\tan^{-1}\left\{ \left[\frac{2e_1e_2}{e_1^2 - e_2^2} \right] \cos\delta \right\} \tag{5.2.6}$$

where a and b are the semi-major and semi-minor axes of the polarization ellipse and ψ is the angle between the x axis and the major axis of the polarization ellipse. The Stokes' parameters are then defined by

$$Q = e_1^2 - e_2^2 = \frac{a^2 - b^2}{a^2 + b^2}\cos(2\psi)I_p \tag{5.2.7}$$

$$U = 2e_1e_2\cos\delta = \frac{a^2 - b^2}{a^2 + b^2}\sin(2\psi)I_p \tag{5.2.8}$$

$$V = 2e_1e_2 \sin \delta = \frac{2ab}{a^2 + b^2} I_p \qquad (5.2.9)$$

where I_p is the intensity of the polarized component of the light. From equations (5.2.7), (5.2.8) and (5.2.9) we have

$$I_p = (Q^2 + U^2 + V^2)^{1/2}. \qquad (5.2.10)$$

The fourth Stokes' parameter, I, is just the total intensity of the partially polarized light

$$I = I_u + I_p \qquad (5.2.11)$$

where I_u is the intensity of the unpolarized component of the radiation. (*Note*: The notation and definitions of the Stokes' parameters can vary. While that given here is probably the commonest usage, a check should always be carried out in individual cases to ensure that a different usage is not being employed.)

The degree of polarization, π, of the radiation is given by

$$\pi = \frac{(Q^2 + U^2 + V^2)^{1/2}}{I} = \frac{I_p}{I} \qquad (5.2.12)$$

while the degree of linear polarization, π_L, and the degree of ellipticity, π_e, are

$$\pi_L = \frac{(Q^2 + U^2)^{1/2}}{I} \qquad (5.2.13)$$

$$\pi_e = \frac{V}{I}. \qquad (5.2.14)$$

When $V = 0$ (i.e. the phase difference, δ, is 0 or π radians) we have linearly polarized radiation. The degree of polarization is then equal to the degree of linear polarization, and is the quantity that is commonly determined experimentally

$$\pi = \pi_L = \frac{I_{max} - I_{min}}{I_{max} + I_{min}} \qquad (5.2.15)$$

where I_{max} and I_{min} are the maximum and minimum intensities that are observed through a polarizer as it is rotated. The value of π_e is positive for right-handed and negative for left-handed radiation.

When several incoherent beams of radiation are mixed, their various Stokes' parameters combine individually by simple addition. A given monochromatic partially elliptically polarized beam of radiation may therefore have been formed in many different ways, and these are indistinguishable by the intensity and polarization measurements alone. It is often customary, therefore, to regard partially elliptically polarized light as formed from two separate components, one of which is unpolarized and the other completely elliptically polarized. The Stokes' parameters of these components are then:

	I	Q	U	V
Unpolarized component	I_u	0	0	0
Elliptically polarized component	I_p	Q	U	V

and the normalized Stokes' parameters for more specific mixtures are:

Type of radiation	Stokes' parameters			
	I/I	Q/I	U/I	V/I
Right-hand circularly polarized (clockwise)	1	0	0	1
Left-hand circularly polarized (anticlockwise)	1	0	0	-1
Linearly polarized at an angle ψ to the x axis	1	$\cos 2\psi$	$\sin 2\psi$	0

The Stokes' parameters are related to more familiar astronomical quantities by

$$\theta = \frac{1}{2}\tan^{-1}\left(\frac{U}{Q}\right) = \psi \tag{5.2.16}$$

$$e = \left\{1 - \tan^2\left[\frac{1}{2}\sin^{-1}\left(\frac{V}{(Q^2 + U^2 + V^2)^{1/2}}\right)\right]\right\}^{1/2} \tag{5.2.17}$$

where θ is the position angle (section 5.1) of the semi-major axis of the polarization ellipse (when the x axis is aligned north–south) and e is the eccentricity of the polarization ellipse and is 1 for linearly polarized radiation and 0 for circularly polarized radiation.

Optical components for polarimetry

Polarimeters usually contain a number of components that are optically active in the sense that they alter the state of polarization of the radiation. They may be grouped under three headings: polarizers, converters and depolarizers. The first produces linearly polarized light, the second converts elliptically polarized light into linearly polarized light, or vice versa, while the last eliminates polarization. Most of the devices rely upon birefringence for their effects, so we must initially discuss some of its properties before looking at the devices themselves.

Birefringence

The difference between a birefringent material and a more normal optical one may best be understood in terms of the behaviour of the Huygens' wavelets.

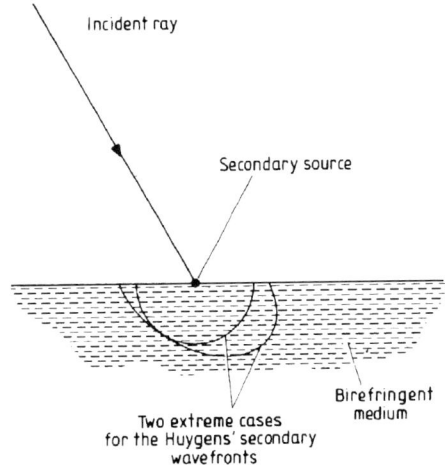

Figure 5.2.2. Huygens' secondary wavelets in a birefringent medium.

In a normal material the refracted ray can be constructed from the incident ray by taking the envelope of the Huygens' wavelets, which spread out from the incident surface with a uniform velocity. In a birefringent material, however, the velocities of the wavelets depend upon the polarization of the radiation, and for at least one component the velocity will also depend on the orientation of the ray with respect to the structure of the material. In some materials it is possible to find a direction for linearly polarized radiation for which the wavelets expand spherically from the point of incidence as in 'normal' optical materials. This ray is then termed the ordinary ray and its behaviour may be described by normal geometrical optics. The ray that is polarized orthogonally to the ordinary ray is then termed the extraordinary ray, and wavelets from its point of incidence spread out elliptically (figure 5.2.2). The velocity of the extraordinary ray thus varies with direction. We may construct the two refracted rays in a birefringent material by taking the envelopes of their wavelets as before (figure 5.2.3). The direction along which the velocities of the ordinary and extraordinary rays are equal is called the optic axis of the material. When the velocity of the extraordinary ray is in general larger than that of the ordinary ray (as illustrated in figures 5.2.2 and 5.2.3), then the birefringence is negative. It is positive when the situation is reversed. The degree of the birefringence may be obtained from the principal extraordinary refractive index, μ_E. This is the refractive index corresponding to the maximum velocity of the extraordinary ray for negative materials, and the minimum velocity for positive materials. It will be obtained from rays travelling perpendicularly to the optic axis of the material. The degree of birefringence, which is often denoted by the symbol J, is then simply the difference between the principal extraordinary

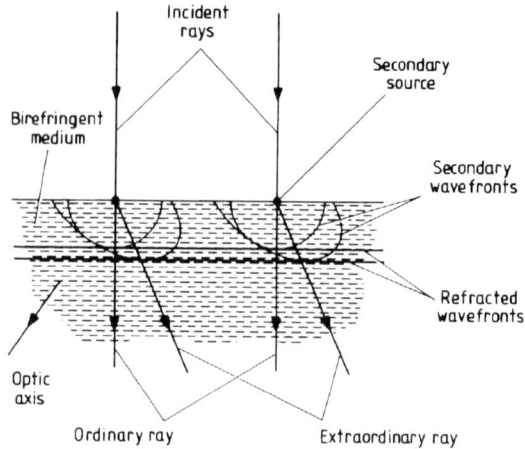

Figure 5.2.3. Formation of ordinary and extraordinary rays in a birefringent medium.

refractive index and the refractive index for the ordinary ray, μ_O:

$$J = \mu_E - \mu_O. \tag{5.2.18}$$

Most crystals exhibit natural birefringence, and this can be introduced into many more and into amorphous substances such as glass by the presence of strain in the material. One of the most commonly encountered birefringent materials is calcite. The cleavage fragments form rhombohedrons and the optic axis then joins opposite blunt corners if all the edges are of equal length (figure 5.2.4). The refractive index of the ordinary ray is 1.658, while the principal extraordinary refractive index is 1.486, giving calcite the very high degree of birefringence of -0.172.

Crystals such as calcite are uniaxial and have only one optic axis. Uniaxial crystals belong to the tetragonal or hexagonal crystallographic systems. Crystals that belong to the cubic system are not usually birefringent, while crystals in the remaining systems—orthorhombic, monoclinic, or triclinic—generally produce biaxial crystals. In the latter case, normal geometrical optics breaks down completely and all rays are extraordinary.

Some crystals such as quartz that are birefringent ($J = +0.009$) are in addition optically active. This is the property whereby the plane of polarization of a beam of radiation is rotated as it passes through the material. Many substances other than crystals, including most solutions of organic chemicals, can be optically active. Looking down a beam of light, against the motion of the photons, a substance is called dextro-rotatory or right handed if the rotation of the plane of vibration is clockwise. The other case is called laevo-rotatory or left handed. Unfortunately and confusingly the opposite convention is also in occasional use.

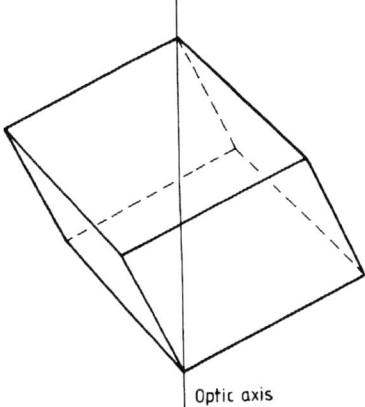

Optic axis

Figure 5.2.4. An equilateral calcite cleavage rhomb and its optic axis.

The description of the behaviour of an optically active substance in terms of the Huygens wavelets is rather more difficult than was the case for birefringence. Incident light is split into two components as previously, but the velocities of both components vary with angle, and in no direction do they become equal. The optic axis therefore has to be taken as the direction along which the difference in velocities is minimized. Additionally the nature of the components changes with angle as well. Along the optic axis they are circularly polarized in opposite senses, while perpendicular to the optic axis they are orthogonally linearly polarized. Between these two extremes the two components are elliptically polarized to varying degrees and in opposite senses (figure 5.2.5).

Polarizers[64]

These are devices that only allow the passage of light that is linearly polarized in some specified direction. There are several varieties that are based upon birefringence, of which the Nicol prism is the best known. This consists of two calcite prisms cemented together by Canada balsam. Since the refractive index of Canada balsam is 1.55, it is possible for the extraordinary ray to be transmitted, while the ordinary ray is totally internally reflected (figure 5.2.6). The Nicol polarizer has the drawbacks of displacing the light beam and of introducing some elliptical polarization into the emergent beam. Its inclined faces also introduce additional light losses by reflection. Various other designs of polarizers have therefore been developed, some of which in fact produce mutually orthogonally polarized beams with an angular separation.

---

[64] Also often known as analysers.

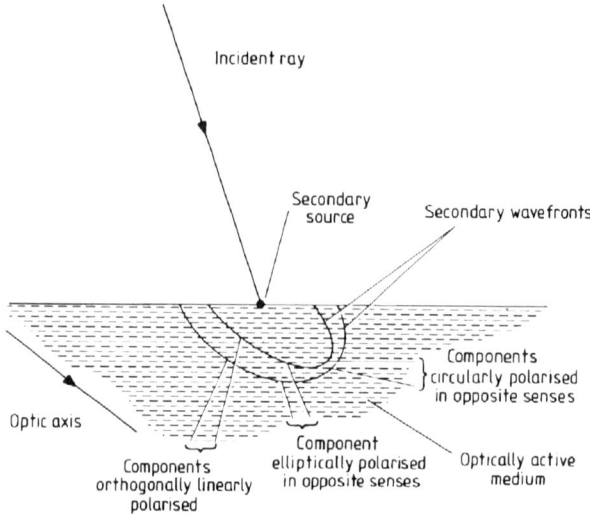

Figure 5.2.5. Huygens' secondary wavelets in an optically active medium.

Examples of several such designs are shown in figure 5.2.7. Magnesium fluoride and quartz can also be used to form polarizers for the visible region, while lithium niobate and sapphire are used in the infrared.

The ubiquitous polarizing sunglasses are based upon another type of polarizer. They employ dichroic crystals that have nearly 100% absorption for one plane of polarization and less than 100% for the other. Generally the dichroism varies with wavelength so that these polarizers are not achromatic. Usually, however, they are sufficiently uniform in their spectral behaviour to be usable over quite wide wavebands. The polarizers are produced commercially in sheet form and contain many small aligned crystals rather than one large one. The two commonly used compounds are polyvinyl alcohol impregnated with iodine and polyvinyl alcohol catalysed

Figure 5.2.6. The Nicol prism.

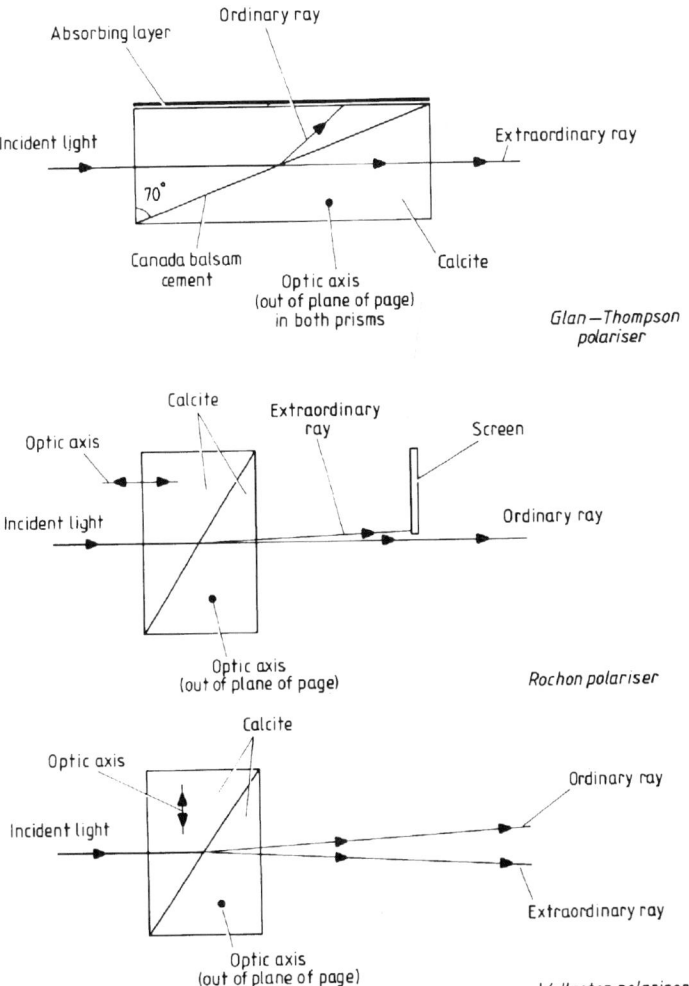

Figure 5.2.7. Examples of birefringence polarizers.

to polyvinylene by hydrogen chloride. The alignment is achieved by stretching the film. The use of microscopic crystals and the existence of the large commercial market means that dichroic polarizers are far cheaper than birefringent polarizers, and so they may be used even when their performance is poorer than that of the birefringent polarizers.

Polarization by reflection can be used to produce a polarizer. A glass plate inclined at the Brewster angle will reflect a totally polarized beam. However, only a small percentage (about 7.5% for crown glass) of the incident energy is reflected. Reflection from the second surface will reinforce

Figure 5.2.8. A reflection polarizer.

the first reflection, however, and then several plates may be stacked together to provide further reflections (figure 5.2.8). The transmitted beam is only partially polarized. However, as the number of plates is increased, the total intensity of the reflected beams (ignoring absorption) will approach half of the incident intensity (figure 5.2.9). Hence the transmitted beam will

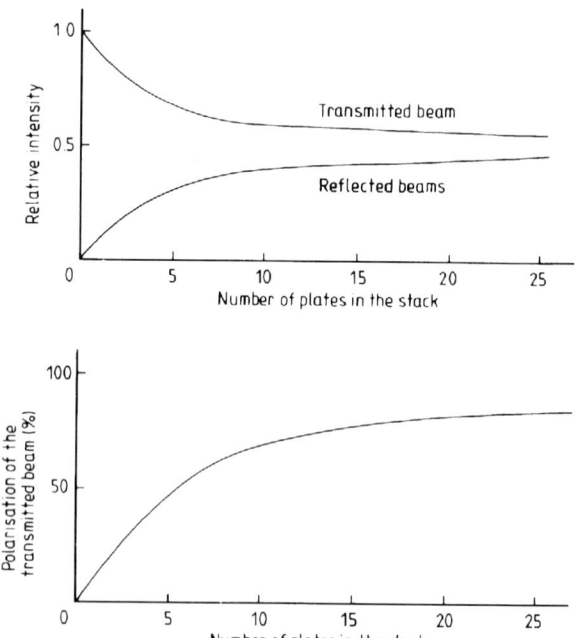

Figure 5.2.9. Properties of reflection polarizers, assuming negligible absorption and a refractive index of 1.5.

approach complete polarization. In practice, therefore, the reflection polarizer is used in transmission, since the problem of recombining the multiple reflected beams is thereby avoided and the beam suffers no angular deviation.

Outside the optical, near-infrared and near-ultraviolet regions, the polarizers tend to be somewhat different in nature. In the radio region, a linear dipole is naturally only sensitive to radiation polarized along its length. In the microwave region and the medium to far infrared, wire grid polarizers are suitable. These, as their name suggests, are grids of electrically conducting wires. Their spacing is about five times their thickness, and they transmit the component of the radiation that is polarized along the length of the wires. They work efficiently for wavelengths longer than their spacing. In the x-ray region, Bragg reflection is polarized (section 1.3) and so a rotating Bragg spectrometer can also act as a linear polarization detector.

The behaviour of a polarizer may be described mathematically by its effect upon the Stokes' parameters of the radiation. This is most easily accomplished by writing the parameters as a column vector. A matrix multiplying the vector on the left may then represent the effect of the polarizer, and also of the other optical components that we discuss later in this section. The technique is sometimes given the name of the Mueller calculus. There is also an alternative, and to some extent complementary, formulation termed the Jones calculus. In the Mueller calculus the effect of passing the beam through several optical components is simply found by successive matrix multiplications of the Stokes' vector. The first optical component's matrix is closest to the original Stokes' vector, and the matrices of subsequent components are positioned successively farther to the left as the beam passes through the optical system. For a perfect polarizer whose transmission axis is at an angle, θ, to the reference direction, we have

$$\begin{bmatrix} I' \\ Q' \\ U' \\ V' \end{bmatrix} = \begin{bmatrix} \frac{1}{2} & \frac{1}{2}\cos 2\theta & \frac{1}{2}\sin 2\theta & 0 \\ \frac{1}{2}\cos 2\theta & \frac{1}{2}\cos^2 2\theta & \frac{1}{2}\cos 2\theta \sin 2\theta & 0 \\ \frac{1}{2}\sin 2\theta & \frac{1}{2}\cos 2\theta \sin 2\theta & \frac{1}{2}\sin^2 2\theta & 0 \\ 0 & 0 & 0 & 0 \end{bmatrix} \begin{bmatrix} I \\ Q \\ U \\ V \end{bmatrix} \qquad (5.2.19)$$

where the primed Stokes' parameters are for the beam after passage through the polarizer and the unprimed ones are for the beam before such passage.

Converters

These are devices that alter the type of polarization and/or its orientation. They are also known as retarders or phase plates. We have seen earlier (equations (5.2.1) and (5.2.2) and figure 5.2.1) that elliptically polarized light may be resolved into two orthogonal linear components with a phase difference. Altering that phase difference will alter the degree of ellipticity (equations (5.2.3) and (5.2.4)) and inclination (equation (5.2.6)) of the ellipse. We have also seen that the velocities of mutually orthogonal linearly polarized beams of radiation will in general differ from each other when the beams pass through a birefringent material. From inspection of figure 5.2.6 it will be seen that if the optic axis is rotated until it is perpendicular to the incident radiation, then the ordinary and extraordinary rays will travel in the same direction. Thus they will pass together through a layer of a birefringent material that is oriented in this way, and will recombine upon emergence, but with an altered phase delay, due to their differing velocities. The phase delay, δ', is given to a first-order approximation by

$$\delta' = \frac{2\pi d}{\lambda} J \tag{5.2.20}$$

where d is the thickness of the material and J is the birefringence of the material. Now let us define the x axis of figure 5.2.1 to be the polarization direction of the extraordinary ray. We then have from equation (5.2.6) the intrinsic phase difference, δ, between the components of the incident radiation

$$\delta = \cos^{-1}\left\{\frac{e_1^2 - e_2^2}{2e_1 e_2} \tan 2\psi\right\}. \tag{5.2.21}$$

The ellipse for the emergent radiation then has a minor axis given by

$$b' = a' \tan\left[\frac{1}{2}\sin^{-1}\left\{\left[\frac{2e_1 e_2}{e_1^2 + e_2^2}\right] \sin(\delta + \delta')\right\}\right] \tag{5.2.22}$$

where the primed quantities are for the emergent beam. So

$$b' = 0 \quad \text{for } \delta + \delta' = 0 \tag{5.2.23}$$

$$b' = a' \quad \text{for } \delta + \delta' = \sin^{-1}\left[\frac{e_1^2 + e_2^2}{2e_1 e_2}\right] \tag{5.2.24}$$

and also

$$\psi = \frac{1}{2}\tan^{-1}\left\{\left[\frac{2e_1 e_2}{e_1^2 - e_2^2}\right] \cos(\delta + \delta')\right\}. \tag{5.2.25}$$

Thus

$$\psi' = -\psi \quad \text{for } \delta' = 180° \tag{5.2.26}$$

and

$$a' = a, \qquad b' = b. \tag{5.2.27}$$

Thus we see that, in general, elliptically polarized radiation may have its degree of ellipticity altered and its inclination changed by passage through a converter. In particular cases it may be converted into linearly polarized or circularly polarized radiation, or its orientation may be reflected about the fast axis of the converter.

In real devices, the value of δ' is chosen to be 90° or 180° and the resulting converters are called quarter-wave plates or half-wave plates respectively, since one beam is delayed with respect to the other by a quarter or a half of a wavelength. The quarter-wave plate is then used to convert elliptically or circularly polarized light into linearly polarized light or vice versa, while the half-wave plate is used to rotate the plane of linearly polarized light.

Many substances can be used to make converters, but mica is probably the commonest because of the ease with which it may be split along its cleavage planes. Plates of the right thickness (about 40 µm) are therefore simple to obtain. Quartz cut parallel to its optic axis can also be used, while for ultraviolet work magnesium fluoride is suitable. Amorphous substances may be stretched or compressed to introduce stress birefringence. It is then possible to change the phase delay in the material by changing the amount of stress. Extremely high acceptance angles and chopping rates can be achieved with low power consumption if a small acoustic transducer at one of its natural frequencies drives the material. This is then commonly called a photoelastic modulator. An electric field can also induce birefringence. Along the direction of the field, the phenomenon is called the Pockels effect, while perpendicular to the field it is called the Kerr effect. Suitable materials abound and may be used to produce quarter- or half-wave plates. Generally these devices are used when rapid switching of the state of birefringence is needed, as for example in Babcock's solar magnetometer (section 5.3). In glass the effects take up to a minute to appear and disappear, but in other substances the delay can be far shorter. The Kerr effect in nitrobenzene, for example, permits switching at up to 1 MHz, and the Pockels effect can be used similarly with ammonium dihydrogen phosphate or potassium dihydrogen phosphate.

All the above converters will normally be chromatic and usable over only a restricted wavelength range. Converters that are more nearly achromatic can be produced which are based upon the phase changes that occur in total internal reflection. The phase difference between components with electric vectors parallel and perpendicular to the plane of incidence is shown in figure 5.2.10. For two of the angles of incidence (approximately 45° and 60° in the example shown) the phase delay is 135°. Two such reflections therefore produce a total delay of 270°, or, as one may also view the situation, an advance of the second component with respect to the first by 90°. The minimum value of the phase difference shown in figure 5.2.10 is equal to 135° when the refractive index is 1.497. There is then only one suitable angle of incidence and this is 51° 47'. For optimum results, the optical design should approach as close to this ideal as possible. A quarter-wave retarder that is nearly achromatic can thus be formed using two total internal reflections at appropriate angles. The precise angles usually have to be determined by trial and error since additional phase changes are produced when the beam interacts with the entrance and exit

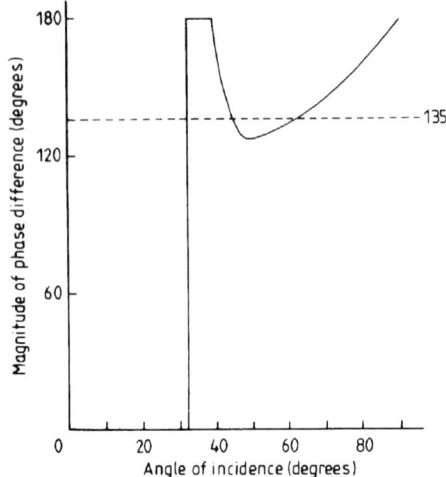

Figure 5.2.10. Differential phase delay upon internal reflection in a medium with a refractive index of 1.6.

faces of the component. Two practical designs—the Fresnel rhomb and the Mooney rhomb—are illustrated in figure 5.2.11. Half-wave retarders can be formed by using two such quarter-wave retarders in succession.

Combining three retarders can make pseudo-quarter- and half-wave plates that are usable over several hundred nanometres in the visual. The Pancharatnam design employs two retarders with the same delay and orientation, together with a third sandwiched between the first two. The inner retarder is rotated with respect to the other two, and possibly has a differing delay. A composite 'half-wave plate' with actual delays in the range $180° \pm 2°$ over wavelengths from 300 to 1200 nm can, for example, be formed from three $180°$ retarders with the centre one oriented at about $60°$ to the two outer ones. Quartz and magnesium fluoride are commonly used materials for the production of such super-achromatic waveplates.

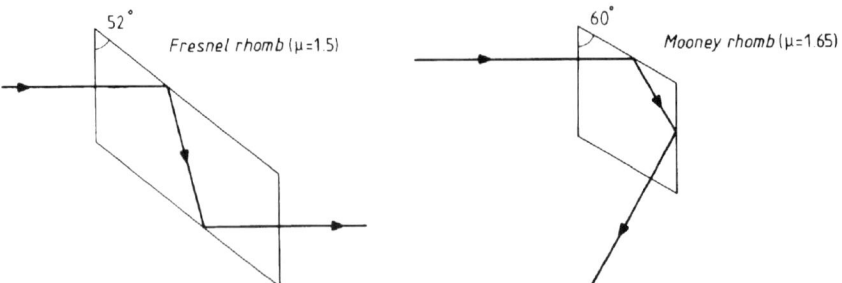

Figure 5.2.11. Quarter-wave retarders using total internal reflection.

The Mueller matrices for converters are

$$\begin{bmatrix} 1 & 0 & 0 & 0 \\ 0 & \cos^2 2\psi & \cos 2\psi \sin 2\psi & -\sin 2\psi \\ 0 & \cos 2\psi \sin 2\psi & \sin^2 2\psi & \cos 2\psi \\ 0 & \sin 2\psi & -\cos 2\psi & 0 \end{bmatrix} \qquad (5.2.28)$$

for a quarter-wave plate, and

$$\begin{bmatrix} 1 & 0 & 0 & 0 \\ 0 & \cos^2 2\psi - \sin^2 2\psi & 2\cos 2\psi \sin 2\psi & 0 \\ 0 & 2\cos 2\psi \sin 2\psi & \sin^2 2\psi - \cos^2 2\psi & 0 \\ 0 & 0 & 0 & -1 \end{bmatrix} \qquad (5.2.29)$$

for a half-wave plate, while in the general case we have

$$\begin{bmatrix} 1 & 0 & 0 & 0 \\ 0 & \cos^2 2\psi + \sin^2 2\psi \cos \delta & (1 - \cos \delta)\cos 2\psi \sin 2\psi & -\sin 2\psi \sin \delta \\ 0 & (1 - \cos \delta)\cos 2\psi \sin 2\psi & \sin^2 2\psi + \cos^2 2\psi \cos \delta & \cos 2\psi \sin \delta \\ 0 & \sin 2\psi \sin \delta & -\cos 2\psi \sin \delta & \cos \delta \end{bmatrix} .$$

$$(5.2.30)$$

Depolarizers

The ideal depolarizer would accept any form of polarized radiation and produce unpolarized radiation. No such device exists, but pseudo-depolarizers can be made. These convert the polarized radiation into radiation that is unpolarized when it is averaged over wavelength, time or area.

A monochromatic depolarizer can be formed from a rotating quarter-wave plate that is in line with a half-wave plate rotating at twice its rate. The emerging beam at any given instant will have some form of elliptical polarization, but this will change rapidly with time, and the output will average to zero polarization over several rotations of the plates.

The Lyot depolarizer averages over wavelength. It consists of two retarders with phase differences very much greater than 360°. The second plate has twice the thickness of the first and its optic axis is rotated by 45° with respect to that of the first. The emergent beam will be polarized at any given wavelength, but the polarization will vary very rapidly with wavelength. In the optical region, averaging over a waveband a few tens of nanometres wide is then sufficient to reduce the net polarization to one per cent of its initial value.

If a retarder is left with a rough surface and immersed in a liquid whose refractive index is the average refractive index of the retarder, then a beam of light will be undisturbed by the roughness of the surface because the hollows will be filled in by the liquid. However, the retarder will vary, on the scale of its roughness, in its effect. Thus the polarization of the emerging beam will vary on the same scale, and a suitable choice for the parameters of the system can lead to an average polarization of zero over the whole beam.

The Mueller matrix of an ideal depolarizer is simply

$$
\begin{bmatrix}
1 & 0 & 0 & 0 \\
0 & 0 & 0 & 0 \\
0 & 0 & 0 & 0 \\
0 & 0 & 0 & 0
\end{bmatrix}. \tag{5.2.31}
$$

Polarimeters

A polarimeter is an instrument that measures the state of polarization, or some aspect of the state of polarization, of a beam of radiation. Ideally the values of all four Stokes' parameters should be determinable, together with their variations with time, space and wavelength. In practice this is rarely possible, at least for astronomical sources. Most of the time only the degree of linear polarization and its direction are found. Astronomical polarimeters normally use CCDs but photographic or photoelectric detectors have been used in the past and may still be encountered.

Photographic polarimeters

At its simplest, photographic polarimetry is just photographic photometry (section 3.2) with the addition of a polarizer into the light path. Polaroid film is most frequently used as the polarizer. Several separate exposures are made through it with the polarizer rotated relative to the image plane by some fixed angle each time. The errors of the photographic photometric method are usually 0.01 to 0.05 magnitudes or more. This technique is therefore of little use for most stellar polarimetry, since the degree of polarization to be measured is normally 1% or less. It is useful, however, for highly polarized extended objects such as the Crab nebula.

The photographic method can be improved considerably by the use of a double-beam polarizer such as the Wollaston prism. Two images are produced for each star in light that is mutually orthogonally linearly polarized. Again, several exposures with a rotation of the polarizer are needed. Multicolour observations can be made by using two Wollaston prisms with an interposed optically active element (figure 5.2.12). The degree of rotation introduced by the latter is generally wavelength dependent. Hence four images result in which one pair has a similar colour balance and mutually orthogonal polarizations, while the second pair are also orthogonally polarized but have a different colour balance from the first pair.

Treanor has devised a photographic method that requires only a single exposure. A polarizer and a slightly inclined plate of glass are rotated in front

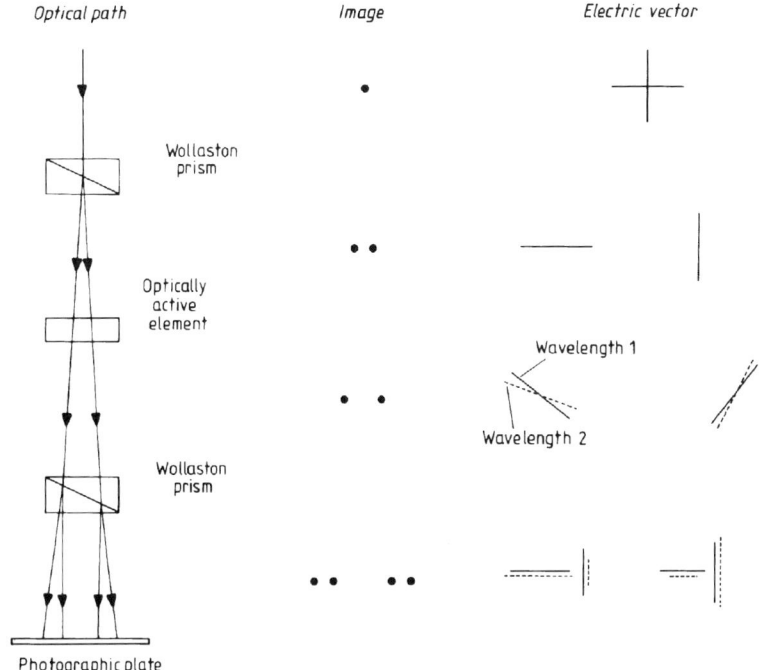

Figure 5.2.12. Öhman's multicolour polarimeter.

of the emulsion. The inclined plate displaces the instantaneous image by a small amount, and its rotation spreads this out into a ring. The polarizer then modulates the intensity around the ring if the original source is polarized. By adding a neutral density filter to cover a segment of the rotor, a photographic intensity calibration can also be included in the system.

Circular polarization can be studied photographically using large flat plastic quarter-wave plates. These are placed before the polarizer and convert the circular (or elliptical) radiation into linearly polarized radiation. The detection of this then proceeds along the lines just discussed. The detection threshold is currently about 10% polarization for significantly accurate measurements.

Photoelectric polarimeters

Most of these devices bear a distinct resemblance to photoelectric photo-meters (section 3.2), and indeed many polarimeters can also function as photometers. The major differences, apart from the components that are needed to measure the polarization, arise from the necessity of reducing or eliminating the instrumentally induced polarization. The instrumental polar-ization originates primarily in inclined planar reflections or in the detector.

Thus Newtonian and Coudé telescopes cannot be used for polarimetry because of their inclined subsidiary mirrors. Inclined mirrors must also be avoided within the polarimeter. Cassegrain and other similar designs should not in principle introduce any net polarization because of their symmetry about the optical axis. However, problems may arise even with these telescopes if mechanical flexure or poor adjustments move the mirrors and/or the polarimeter from the line of symmetry.

Many detectors, especially photomultipliers (section 1.1), are sensitive to polarization, and furthermore this sensitivity may vary over the surface of the detector. If the detector is not shielded from the Earth's magnetic field, the sensitivity may vary with the position of the telescope as well. In addition, although an ideal Cassegrain system will produce no net polarization, this sensitivity to polarization arises because rays on opposite sides of the optical axis are polarized in opposite but equal ways. The detector non-uniformity may then result in incomplete cancellation of the telescope polarization. Thus a Fabry lens and a depolarizer immediately before the detector are almost always essential components of a polarimeter.

Alternatively the instrumental polarization can be reversed using a Bowen compensator. This comprises two retarders with delays of about $\lambda/8$ that may be rotated with respect to the light beam and to each other until their polarization effects are equal and opposite to those of the instrument. Small amounts of instrumental polarization can be corrected by the insertion of a tilted glass plate into the light beam to introduce opposing polarization. In practice an unpolarized source is observed, and the plate tilted and rotated until the observed polarization is minimized or eliminated.

There are many detailed designs of polarimeter, but we may group them into single- and double-channel devices, with the former further subdivided into discrete and continuous systems. The single-channel discrete polarimeter is the basis of most of the other designs, and its main components are shown in figure 5.2.13, although the order is not critical for some of them. Since stellar polarizations are usually of the order of 1% or less, it is also usually necessary to incorporate a chopper and use a phase-sensitive detector plus integration in order to obtain sufficiently accurate measurements. Switching between the sky and background is also vital since the background radiation is very likely to be polarized. The polarizer is rotated either in steps or continuously, but at a rate which is slow compared with the chopping and integration times. The output for a linearly polarized source is then modulated with the position of the polarizer.

The continuous single-beam systems use a rapidly rotating polarizer. This gives an oscillating output for a linearly polarized source, and this may be fed directly into a computer or into a phase-sensitive detector that is driven at the same frequency. Alternatively a quarter-wave plate and a photoelastic modulator can be used. The quarter-wave plate produces circularly polarized light, and this is then converted back into linearly polarized

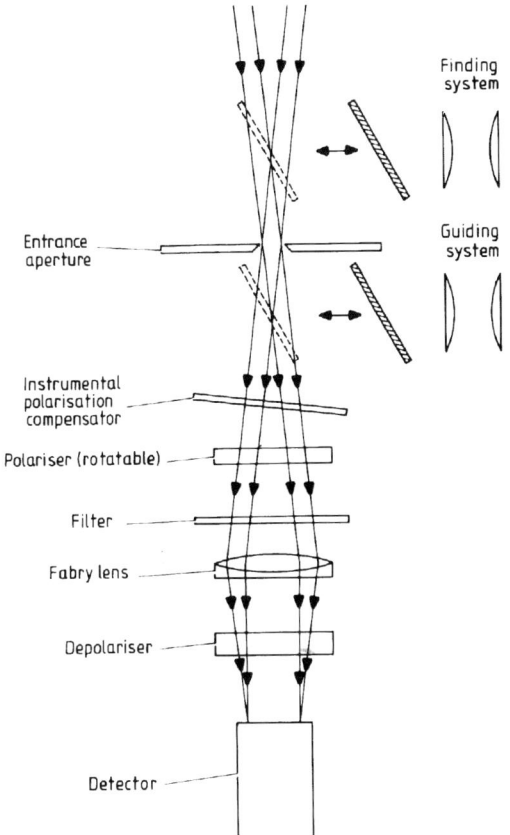

Entrance
aperture

Instrumental
polarisation
compensator

Polariser (rotatable)

Filter

Fabry lens

Depolariser

Detector

Finding
system

Guiding
system

Figure 5.2.13. The basic polarimeter: schematic design and optical paths.

light by the photoelastic modulator but with its plane of polarization rotated by 90° between alternate peaks of its cycle. A polarizer that is stationary will then introduce an intensity variation with time that may be detected and analysed as before. The output of the phase-sensitive detector in systems such as these is directly proportional to the degree of polarization, with unpolarized sources giving zero output. They may also be used as photo-meters to determine the total intensity of the source, by inserting an additional fixed polarizer before the entrance aperture, to give in effect a source with 100% polarization and half the intensity of the actual source (provided that the intrinsic polarization is small). An early polarimeter system that is related to these is due to Dolfus. It uses a quarter-wave plate to convert the linearly polarized component of the radiation into circularly polarized light. A rotating disc that contains alternate sectors made from zero- and half-wave plates then chops the beam. The circular polarization alternates in its direction but remains circularly polarized. Another quarter-wave plate converts the circularly polarized light back to linearly polarized light, but the plane of its polarization switches through 90° at the chopping frequency. Finally a polarizer is used to eliminate one of the linearly polarized components, so that the detector produces a square wave output.

The continuous single-beam polarimeter determines both linear polarization components quasi-simultaneously, but it still requires a separate observation to acquire the total intensity of the source. In double-beam devices, the total intensity is found with each observation, but several observations are required to determine the polarization. These instruments are essentially similar to that shown in figure 5.2.13, but a double-beam polarizer such as a Wollaston prism is used in place of the single-beam polarizer shown there. Two detectors then observe both beams simultaneously. The polarizer is again rotated either in steps or continuously but slowly in order to measure the polarization. Alternatively a half-wave plate can be rotated above a fixed Wollaston prism. The effects of scintillation and other atmospheric changes are much reduced by the use of dual beam instruments.

The current state of the art is represented by a polarimeter designed by James Hough at the University of Hertfordshire to detect extra-solar planets. It will use a 20 kHz photoelastic modulator and a Wollaston prism to detect the ordinary and extraordinary rays simultaneously using avalanche photodiodes and have a separate sky channel. The whole instrument will be rotated through 45° to measure the Q and U Stokes' parameters. Its sensitivity should be 1 part in 10^6 or better, and on an 8 m telescope it should be able to detect the polarization induced into the light from a star by a Jupiter-sized planet for stars brighter than 6.5^m.

CCD and other array detector polarimeters

Again many designs are possible, but the dual-beam polarizer using a rotating wave plate and a Wollaston prism is in widespread use. For extended sources, the polarized images are detected directly by a CCD or infrared detector array. To avoid overlap, a mask is used at the focal plane of the telescope comprising linear obscuring bars and gaps of equal width (sometimes known as a comb dekker). The image from one of the beams then occupies the space occupied by the 'image' of the mask. Clearly two sets of observations, shifted by the width of an obscuring bar, are then needed to show the complete extended object. In the ultraviolet little work has been undertaken to date. The solar maximum mission satellite, however, did include an ultraviolet spectrometer into which a magnesium fluoride retarder could be inserted, so that some measurements of the polarization in strong spectrum lines could be made. The Wisconsin Ultra Violet Photo-Polarimeter Experiment (WUPPE) was flown on the space shuttle in 1993 and the Faint Object Camera (FOC) on board the Hubble space telescope can be used for polarimetry in the ultraviolet.

A few polarimeter designs have been produced, particularly for infrared work, in which a dichroic mirror is used. These are mirrors that reflect at one wavelength and are transparent at another. They are not related to the other form of dichroism that was discussed earlier in this section. A thinly plated gold mirror reflects infrared radiation but transmits visible radiation, for example. Continuous guiding on the object being studied is therefore possible, providing that it is emitting visible radiation, through the use of such a mirror in the polarimeter. Unfortunately the reflection introduces instrumental polarization, but this can be reduced by incorporating a second 90° reflection whose plane is orthogonal to that of the first into the system.

Spectropolarimetry

Spectropolarimetry that provides information on the variation of polarization with wavelength (see also Öhman's multicolour polarimeter, above) can be realized by several methods. Almost any spectroscope (section 4.2) suited for long slit spectroscopy can be adapted to spectropolarimetry. A quarter- or half-wave plate and a block of calcite are placed before the slit. The calcite is oriented so that the ordinary and extraordinary images lie along the length of the slit. Successive exposures are then obtained with the wave plate rotated between each exposure. For linear spectropolarimetry four positions separated by 22.5° are needed for the half-wave plate, while for circular spectropolarimetry just two orthogonal positions are required for the quarter-wave plate.

Data reduction and analysis

The output of a polarimeter is usually in the form of a series of intensity measurements for varying angles of the polarizer. These must be corrected for instrumental polarization if this has not already been removed or has only been incompletely removed by the use of an inclined plate. The atmospheric contribution to the polarization must then be removed by comparison of the observations of the star and its background. Other photometric corrections, such as for atmospheric extinction (section 3.2) must also be made, especially if the observations at differing orientations of the polarizer are made at significantly different times so that the position of the source in the sky has changed.

The normalized Stokes' parameters are then obtained from

$$\frac{Q}{I} = \frac{I(0) - I(90)}{I(0) + I(90)} \tag{5.2.32}$$

and

$$\frac{U}{I} = \frac{I(45) - I(135)}{I(45) + I(135)} \tag{5.2.33}$$

where $I(\theta)$ is the intensity at a position angle θ for the polarizer. Since for most astronomical sources the elliptically polarized component of the radiation is 1% or less of the linearly polarized component, the degree of linear polarization, π_{L}, can then be found via equation (5.2.13), or more simply from equation (5.2.15) if sufficient observations exist to determine $I_{\max}$ and $I_{\min}$ with adequate precision. The position angle of the polarization may be

Figure 5.2.14. Flowchart of polarimetric analysis.

found from equation (5.2.16), or again directly from the observations if these are sufficiently closely spaced. The fourth Stokes' parameter, V, is rarely determined in astronomical polarimetry. However, the addition of a quarter-wave plate to the optical train of the polarimeter is sufficient to enable it to be found if required (section 5.4 for example). The reason for the lack of interest is, as already remarked, the very low levels of circular polarization that are usually found. The first Stokes' parameter, I, is simply the intensity of the source, and so if it is not measured directly as a part of the experimental procedure, can simply be obtained from

$$I = I(\theta) + I(\theta + 90°). \tag{5.2.34}$$

A flowchart to illustrate the steps in an observation and analysis sequence to determine all four Stokes' parameters is shown in figure 5.2.14.

Exercises

5.2.1 Obtain equation (5.2.6) from equations (5.2.1) and (5.2.2).

5.2.2 Show, using the Mueller calculus, that the effect of two ideal half-wave plates upon a beam of radiation of any form of polarization is zero.

5.3 Solar studies

Introduction

The Sun is a reasonably typical star and as such it can be studied by many of the techniques discussed elsewhere in this book. In one major way, however, the Sun is quite different from all other stars, and that is that it may easily be studied using an angular resolution smaller than its diameter. This, combined with the vastly greater amount of energy that is available compared with that from any other star, has led to the development of many specialized instruments and observing techniques. Many of these have already been mentioned or discussed in some detail in other sections. The main such techniques include: solar cosmic ray and neutron detection (section 1.4), neutrino detection (section 1.5), radio observations (sections 1.2 and 2.5), x-ray and gamma ray observations (section 1.3), adaptive optics (section 2.5), radar observations (section 2.8), magnetometry (section 5.4) and spectroscopy and prominence spectroscopes (section 4.2). The other remaining important specialist instrumentation and techniques for solar work are covered in this section.

The warning on the next page may be quantified by the European Community directive, EN 167:

For safe viewing of the Sun with the unaided eye, filters should have a minimum size of 35 mm × 115 mm so that both eyes are covered and have

---

WARNING

Studying the Sun can be dangerous. Permanent eye damage or even blindness can easily occur through looking directly at the Sun with the naked eye. Smoked glass, exposed film, CDs, space blankets, aluminized helium balloons, crossed polaroid filters, sunglasses, etc. are not safe; *ONLY* filters made and sold specifically for viewing the Sun should ever be used.

Herschel wedges (see below) are no longer recommended except if used on very small telescopes, Also, care must be taken to ensure that the rejected solar radiation does not cause burns or start fires.

Eyepiece projection is safe from the point of view of personal injury, but can damage the eyepiece and/or the telescope, especially with the shorter focal ratio instruments.

This warning should be carefully noted by all intending solar observers, and passed on to inexperienced persons with whom the observer may be working. In particular, the dangers of solar observing should be made quite clear to any groups of laymen touring the observatory, because the temptation to try looking at the Sun for themselves is very great, especially when there is an eclipse.

---

maximum transmissions of

>0.003%—ultraviolet (280–380 nm)
>0.0032%—visible (380–780 nm)
>0.027%—near infrared (780–1400 nm).

Although these maxima are safe with regard to damage to the eye, the Sun may still be uncomfortably bright, and a transmittance of 0.0003% may found to be better. Since the surface brightness of the Sun is not increased by the use of a telescope (equation (1.1.68) *et seq.*), the same limits are safe for telescopic viewing, although the larger angular size of the image may increase the total energy entering the eye to the point of discomfort, so filters should generally be denser than this directive suggests. A transmission of 0.003% corresponds to an optical density for the filter of 4.5. Full aperture filters (see below) sold for solar observing generally have optical densities of 5 to 6.

Solar telescopes

Almost any telescope may be used to image the Sun. While specialized telescopes are often built for professional work (see later discussion), small conventional telescopes are frequently used by amateurs or for teaching or training purposes. Since it is dangerous to look at the Sun through an unadapted telescope (see the warning above), special methods must be used. By far the best approach is to use a full aperture filter sold by a reputable supplier specifically for solar observing (see below), but eyepiece projection may also be possible. In this the image is projected on to a sheet of white card held behind the eyepiece. This method may be preferred when demonstrating to a group of people, since all can see the image at the same time, and they are looking away from the Sun itself in order to see its image. However, telescopes such as Schmidt–Cassegrains and Maksutovs have short focal length primary mirrors, and these can concentrate the light internally sufficiently to damage the instrument. The use of a telescope for solar projection may well invalidate the manufacturer's guarantee, and the manufacturer should always be consulted on this in advance of using the telescope for solar work. Even for other telescope designs, the heat passing through the eyepiece may be sufficient to damage it, especially with the more expensive multi-component types. The Herschel wedge, in the past a common telescope accessory, is no longer recommended (see below). Filters placed at the eyepiece end of the telescope, even if of adequate optical density, should never be used. They can become very hot and shatter, and the unfiltered image within the telescope can cause damage to the instrument.

The Herschel wedge or solar diagonal is shown in figure 5.3.1. It is a thin prism with unsilvered faces. The first face is inclined at 45° to the optical axis and thus reflects about 5% of the solar radiation into an eyepiece in the normal star diagonal position. The second face also reflects about 5% of

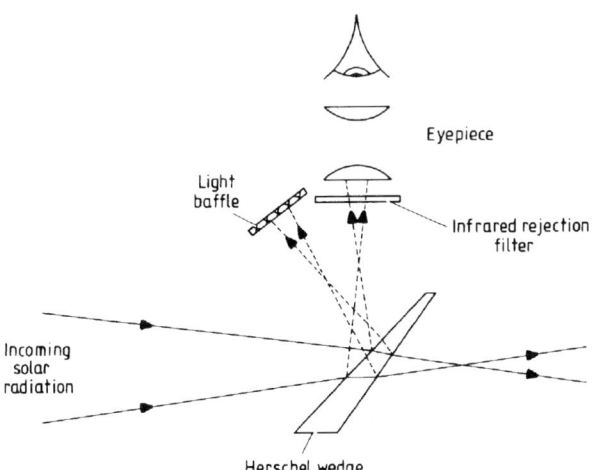

Figure 5.3.1. Optical paths in a Herschel wedge.

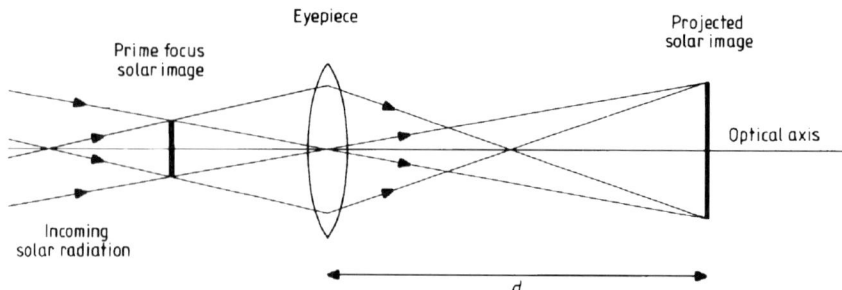

Figure 5.3.2. Projection of the solar image.

the radiation, but its inclination to the optical axis is not 45°, and so it is easy
to arrange for this radiation to be intercepted by a baffle before it reaches the
eyepiece. The remaining radiation passes through the prism and can be
absorbed in a heat trap, or more commonly (and dangerously) just
allowed to emerge as an intense beam of radiation at the rear end of the telescope. In order to reduce the solar intensity to less than the recommended
maximum, the wedge must be used on a very small telescope at a high
magnification—a 50 mm telescope would need a minimum magnification
of ×300 to be safe, for example. The device must also incorporate a separate
infrared filter. Since modern 'small' telescopes range from 100 to 300 mm or
more in aperture, they are much too large to use with a Herschel wedge. Even
if a larger telescope were to be stopped down to 50 mm (say), the Herschel
wedge still has the disadvantage of producing a real image inside the
telescope that may damage the telescope and/or invalidate its guarantee.
The recommendation now, therefore, is that Herschel wedges should not
be used for solar observing. If you have a Herschel wedge already, then it
would be best to get the front surface of the wedge aluminized and then
use it as a star diagonal.

   The principle of eyepiece projection is shown in figure 5.3.2, and it is the
same as the eyepiece projection method of imaging (section 2.2). The eyepiece
is moved outwards from its normal position for visual use, and a real image
of the Sun is obtained that may easily be projected on to a sheet of white
paper or cardboard etc. for viewing purposes. From equation (2.2.9), we
have the effective focal length (EFL) of the telescope

$$\mathrm{EFL} = \frac{Fd}{e} \qquad (5.3.1)$$

where F is the focal length of the objective, e is the focal length of the
eyepiece and d is the projection distance (see figure 5.3.2). The final image
size, D, for the Sun is then simply

$$D = \frac{0.0093Fd}{e}. \qquad (5.3.2)$$

There are two drawbacks to this method. The first is that a real image of the Sun is formed at the prime focus of the objective. If this is larger than the eyepiece acceptance area, or if this image of the Sun is intercepted by a part of the telescope's structure during acquisition, then structural damage may occur through the intense heat in the beam. This is especially a problem for modern short focal ratio Schmidt–Cassegrain and Maksutov telescopes. The second drawback is that the eyepiece lenses will absorb a small fraction of the solar energy, and this may stress the glass to the point at which it may fracture. Even if this does not occur, the heating will distort the lenses and reduce the quality of the final image. The technique is thus normally limited to telescopes with apertures of 75 mm or less. Larger instruments will need to be stopped down to about this area if they are to be used. As noted above, the use of a commercially produced telescope for solar observation by eyepiece projection may invalidate its guarantee, even if the telescope is not damaged, and the manufacturer should always be consulted before using the telescope in this fashion.

Full aperture filters are produced commercially for use on small telescopes. They cover the whole aperture of the telescope as their name suggests, and may be made sufficiently opaque for use on telescopes up to 0.5 m in diameter. They are the preferred approach to converting a telescope to solar observation. One of the earliest types to be produced, and which is still available today, consists of a very thin Mylar film that has been coated with a reflective layer of aluminium, stainless steel etc. The film is so thin that very little degradation of the image occurs; however, this also makes the film liable to damage, and its useful life when in regular use can be quite short. Such filters are produced with single or double layers of reflective coating. The double layer is to be preferred since any pin-holes in one layer are likely to be covered by the second layer and vice versa. Reflective coatings on optically flat glass substrates have become available recently, though they are usually somewhat more expensive than the coated plastic filters. Finally, the filter may simply be a thin sheet of absorbing plastic. (Do not be tempted to use any black plastic that may be lying around—it may not be sufficiently opaque in the ultraviolet or infrared. Always purchase a purpose-made filter from a reliable supplier.) An additional note of warning—do not forget to put a full-aperture filter on the guide telescope (or blank it off with an opaque screen). Without such a filter the guide telescope must NOT be used to align the telescope. Instead, the telescope's shadow should be circularized. This is usually sufficient to bring the solar image into the field of view of the (filtered) main telescope.

Larger instruments built specifically for solar observing come in several varieties. They are designed to try to overcome the major problem of the solar observer (turbulence within and outside the instrument) while taking advantage of the observational opportunity afforded by the plentiful supply of radiation. The most intractable of the problems is that of external turbulence. This arises from the differential heating of the observatory, its surroundings and the atmosphere that leads to the generation of convection currents. The effect reduces with height above the ground, and so one solution is to use a tower telescope. This involves a long focal length telescope that is fixed vertically with its objective supported on a tower at a height of several tens of metres. The solar radiation is fed into it using a coelostat (section 1.1) or heliostat. The latter uses a single plane mirror rather than the double mirror of the coelostat, but has the disadvantage that the field of view rotates as the instrument tracks the Sun. The 150 foot (46 m) tower telescope on Mount Wilson for example was constructed by George Ellery Hale in 1911 with a 0.3 m lens and continues in use to this day.

The other main approach to the turbulence problem (the two approaches are not necessarily mutually exclusive) is to try to reduce the extent of the turbulence. The planting of low-growing tree or bush cover of the ground is one method of doing this, and another is to surround the observatory by water. The Big Bear Solar Observatory is thus built on an artificial island in Big Bear Lake, California, and houses several solar telescopes including a 0.65 m reflector. Painting the telescope and observatory with titanium dioxide-based white paint will help to reduce heating since this reflects the solar incoming radiation, but allows the long-wave infrared radiation from the building to be radiated away. A gentle breeze also helps to clear away local turbulence and some tower telescopes such as the 0.45 m Dutch Open Solar Telescope on La Palma are completely exposed to the elements to facilitate the effect of the wind.

Within the telescope, the solar energy can create convection currents, especially from heating the objective. Sometimes these can be minimized by careful design of the instrument, or the telescope may be sealed and filled with helium whose refractive index is only 1.000036 compared with 1.000293 for air and whose high conductivity helps to keep the components cool and minimize convection. But the ultimate solution to them is to evacuate the optical system to a pressure of a few millibars, so that all the light paths after the objective, or a window in front of the objective, are effectively in a vacuum. Most modern solar telescopes are evacuated, such as the 0.7 m German Vacuum Tower Telescope (VTT) and the 0.9 m French THEMIS (Télescope Héliographique pour l'Etude du Magnétisme et des Instabilités Solaires), both on Tenerife. It is impractical, however, to evacuate larger instruments, such as the proposed 4 m Advanced Technology Solar Telescope (ATST) planned for 2010. The latter will use adaptive optics to overcome turbulence, and its large size is needed in order to achieve high angular ($0.1''$) and time (a few seconds) resolution observations of the Sun.

Optically solar telescopes are fairly conventional (section 1.1) except that very long focal lengths can be used to give large image scales since there is plenty of available light. It is generally undesirable to fold the light paths since the extra reflections will introduce extra scattering and distortion into the image. Thus the instruments are often very cumbersome, and many are fixed in position and use a coelostat to acquire the solar radiation.

Conventional large optical telescopes can be used for solar infrared observations if they are fitted with a suitable mask. The mask covers the whole of the aperture of the telescope and protects the instrument from the visible solar radiation. It also contains one or more apertures filled with infrared transmission filters. The resulting combination provides high-resolution infrared images without running the normal risk of thermal damage when a large telescope is pointed towards the Sun.

Smaller instruments may conveniently be mounted on a solar spar. This is an equatorial mounting that is driven so that it tracks the Sun, and which has a general-purpose mounting bracket in the place normally reserved for a telescope. Special equipment may then be attached as desired. Often several separate instruments will be mounted on the spar together and may be in use simultaneously.

Numerous spacecraft, including many manned missions, have carried instruments for solar observing. Among the more recent spacecraft whose primary mission was aimed at the Sun, we may pick out as examples the Solar Maximum Mission (SMM), Ulysses, Yokhoh and SOHO (SOlar and Heliospheric Observatory). SMM operated from 1980 to 1989 and observed the Sun from the ultraviolet through to gamma rays, as well as monitoring the total solar luminosity. Ulysses (1992 to date) has an orbit that takes it nearly over the solar poles enabling high solar latitude regions to be observed clearly for the first time. As well as monitoring solar x-rays, Ulysses also studies the solar wind. Yokhoh (1991–2002) was designed to detect high-energy radiation from flares and the corona, while SOHO (1995 to date) monitors solar oscillations (see below) and the solar wind. For the future, the STEREO mission plans to operate two spacecraft simultaneously, tens of millions of kilometres apart, so that three-dimensional images can be obtained of events such as coronal mass ejections.

Spectrohelioscope

This is a monochromator that is adapted to provide an image of the whole or a substantial part of the Sun. It operates as a normal spectroscope, but with a second slit at the focal plane of the spectrum that is aligned with the position of the desired wavelength in the spectrum (figure 5.3.3). The spectroscope's entrance slit is then oscillated so that it scans across the solar image. As the entrance slit moves, so also will the position of the spectrum. Hence the second slit must move in sympathy with the first if it is to continue to isolate the required wavelength. Some mechanical arrangements for ensuring that the slits remain in their correct mutual alignments are shown in figure 5.3.4. In the first of these the slit's movements are in antiphase with each other, and this is probably the commonest arrangement. In the second

Figure 5.3.3. Principle of the spectrohelioscope.

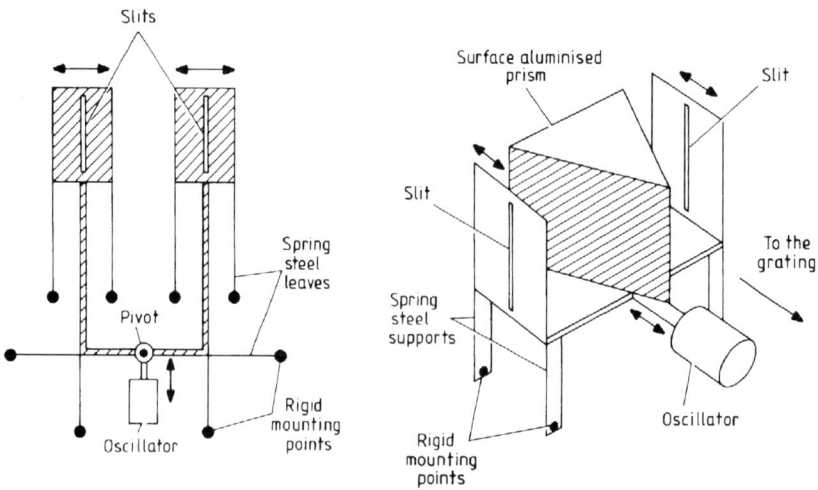

Figure 5.3.4. Mechanical arrangements for co-moving the two slits of a spectrohelioscope.

system the slits are in phase and the mechanical arrangements are far simpler. As an alternative to moving the slits, rotating prisms can be used to scan the beam over the grating and the image plane. If the frequency of the oscillation is higher than 15 to 20 Hz, then the monochromatic image may be viewed directly by eye. Alternatively, an image may be taken and a spectroheliogram produced.

Usually the wavelength that is isolated in a spectrohelioscope is chosen to be that of a strong absorption line, so that particular chromospheric features such as prominences, plages, mottling etc. are highlighted. In the visual region birefringent filters (see below) can be used in place of spectrohelioscopes, but for ultraviolet imaging and scanning, satellite-based spectrohelioscopes are used exclusively.

Narrow band filters

The spectrohelioscope produces an image of the Sun over a very narrow range of wavelengths. However, a similar result can be obtained by the use of a very narrow band filter in the optical path of a normal telescope. Since the bandwidth of the filter must lie in the region of 0.01 to 0.5 nm if the desired solar features are to be visible, normal dye filters or interference filters (section 4.1) are not suitable. Instead a filter based upon a birefringent material (section 5.2) has been developed by Bernard Lyot. It has various names—quartz monochromator, Lyot monochromator, or birefringent filter being among the commonest.

Its operational principle is based upon a slab of quartz or other birefrin-gent material that has been cut parallel to its optic axis. As we saw in section 5.2, the extraordinary ray will then travel more slowly than the ordinary ray and so the two rays will emerge from the slab with a phase difference (figure 5.3.5). The rays then pass through a sheet of polaroid film whose axis is midway between the directions of polarization of the ordinary and extraordinary rays. Only the components of each ray which lie along the polaroid's axis are transmitted by it. Thus the rays emerging from the polaroid film have parallel polarization directions and a constant phase difference and so they can mutually interfere. If the original electric vectors of the radiation, E_o and E_e, along the directions of the ordinary and extra-ordinary rays' polarizations respectively, are given at some point by

$$E_o = E_e = a\cos(2\pi\nu t) \qquad (5.3.3)$$

then after passage through the quartz, we will have at some point

$$E_o = a\cos(2\pi\nu t) \qquad (5.3.4)$$

and

$$E_e = a\cos(2\pi\nu t + \delta) \qquad (5.3.5)$$

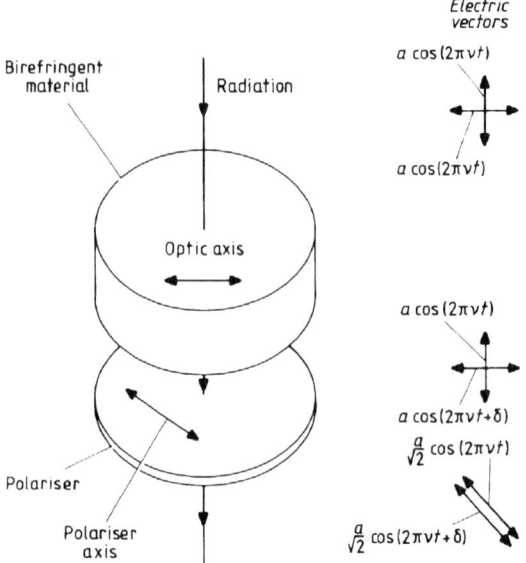

Figure 5.3.5. Basic unit of a birefringent monochromator.

where δ is the phase difference between the two rays. After passage through the polaroid, we then have

$$E_{45,o} = \frac{a}{\sqrt{2}} \cos(2\pi\nu t) \tag{5.3.6}$$

$$E_{45,e} = \frac{a}{\sqrt{2}} \cos(2\pi\nu t + \delta) \tag{5.3.7}$$

where $E_{45,x}$ is the component of the electric vector along the polaroid's axis for the xth component. Thus the total electric vector of the emerging radiation, E_{45}, will be given by

$$E_{45} = \frac{a}{\sqrt{2}} [\cos(2\pi\nu t) + \cos(2\pi\nu t + \delta)] \tag{5.3.8}$$

and so the emergent intensity of the radiation, I_{45}, is

$$I_{45} = a^2 \quad \text{for } \delta = 2n\pi \tag{5.3.9}$$

$$I_{45} = 0 \quad \text{for } \delta = (2n + 1)\pi \tag{5.3.10}$$

where n is an integer. Now

$$\delta = \frac{2\pi c \, \Delta t}{\lambda} \tag{5.3.11}$$

where Δt is the time delay introduced between the ordinary and extra-ordinary rays by the material. If v_o and v_e are the velocities of the ordinary and extraordinary rays in the material then

$$\Delta t = \frac{T}{v_o} - \frac{T}{v_e} \tag{5.3.12}$$

$$= T\left(\frac{v_e - v_o}{v_e v_o}\right) \tag{5.3.13}$$

$$= \frac{TJ}{c} \tag{5.3.14}$$

where J is the birefringence of the material (section 5.2) and T is the thickness of the material. Thus

$$\delta = \frac{\left[2\pi c\left(\frac{TJ}{c}\right)\right]}{\lambda} \tag{5.3.15}$$

$$= \frac{2\pi TJ}{\lambda}. \tag{5.3.16}$$

The emergent ray therefore reaches a maximum intensity at wavelengths, λ_{max}, given by

$$\lambda_{max} = \frac{TJ}{n} \tag{5.3.17}$$

and is zero at wavelengths, λ_{min}, given by

$$\lambda_{min} = \frac{2TJ}{2n + 1} \tag{5.3.18}$$

(see figure 5.3.6).

Now if we require the eventual filter to have a whole bandwidth of $\Delta\lambda$ centred on λ_c, then the parameters of the basic unit must be such that one of the maxima in figure 5.3.6 coincides with λ_c, and the width of one of the

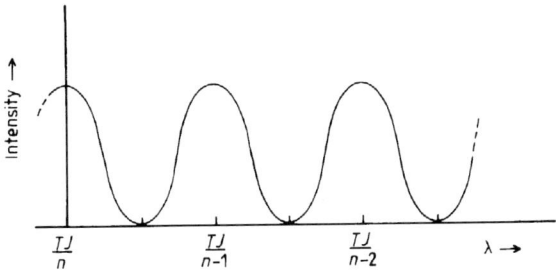

Figure 5.3.6. Spectrum of the emerging radiation from a basic unit of a birefringent monochromator.

fringes is $\Delta\lambda$; that is

$$\lambda_c = \frac{TJ}{n_c} \tag{5.3.19}$$

$$\Delta\lambda = TJ\left[\frac{1}{n_c - \frac{1}{2}} - \frac{1}{n_c + \frac{1}{2}}\right]. \tag{5.3.20}$$

Since selection of the material fixes J, and n_c is deviously related to T, the only truly free parameter of the system is T, and thus for a given filter we have

$$T = \frac{\lambda_c}{2J}\left[\frac{\lambda_c}{\Delta\lambda} + \left(\frac{\lambda_c^2}{\Delta\lambda^2} + 1\right)^{1/2}\right] \tag{5.3.21}$$

and

$$n_c = \frac{TJ}{\lambda_c}. \tag{5.3.22}$$

Normally, however,

$$\lambda_c \gg \Delta\lambda \tag{5.3.23}$$

and quartz is the birefringent material, for which in the visible

$$J = +0.0092 \tag{5.3.24}$$

so that

$$T \approx \frac{109\lambda_c^2}{\Delta\lambda}. \tag{5.3.25}$$

Thus for a quartz filter to isolate the Hα line at 656.2 nm with a bandwidth of 0.1 nm, the thickness of the quartz plate should be about 470 mm.

   Now from just one such basic unit, the emergent spectrum will contain many closely spaced fringes as shown in figure 5.3.5. But if we now combine it with a second basic unit oriented at 45° to the first, and whose central frequency is still λ_c, but whose bandwidth is $2\Delta\lambda$, then the final output will be suppressed at alternate maxima (figure 5.3.7). From equation (5.3.25) we see that the thickness required for the second unit, for it to behave in this way, is just half the thickness of the first unit. Further basic units may be added, with thicknesses 1/4, 1/8, 1/16 etc. that of the first, whose transmissions continue to be centred upon λ_c, but whose bandwidths are 4, 8, 16 etc. times that of the original unit. These continue to suppress additional unwanted maxima. With six such units, the final spectrum has only a few narrow maxima that are separated from λ by multiples of $32\Delta\lambda$ (figure 5.3.8). At this stage the last remaining unwanted maxima are sufficiently well separated from λ_c for them to be eliminated by conventional dye or interference filters, so that just the desired transmission curve remains. One further refinement to the filter is required and that is to ensure that the

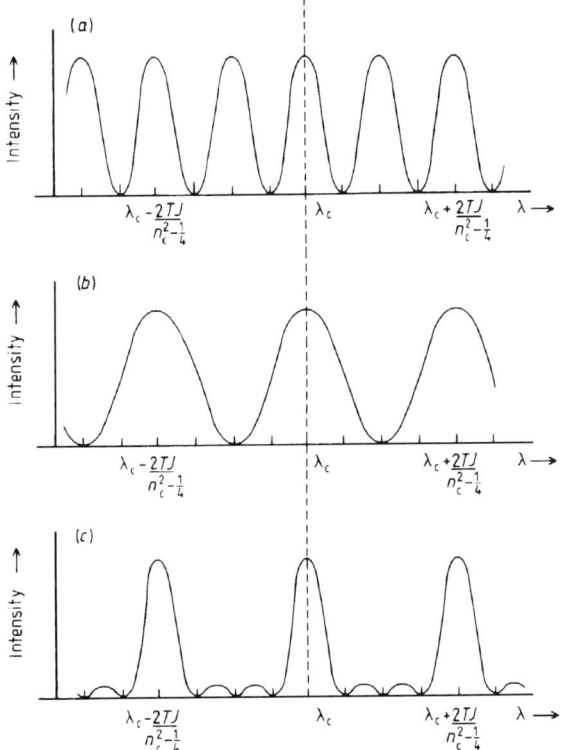

Figure 5.3.7. Transmission curves of birefringent filter basic units. (a) Basic unit of thickness T; (b) basic unit of thickness $T/2$; (c) combination of the units in (a) and (b).

initial intensities of the ordinary and extraordinary rays are equal, since this was assumed in obtaining equation (5.3.3). This is easily accomplished, however, by placing an additional sheet of polaroid before the first unit that has its transmission axis at $45°$ to the optic axis of that first unit. The complete unit is shown in figure 5.3.9. Neglecting absorption and other

Figure 5.3.8. Transmission curve of a birefringent filter comprising six basic units with a maximum thickness T.

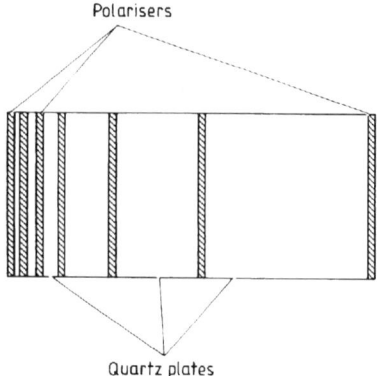

Figure 5.3.9. Six-element quartz birefringent filter.

losses, the peak transmitted intensity is half the incident intensity because of the use of the first polaroid. The whole filter, though, uses such a depth of quartz that its actual transmission is a few per cent. The properties of quartz are temperature dependent and so the whole unit must be enclosed and its temperature controlled to a fraction of a degree when it is in use. The temperature dependence allows the filter's central wavelength to be tuned slightly by changing the operating temperature. Much larger changes in wavelength may be encompassed by splitting each retarder into two opposed wedges. Their combined thicknesses can then be varied by displacing one set of wedges with respect to the other in a direction perpendicular to the axis of the filter. The wavelength may also be decreased by slightly tilting the filter to the direction of the incoming radiation, so increasing the effective thicknesses of the quartz plates. For the same reason, the beam of radiation must be collimated for its passage through the filter, or the bandwidth will be increased.

A closely related filter is due to Solč.[65] It uses only two polarizers placed before and after the retarders. All the retarders are of the same thickness, and they have their optic axes alternately oriented with respect to the axis of the first polarizer in clockwise and anticlockwise directions at some specified angle. The output consists of a series of transmission fringes whose spacing in wavelength terms increases with wavelength. Two such units of differing thicknesses can then be combined so that only at the desired wavelength do the transmission peaks coincide.

Another device for observing the Sun over a very narrow wavelength range is the Magneto-Optical Filter (MOF). This, however, can only operate at the wavelengths of strong absorption lines due to gases. It comprises two polarizers on either side of a gas cell. The polarizers are

[65] Pronounced 'Sholts'.

oriented orthogonally to each other if they are linear or are left- and right-handed if circular in nature. In either case, no light is transmitted through the system. A magnetic field is then applied to the gas cell, inducing the Zeeman effect. The Zeeman components of the lines produced by the gas are then linearly and/or circularly polarized (section 5.4) and so permit the partial transmission of light at their wavelengths through the whole system. The gases currently used in MOFs are vapours of sodium or potassium.

Relatively inexpensive Hα filters can be made as solid Fabry–Pérot etalons (section 4.1). A thin fused-silica spacer between two optically flat dielectric mirrors is used together with a blocking filter. Peak transmissions of several tens of percent and bandwidths of better than a tenth of a nanometre can be achieved in this way.

Coronagraph

This instrument enables observations of the corona to be made at times other than during solar eclipses. It does this by producing an artificial eclipse. The principle is very simple: an occulting disc at the prime focus of a telescope obscures the photospheric image while allowing that of the corona to pass by. The practice, however, is considerably more complex, since scattered and/or diffracted light etc. in the instrument and the atmosphere can still be several orders of magnitude brighter than the corona. Extreme precautions have therefore to be taken in the design and operation of the instrument in order to minimize this extraneous light.

The most critical of these precautions lies in the structure of the objective. A single simple lens objective is used in order to minimize the number of surfaces involved and it is formed from a glass blank that is as free from bubbles, striae and other imperfections as possible. The surfaces of the lens are polished with extreme care in order to eliminate all scratches and other surface markings. In use, they are kept dust free by being tightly sealed when not in operation, and by the addition of a very long tube lined with grease, in front of the objective to act as a dust trap.

The occulting disc is a polished metal cone or an inclined mirror so that the photospheric radiation may be safely reflected away to a separate light and heat trap. Diffraction from the edges of the objective is eliminated by imaging the objective with a Fabry lens after the occulting disc, and by using a stop that is slightly smaller than the image of the objective to remove the edge effects. Alternatively the objective can be apodized (section 2.5); its transparency decreases from its centre to its edge in a Gaussian fashion. This leads to full suppression of the diffraction halo although with some loss in resolution. A second occulting disc before the final objective removes the effects of multiple reflections within the first objective. The final image of the corona is produced by this second objective,

Figure 5.3.10. Schematic optical arrangement of a coronagraph.

and this is placed after the diffraction stop. The full system is shown in figure 5.3.10.

The atmospheric scattering can only be reduced by a suitable choice of observing site. Thus the early coronagraphs are to be found at high-altitude observatories, while more recently they have been mounted on spacecraft in order to eliminate the problem of the Earth's atmosphere entirely.

The use of a simple lens results in a chromatic prime focus image. A filter must therefore normally be added to the system. This is desirable in any case since the coronal spectrum is largely composed of emission lines super-imposed upon a diluted solar photospheric spectrum. Selection of a narrow band filter that is centred upon a strong coronal emission line there-fore improves considerably the contrast of the final image. White light or wideband imaging of the corona is only possible using Earth-based instru-ments on rare occasions, and it can only be attempted under absolutely optimum observing conditions. Satellite-borne instruments may be so used more routinely.

Improvements to the basic coronagraph may be justified for balloon or satellite-borne instruments, since the sky background is then less than the scattered light in a more normal device, and have taken two different forms. A reflecting objective can be used. This is formed from an uncoated off-axis parabola. Most of the light passes through the objective and is absorbed. Bubbles and striations in the glass are of much less importance since they cause scattering primarily in the forward direction. The mirror is uncoated since metallic films are sufficiently irregular to a cause a con-siderable amount of scattering. In other respects, the coronagraph then follows the layout of figure 5.3.10. The second approach to the improvement of coronagraphs is quite different and consists simply of producing the artificial eclipse outside the instrument rather than at its prime focus. An occulting disc is placed well in front of the first objective of an otherwise fairly conventional coronagraph. The disc must be large enough to ensure that the first objective lies entirely within its umbral shadow. The inner parts of the corona will therefore be badly affected by vignetting. However, this is of little importance since these are the brightest portions of the corona, and it may even be advantageous since it will reduce the dynamic range that must be covered by the detector. A simple disc produces an image with a bright central spot due to diffraction, but this can be eliminated by using a disc whose edge is formed into a zig-zag pattern of sharp teeth, or by the use of multiple occulting discs. By such means the instrumentally scattered light can be reduced to 10^{-4} of that of a basic coronagraph.

The final image in a coronagraph may be imaged directly, but more commonly is fed to a spectroscope, photometer or other ancillary instrument. From the Earth the corona may normally only be detected out to about one solar radius, but satellite-based coronagraphs have been used successfully out to six solar radii or more.

Devices similar to the coronagraph are also sometimes carried on spacecraft so that they may observe the atmospheres of planets, while shielding the detector from the radiation from the planet's surface. The contrast is generally far smaller than in the solar case so that such planetary 'coronagraphs' can be far simpler in design.

Terrestrial and spacecraft-borne stellar coronagraphs have also recently been constructed. These are designed to allow the detection of faint objects close to much brighter objects, such as stellar companions and accretion discs, by eliminating the bright stellar image in a similar manner to the removal of the solar image in a solar coronagraph. With the addition of suitable apodization stops (section 4.1) and adaptive optics (section 1.1) it may even be possible to detect directly extra-solar planets. The Coronagraphic Imager with Adaptive Optics (CIAO) for the 8 m Subaru telescope, for example, is a near infrared coronagraph using a 1024×1024 indium antimonide array detector. It reduces the intensity by a factor of 30 to 40 within a few tenths of an arc second of the central object. It can also be used for direct imaging, spectroscopy and polarimetry. The nulling interferometer (section 2.5) similarly reduces the intensity of the central object.

Pyrheliometer

This is an instrument intended to measure the total flux of solar radiation at all wavelengths. In practice, current devices measure the energy from the microwave to the soft x-ray region. Modern pyrheliometers are active cavity radiometers. The radiation is absorbed within a conical cavity. The cavity is held in contact with a heat sink and maintained at a temperature about 1 K higher than the heat sink by a small heater. The difference between the power used by the heater when the cavity is exposed to the Sun and that when a shutter closes it off provides the measure of the solar energy.

Solar oscillations

Whole-body vibrations of the Sun reveal much about its inner structure. They are small-scale effects and their study requires very precise measurements of the velocities of parts of the solar surface. The resonance scattering spectrometer originated by the solar group at the University of Birmingham is capable of detecting motions of a few tens of millimetres per second. It operates by passing the 770 nm potassium line from the Sun through a container of heated potassium vapour. A magnetic field is directed through the vapour and the solar radiation is circularly polarized. Depending upon the direction of the circular polarization either the light from the long-wave wing (I_L) or the short-wave wing (I_S) of the line is resonantly scattered. By switching the direction of the circular polarization rapidly, the two intensities may be measured nearly simultaneously. The line of sight velocity is then given by

$$v = k \frac{I_S - I_L}{I_S + I_L} \qquad (5.3.26)$$

where k is a constant with a value around $3 \, \text{km s}^{-1}$.

BISON (Birmingham–Izaña Solar Oscillations Network) has six automated observatories around the world so that almost continuous observations of the Sun can be maintained. It measures the average radial velocity over the whole solar surface using resonance scattering spectrometers. The GONG project (Global Oscillation Network Group) likewise has six observatories, and measures radial velocities on angular scales down to $8''$, using Michelson interferometers to monitor the position of the nickel 676.8 nm spectrum line. For the SOHO spacecraft, a resonance scattering device based upon sodium D line detects motions to better than $1 \, \text{mm s}^{-1}$.

Other solar observing methods

Slitless spectroscopes (section 4.2) are of considerable importance in observing regions of the Sun whose spectra consist primarily of emission lines. They are simply spectroscopes in which the entrance aperture is large enough to accept a significant proportion of the solar image. The resulting spectrum is therefore a series of monochromatic images of this part of the Sun in the light of each of the emission lines. They have the advantage of greatly improved speed over a slit spectroscope combined with the obvious ability to obtain spectra simultaneously from different parts of the Sun. Their best known application is during solar eclipses, when the 'flash spectrum' of the chromosphere may be obtained in the few seconds immediately after the start of totality or just before its end. More recently, they have found widespread use as spacecraft-borne instrumentation for observing solar flares in the ultraviolet part of the spectrum.

One specialized satellite-borne instrument based upon slitless spectroscopes is the Ly α camera. The Lyman α line is a strong emission line. If the image of the whole disc of the Sun from an ultraviolet telescope enters a slitless spectroscope, then the resulting Ly α image may be isolated from the rest of the spectrum by a simple diaphragm, with very little contamination from other wavelengths. A second identical slitless spectroscope whose entrance aperture is this diaphragm, and whose dispersion is perpendicular to that of the first spectroscope, will then provide a stigmatic spectroheliogram at a wavelength of 121.6 nm.

In the radio region, solar observations tend to be undertaken by fairly conventional equipment although there are now a few dedicated solar radio telescopes. One exception to this, however, is the use of multiplexed receivers to provide a quasi-instantaneous radio spectrum of the Sun. This is of particular value for the study of solar radio bursts, since their emitted bandwidth may be quite narrow, and their wavelength drifts rapidly with time. These radio spectroscopes and the acousto-optical radio spectroscope were discussed more fully in section 1.2. The data from them are usually presented as a frequency/time plot, from whence the characteristic behaviour patterns of different types of solar bursts may easily be recognized.

This account of highly specialized instrumentation and techniques for solar observing could be extended almost indefinitely, and might encompass all the equipment designed for parallax observations, solar radius determinations, oblateness determinations, eclipse work, tree rings and ^{14}C determination and so on. However, at least in the author's opinion, these are becoming too specialized for inclusion in a general book like this, and so the reader is referred to more restricted texts and the scientific journals for further information upon them.

Exercise

5.3.1 Calculate the maximum and minimum thicknesses of the elements required for an Hα birefringent filter based upon calcite, if its whole bandwidth is to be 0.05 nm, and it is to be used in conjunction with an interference filter whose whole bandwidth is 3 nm. The birefringence of calcite is −0.172.

5.4 Magnetometry

Background

The measurement of astronomical magnetic fields is accomplished in two quite separate ways. The first is direct measurement by means of apparatus carried by spacecraft, while the second is indirect and is based upon the Zeeman effect of a magnetic field upon spectrum lines (or more strictly upon the inverse Zeeman effect, since it is usually applied to absorption lines).

Zeeman effect

The Zeeman effect describes the change in the structure of the emission lines in a spectrum when the emitting object is in a magnetic field. The simplest change arises for singlet lines, that is lines arising from transitions between levels with a multiplicity of one, or a total spin quantum number, M_s, of zero for each level. For these lines the effect is called the normal Zeeman effect. The line splits into two or three components depending on whether the line of sight is along or perpendicular to the magnetic field lines. An appreciation of the basis of the normal Zeeman effect may be obtained from a classical approach to the problem. If we imagine an electron in an orbit around an atom, then its motion may be resolved into three simple harmonic motions along the three coordinate axes. Each of these in turn we may imagine to be the sum of two equal but opposite circular motions (figure 5.4.1). If we now imagine a magnetic field applied along the z axis, then these various motions may be modified by it. The simple harmonic motion along the z axis will be unchanged since it lies along the direction of the magnetic field. The simple harmonic motions along the x and y axes, however, are cutting across the magnetic field and so will be altered. We may best see how their motion changes by considering the effect upon their circular components. When the magnetic field is applied, the radii of the circular motions remain unchanged, but their frequencies alter. If ν is the original frequency of the circular motion, and H is the magnetic field strength, then the new frequencies of the two resolved components ν^+

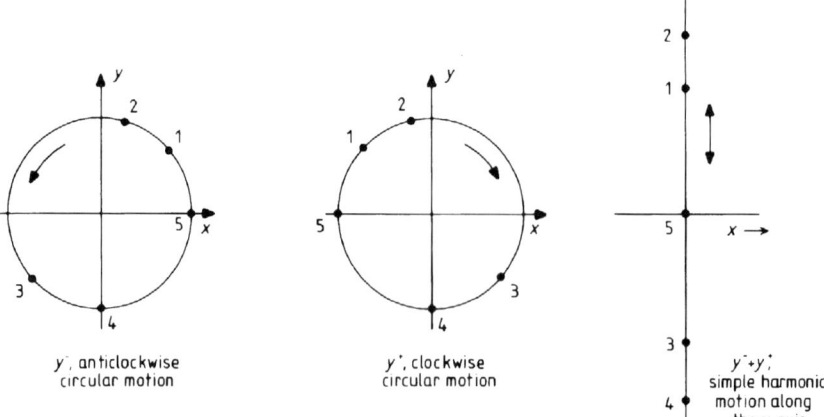

Figure 5.4.1. Resolution of simple harmonic motion along the y axis into two equal but opposite circular motions.

and ν^- are

$$\nu^+ = \nu + \Delta\nu \qquad (5.4.1)$$

$$\nu^- = \nu - \Delta\nu \qquad (5.4.2)$$

where

$$\Delta\nu = \frac{eH}{4\pi m_e c} \qquad (5.4.3)$$

$$= 1.40 \times 10^{10} H \quad \mathrm{Hz\,T^{-1}}. \qquad (5.4.4)$$

Thus we may combine the two higher frequency components arising from the x and y simple harmonic motions to give a single elliptical motion in the xy plane at a frequency of $\nu + \Delta\nu$. Similarly we may combine the lower frequency components to give another elliptical motion in the xy plane at a frequency of $\nu - \Delta\nu$. Thus the electron's motion may be resolved, when it is in the presence of a magnetic field along the z axis, into two elliptical motions in the xy plane, plus simple harmonic motion along the z axis (figure 5.4.2), the frequencies being $\nu + \Delta\nu$, $\nu - \Delta\nu$ and ν, respectively. Now if we imagine looking at such a system, then only those components that have some motion across the line of sight will be able to emit light towards the observer, since light propagates in a direction perpendicular to its electric vector. Hence looking along the z axis (i.e. along the

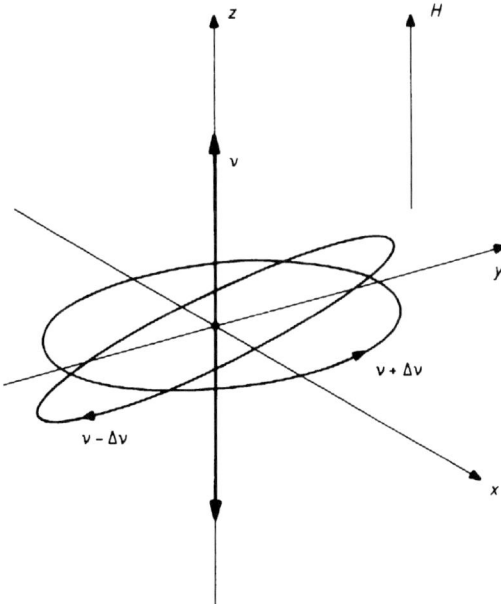

Figure 5.4.2. Components of electron orbital motion in the presence of a magnetic field.

Figure 5.4.3. The normal Zeeman effect.

magnetic field lines) only emission from the two elliptical components of the electron's motion will be visible. Since the final spectrum contains contributions from many atoms, these will average out to two circularly polarized emissions, shifted by $\Delta\nu$ from the normal frequency (figure 5.4.3), one with clockwise polarization, and the other with anticlockwise polarization. Lines of sight perpendicular to the magnetic field direction, that is within the xy plane, will in general view the two elliptical motions as two collinear simple harmonic motions, while the z axis motion will remain as simple harmonic motion orthogonal to the first two motions. Again the spectrum is the average of many atoms, and so will comprise three linearly polarized lines. The first of these is at the normal frequency of the line and is polarized parallel to the field direction. It arises from the z axis motion. The other two lines are polarized at right angles to the first and are shifted in frequency by $\Delta\nu$ (figure 5.4.3) from the normal position of the line. When observing the source along the line of the magnetic field, the two spectrum lines have equal intensities, while when observing perpendicular to the magnetic field, the central line has twice the intensity of either of the other components. Thus imagining the magnetic field progressively reducing then, as the components re-mix, an unpolarized line results, as one would expect. This pattern of behaviour for a spectrum line originating in a magnetic field is termed the normal Zeeman effect.

In astronomy, absorption lines, rather than emission lines, are the main area of interest, and the inverse Zeeman effect describes their behaviour. This, however, is precisely the same as the Zeeman effect except that emission processes are replaced by their inverse absorption processes. The above analysis may therefore be equally well applied to describe the behaviour of absorption lines. The one major difference from the emission line case is that the observed radiation remaining at the wavelength of one of the lines is preferentially polarized in the opposite sense to that of the Zeeman

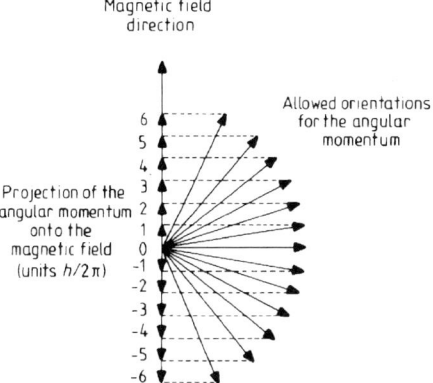

Figure 5.4.4. Space quantization for an atom with $J = 6$.

component, since the Zeeman component is being subtracted from unpolarized radiation.

If the spectrum line does not originate from a transition between singlet levels (i.e. $M_s \neq 0$), then the effect of a magnetic field is more complex. The resulting behaviour is known as the anomalous Zeeman effect, but this is something of a misnomer since it is anomalous only in the sense that it does not have a classical explanation. Quantum mechanics describes the effect completely. The orientation of an atom in a magnetic field is quantized. The angular momentum of the atom is given by

$$[J(J+1)] \frac{h}{2\pi} \tag{5.4.5}$$

where J is the inner quantum number, and its space quantization is such that the projection of the angular momentum on to the magnetic field direction must be an integer multiple of $h/2\pi$ when J is an integer, or a half-integer multiple of $h/2\pi$ when J is a half-integer. Thus there are always $(2J+1)$ possible quantized states (figure 5.4.4). Each state may be described by a total magnetic quantum number, M, which for a given level can take all integer values from $-J$ to $+J$ when J is an integer, or all half-integer values over the same range when J is a half-integer. In the absence of a magnetic field, electrons in these states all have the same energy (i.e. the states are degenerate), and the set of states forms a level. Once a magnetic field is present, however, electrons in different states have different energies, with the energy change from the normal energy of the level, ΔE, being given by

$$\Delta E = \frac{eh}{4\pi m_e c} MgH \tag{5.4.6}$$

where g is the Landé factor, given by

$$g = 1 + \frac{J(J+1) + M_s(M_s+1) + L(L+1)}{2J(J+1)} \tag{5.4.7}$$

where L is the total azimuthal quantum number. Thus the change in the frequency, $\Delta\nu$, for a transition to or from the state is

$$\Delta\nu = \frac{e}{4\pi m_e c} MgH \tag{5.4.8}$$

$$= 1.40 \times 10^{10} MgH \quad \mathrm{Hz\,T^{-1}}. \tag{5.4.9}$$

Now for transitions between such states we have the selection rule that M can change by 0 or ± 1 only. Thus we may understand the normal Zeeman effect in the quantum mechanical case simply from the allowed transitions (figure 5.4.5) plus the fact that the splitting of each level is by the same amount since the singlet levels

$$M_s = 0 \tag{5.4.10}$$

Figure 5.4.5. Quantum mechanical explanation of the normal Zeeman effect.

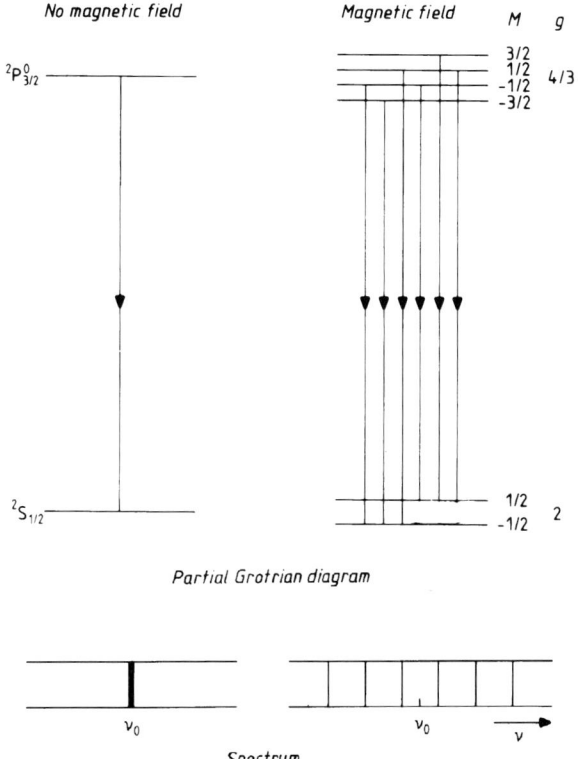

Partial Grotrian diagram

Spectrum

Figure 5.4.6. Quantum mechanical explanation of the anomalous Zeeman effect.

so that

$$J = L \tag{5.4.11}$$

and so

$$g = 1. \tag{5.4.12}$$

Each of the normal Zeeman components for the example shown in figure 5.4.5 is therefore triply degenerate, and only three lines result from the nine possible transitions. When M_s is not zero, g will in general be different for the two levels and the degeneracy will cease. All transitions will then produce separate lines, and hence we get the anomalous Zeeman effect (figure 5.4.6). Only one of many possible different patterns is shown in the figure; the details of any individual pattern will depend upon the individual properties of the levels involved.

As the magnetic field strength increases, the pattern changes almost back to that of the normal Zeeman effect. This change is known as the Paschen–Back effect. It arises as the magnetic field becomes strong enough to decouple L and M_s from each other. They then no longer couple together to form J, which then couples with the magnetic field, as just described, but couple separately and independently to the magnetic field. The pattern of spectrum lines is then that of the normal Zeeman effect, but with each component of the pattern formed from a narrow doublet, triplet etc, accordingly as the original transition was between doublet, triplet etc. levels. Field strengths of around 0.5 T or more are usually necessary for the complete development of the Paschen–Back effect.

At very strong magnetic field strengths ($>10^3$ T), the quadratic Zeeman effect will predominate. This displaces the spectrum lines to higher frequencies by an amount $\Delta\nu$, given by

$$\Delta\nu = \frac{\varepsilon_0 h^3}{8\pi^2 m_e^3 c^2 e^2 \mu_0} n^4 (1 + M^2) H^2 \qquad (4.4.13)$$

$$= (1.489 \times 10^4) n^4 (1 + M^2) H^2 \quad \text{Hz} \qquad (4.4.14)$$

where n is the principal quantum number.

Magnetometers

Among the direct measuring devices, the commonest type is the flux gate magnetometer illustrated in figure 5.4.7. Two magnetically soft cores have windings coiled around them as shown. Winding A is driven by an alternating current. Winding B has equal and opposite currents induced in its two coils by the alternating magnetic fields of the cores so long as there is no external magnetic field, and so there is no output. The presence of an external magnetic field introduces an imbalance into the currents, so that there is a net alternating current output. This may then easily be detected and calibrated in terms of the external field strength. Spacecraft usually carry three such magnetometers oriented orthogonally to each other so that all the components of the external magnetic field may be measured.

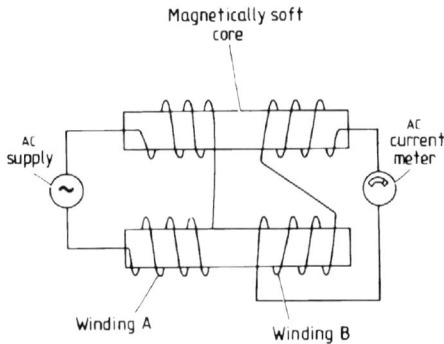

Figure 5.4.7. Flux gate magnetometer.

Another type of magnetometer that is often used on spacecraft is based upon atoms in a gas oscillating at their Larmor frequency. It is often used for calibrating flux gate magnetometers in orbit. The vector helium magnetometer operates by detecting the effect of a magnetic field upon the population of a metastable state of helium. A cell filled with helium is illuminated by 1.08 μm radiation that pumps the electrons into the metastable state. The efficiency of the optical pump is affected by the magnetic field, and the population of that state is monitored by observing the absorption of the radiation. A variable artificially generated magnetic field is swept through the cell until the external magnetic field is nullified. The strength and direction of the artificial field are then equal and opposite to the external field.

Any magnetometer on board a spacecraft usually has to be deployed at the end of a long boom after launch in order to remove it from the magnetic effects of the spacecraft itself. An alternative technique altogether is to look at the electron flux. Since electron paths are modified by the magnetic field, their distribution can be interpreted in terms of the local magnetic field strength. This enables the magnetic field to be detected with moderate accuracy over a large volume, in comparison with the direct methods that give very high accuracies, but only in the immediate vicinity of the spacecraft.

Most of the successful work in magnetometry based upon the Zeeman effect has been undertaken by Harold and Horace Babcock or uses instruments based upon their designs. For stars, the magnetic field strength along the line of sight may be determined via the longitudinal Zeeman effect. The procedure is made very much more difficult than in the laboratory by the widths of the spectrum lines. For a magnetic field of 1 T, the change in the frequency of the line is 1.4×10^{10} Hz (equation (5.4.4)), which for lines near 500 nm corresponds to a separation for the components in wavelength terms of only 0.02 nm. This is smaller than the normal linewidth for most stars. Thus even strong magnetic fields do not cause the spectrum lines to split into separate components, but merely to become somewhat broader than normal. The technique is saved by the opposite senses of circular polarization of the components that enables them to be separated from each other. Babcock's differential analyser therefore consists of a quarter-wave plate (section 5.2) followed by a doubly refracting calcite crystal. This is placed immediately before the entrance aperture of a high dispersion spectroscope (figure 5.4.8). The quarter-wave plate converts the two circularly polarized components into two mutually perpendicular linearly polarized components. The calcite is suitably oriented so that one of these components is the ordinary ray and the other the extraordinary ray. The components are therefore separated by the calcite into two beams. These are both accepted by the spectroscope slit and so two spectra are formed in close juxtaposition, with the lines of each slightly shifted with respect to the other due to the Zeeman splitting. This shift may then be measured and translated back to

Figure 5.4.8. Babcock's differential analyser.

give the longitudinal magnetic field intensity. A lower limit on the magnetic field intensity of about 0.02 T is detectable by this method, while the strongest stellar fields found have strengths of a few teslas. By convention the magnetic field is positive when it is directed towards the observer.

On the Sun much greater sensitivity is possible, and fields as weak as 10^{-5} T can be studied. George Ellery Hale first detected magnetic fields in sunspots in 1908, but most of the modern methods and apparatus are again due to the Babcocks. Their method relies upon the slight shift in the line position between the two circularly polarized components, which causes an exaggerated difference in their relative intensities in the wings of the lines (figure 5.4.9). The solar light is passed through a differential analyser

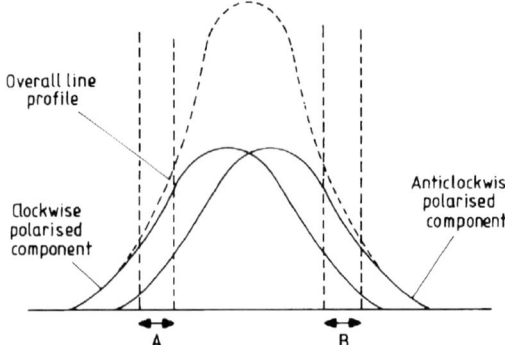

Figure 15.4.9. Longitudinal Zeeman components.

as before. This time, however, the quarter-wave plate is composed of ammonium dihydrogen phosphate, and this requires an electric potential of some 9 kV across it, in order to make it sufficiently birefringent to work. Furthermore, only one of the beams emerging from the differential analyser is fed to the spectroscope. Originally a pair of photomultipliers that accepted the spectral regions labelled A and B in figure 5.4.9 detected the line; now CCDs or other array detectors are used. Thus the wing intensities of, say, the clockwise-polarized component are detected when a positive voltage is applied to the quarter-wave plate, while those of the anticlockwise component are detected when a negative voltage is applied. The voltage is switched rapidly between the two states, and the outputs are detected in phase with the switching. Since all noise except that in phase with the switching is automatically eliminated, and the latter may be reduced by integration and/or the use of a phase-sensitive detector (section 3.2), the technique is very sensitive. The entrance aperture to the apparatus can be scanned across the solar disc so that a magnetogram can be built up with a typical resolution of a few seconds of arc and of a few microtesla. Specialized CCD detectors can be used which have every alternate row of pixels covered over. The accumulating charges in the pixels are switched between an exposed row and a covered row in phase with the switching between the polarized components. Thus, at the end of the exposure, alternate rows of pixels contain the intensities of the clockwise and anti-clockwise components. Similar instruments have been devised for stellar observations, but so far are not as sensitive as the previously mentioned system.

Several other types of magnetometer have been devised for solar work. For example, at the higher field strengths, Leighton's method provides an interesting pictorial representation of the magnetic patterns. Two spectroheliograms are obtained in opposite wings of a suitable absorption line, and through a differential analyser that is arranged so that it only allows the passage into the system of the stronger circularly polarized component in each case. A composite print is then made of both spectroheliograms, with one as a negative, and the other as a positive. Areas with magnetic fields less than about 2 mT are seen as grey, while areas with stronger magnetic fields show up light or dark according to their polarity. More modern devices use video recording but the basic technique remains unaltered. Vector-imaging magnetographs image the Sun over a very narrow wavelength range, and then rapidly step that image through a magnetically sensitive spectrum line. This enables both the direction and strength of the magnetic field to be determined over the selected region. Usually this region is just a small part of the whole solar disc because of the enormous amounts of data that such instruments can generate.

In the radio region, the Zeeman effect also affects lines such as the 1.42 GHz emission from hydrogen. Their Zeeman splitting is found in a similar manner to Babcock's solar method—by observing in the steepest part of the wings. Interstellar magnetic fields of about 10^{-9} T are detectable in this way.

The very strongest fields, up to 10^8 T, are expected to exist in white dwarfs and neutron stars. However, there is usually little spectral structure in such objects from which the fields might be detected. The quadratic Zeeman effect, however, may then be sufficient for circular polarization of the continuum to be distinguished at points where it is varying rapidly in intensity. Field strengths of upwards of 1000 T are found in this way.

A few devices have been constructed to detect electric fields via the Stark effect upon the Paschen hydrogen lines. They are basically similar to magnetometers and for solar studies can reach sensitivities of 500 V m^{-1}.

Data reduction and analysis

Once a flux gate magnetometer has been calibrated, there is little involved in the analysis of its results except to convert them to field strength and direction. However, it may be necessary to correct the readings for influence from the spacecraft, and for the effects of the solar wind and the solar or interplanetary magnetic fields. Measurements with accuracies as high as 10^{-11} T are possible when these corrections are reliably known.

With the first of Babcock's methods, in which two spectra are produced side by side, the spectra are simply measured as if for radial velocity. The longitudinal magnetic field strength can then be obtained via equation (5.4.9). The photoelectric method's data are rather more complex to analyse. No general method can be given since it will depend in detail upon the experimental apparatus and method and upon the properties of the individual line (or continuum) being observed.

With strong solar magnetic fields, such as those found in sunspots, information may be obtainable in addition to the longitudinal field strength. The line in such a case may be split into two or three components, and when these are viewed through a circular analyser (i.e. an analyser which passes only one circularly polarized component) the relative intensities of the three lines of the normal Zeeman effect are given by Seare's equations

$$I_v = \tfrac{1}{4}(1 \pm \cos\theta)^2 I \qquad (5.4.15)$$

$$I_c = \tfrac{1}{2}(\sin^2\theta)I \qquad (5.4.16)$$

$$I_r = \tfrac{1}{4}(1 \mp \cos\theta)^2 I \qquad (5.4.17)$$

where I_v, I_c and I_r are the intensities of the high frequency, central and low frequency components of the triplet, I is the total intensity of all the components and θ is the angle of the magnetic field's axis to the line of sight. Thus the direction of the magnetic field as well as its total strength may be found.

5.5   Computers and the World Wide Web

Introduction

Computers and their applications are essential to just about every aspect of astrophysics, and indeed to most of the rest of science and life in general. Many of the uses of computers, such as controlling telescopes, processing data etc. have already been mentioned within other sections of this book, but two promising applications deserve sections to themselves. These are the availability of digital sky surveys, and the related development of 'virtual' observatories.

A major concern of astronomy, dating back to Hipparchus and earlier, has always been the cataloguing or listing of objects in the sky with their positions and properties. The quantum changes that have occurred in the past few years are that many of these catalogues are now on open access through the World Wide Web, and personal computers affordable by individuals are now powerful enough to enable them to process the vast amounts of data present in the catalogues. The possibility of conducting real research is therefore open to anyone with the interest and some spare time, no longer just to professional scientists. Of course there are still problems, the main one being data links. Some of the surveys contain a terabyte (10^{12} bytes) of data, and using a 1 MHz broadband connection, this would take three to four months to download.[66] Data archives such as SIMBAD and MAST (see below) contain 10 Tbytes at the time of writing and are being added to at a rate of 5 or more Tbytes per year. The cumulative information content within astronomy is now hundreds of Tbytes, and one proposal alone (the Large Synoptic Survey Telescope) could be generating 5 Tbytes of data *per day* within a few years. Fortunately, for many purposes, it is not necessary to download a whole survey in order to work with it; a small subset is sufficient.

Digital sky surveys

We have already seen (section 5.1) that the Hipparcos, Tycho and other astrometric catalogues are available via the web. Other large surveys presented on the web include the USNO's A2.0 catalogue containing 526 million entries, the Hubble space telescope second Guide Star Catalogue (GSC2) containing 500 million objects and the Two-Micron All Sky Survey (2MASS) containing 300 million entries. The Sloan Digital Sky Survey (SDSS) and the Palomar Digital Sky Survey (DPOSS) are currently

---

[66] At the time of writing, it would also need some ten top of the range PCs in which to store it, but that is likely to cease to be a problem soon if the 18-month doubling of performance rule for computers continues for a few more years. The speed of data transmission seems likely to be a more intractable problem.

being conducted and will eventually have data on 100 million and 2000 million objects respectively. Not all digital sky surveys are the prerogative of the professional astronomer. TASS (The Amateur all-Sky Survey, http://www. tass-survey.org/) uses small cameras and CCD detectors to record some 200 Mbytes of data per night while searching for comets and variable stars. A list of the surveys and catalogues available on-line, that currently contains over 3600 entries, is given by VizieR at the CDS (see below).

Many of these surveys have been gathered together to be available through single sites[67] such as CDS[68] (Centre de Données astronomiques de Strasbourg) that operates SIMBAD (Set of Identifications, Measurements and Bibliography for Astronomical Data, http://simbad.u-strasbg.fr/Simbad), NSSDC (National Space Science Data Centre, http://nssdc.gsfc.nasa.gov) and MAST (Multimission Archive at the Space Telescope science institute, http://archive.stsci.edu). Anyone who is interested can easily look up the details on an individual object within these catalogues via the web. However, if more than a few tens of objects are needed, then this approach becomes cumbersome and time consuming, and larger projects therefore need more powerful search and processing software. Some of the sites provide this on an individual basis, but with little consistency in what is available and how to use it. Recently, therefore, virtual observatories have started to come on-stream providing much greater software and other support, though at present their use is limited to accredited astronomers.

Virtual observatories

Virtual observatories are interfaces whereby huge data sets such as the digital sky surveys, and others, such as collections of spectra from the IUE (International Ultra-violet Explorer) and the HST (Hubble Space Telescope) spacecraft, results from other spacecraft such as ROSAT, ISO, Chandra etc., interferometric data from radio and optical telescopes and so on, can be handled and 'mined' for new information. The virtual observatory is a set of data archives, software tools, hardware and staff that enables data in archives to be found and processed in many different ways. Among the types of function that a virtual observatory can provide there are

> Standardizing the formats of differing data sets
> Finding different observations of a given object, or a set of objects, and sorting through them
> Comparing and combining the information about objects available in various catalogues and other data bases

[67] The ADS (Astronomical Data Centre) has recently been closed, and its facilities made available through the CDS.
[68] The CDS also archives bibliographical information, enabling searches for publications relating to objects of interest to be conducted.

Comparing and combining observations taken at different times
Combining archive data with new data obtained from telescopes
Correlating data from different sources
Classifying objects
Performing statistical analyses
Measuring images
Image processing

while the types of scientific study that are possible include

Multi-wavelength studies
Large statistical studies
Searches for changing, moving, rare, unusual or new objects and types
of object
Combining data in new ways leading to unexpected discoveries.

Virtual observatories are just starting to come into being at the time of writing. AstroVirtel started operating in 2000. It is a virtual observatory operated by the ST-ECF (Space Telescope European Coordinating Facility) on behalf of the ESO (European Southern Observatory) and ESA (European Space Agency). It contains in excess of 10 Tbytes of data from the HST and the ESO ground-based telescopes and may be accessed at http://ecf.hq. eso.org/astrovirtel/. Proposals to use AstroVirtel have to be put forward in a similar way to applying to use any major telescope, and currently six proposals are funded each year. The Astrophysical Virtual Observatory (AVO, http://cdsweb.u-strasbg.fr/avo.htx) is currently being developed to succeed AstroVirtel, as are the UK's AstroGrid project (http://www.astrogrid.org/) and, in the USA, the National Virtual Observatory (NVO, http://www.us-vo.org/) with other possible projects in Australia, Germany and Japan. It seems likely that any fifth edition of this book will see virtual observatories as a well established and widely used tool for astrophysicists.

Appendix 1

North polar sequence

Star number	Position RA$_{2000}$			Dec$_{2000}$			Photographic magnitude
	h	m	s	°	′	″	
1	17	32	14	86	35	22	4.48
2	10	20	12	85	51	20	5.28
3	11	30	09	87	01	41	5.81
4	17	30	40	86	58	19	5.99
5	01	32	07	89	00	56	6.49
6	09	22	30	88	34	27	7.11
7	11	21	38	87	38	25	7.31
8	03	38	26	89	06	32	8.23
9	04	04	56	88	55	35	8.83
10	10	02	19	89	34	58	9.02
11	11	51	33	88	55	55	9.55
12	12	55	56	88	57	41	9.86
13	13	15	16	89	09	59	10.30
14	11	51	55	89	35	23	10.65
15	10	52	47	89	18	09	11.08
16	13	38	38	89	36	04	11.40
17	13	59	18	89	21	58	11.63
18	13	45	57	89	26	23	12.06
19	12	48	23	89	25	09	12.42
20	12	30	10	89	28	46	12.73
21	13	28	36	89	21	28	12.96
22	13	27	05	89	25	18	13.19
23	12	10	16	89	26	55	13.34
24	13	05	08	89	25	18	13.64
25	12	50	05	89	24	23	13.87

Star number	Position						Photographic magnitude
	RA$_{2000}$			Dec$_{2000}$			
	h	m	s	°	′	″	
26	12	35	37	89	26	49	14.33
27	12	02	46	89	30	20	14.69
28	11	51	17	89	19	58	15.11
29	11	48	31	89	21	17	15.66
30	11	54	21	89	19	57	16.04
31	12	22	44	89	24	34	16.22
32	11	47	51	89	27	07	16.62
33	12	44	01	89	26	12	16.96
34	12	20	20	89	24	31	17.25
35	11	59	34	89	26	23	17.50
36	12	03	00	89	26	44	17.75
37	12	17	13	89	29	03	17.94
38	12	20	21	89	27	26	18.31
39	12	10	35	89	28	47	18.67
40	12	00	45	89	27	37	19.03
41	12	08	02	89	27	51	19.28
42	11	54	30	89	24	39	19.68
43	12	12	48	89	28	01	20.22
44	12	11	33	89	27	57	20.44
45	12	13	21	89	28	24	20.74
46	11	55	42	89	26	16	21.10

Figure AI.1 gives a finder chart for the fainter stars in this sequence. The data for the table and the finder chart come from Leavitt H S *Annals of the Harvard College Observatory* **71**(3) 49–52.

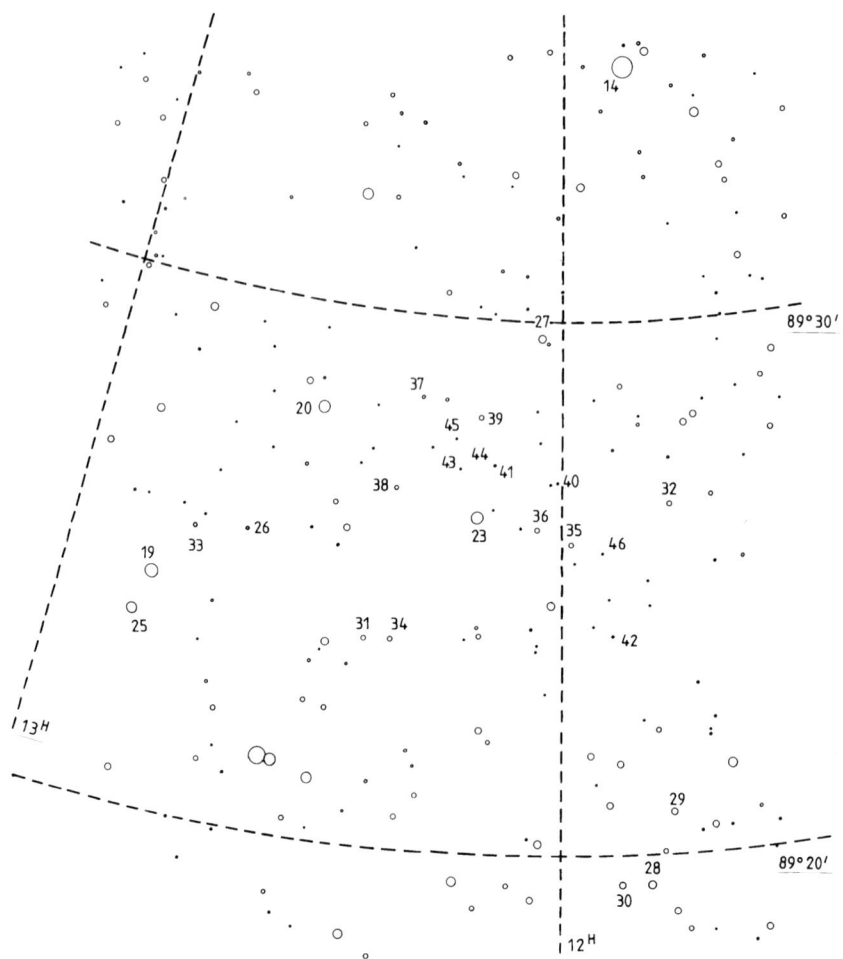

Figure AI.1. Finder chart for faint stars in the north polar sequence (epoch 2000.0).

Appendix 2

Julian date

The Julian date is the number of days elapsed since noon on 24 November 4714 BC (on the Gregorian Calendar) or since noon on 1 January 4713 BC (on the Julian Calendar). The modified Julian date is a variation of the Julian date that starts at midnight on 17 November 1858. The modified Julian date is thus the Julian date minus 2 400 000.5 days

Date (1 January at noon Gregorian reckoning)	Julian day number	Date (1 January at noon Gregorian reckoning)	Julian day number
2050	2 469 808.0	1600	2 305 448.0
2025	2 460 677.0	1200	2 159 351.0
2000	2 451 545.0	800	2 013 254.0
1975	2 442 414.0	400	1 867 157.0
1950	2 433 283.0	0	1 721 060.0
1925	2 424 152.0	400 BC	1 574 963.0
1900	2 415 021.0	800 BC	1 428 866.0
1875	2 405 890.0	1200 BC	1 282 769.0
1850	2 396 759.0	1600 BC	1 136 672.0
1825	2 387 628.0	2000 BC	990 575.0
1800	2 378 497.0	2400 BC	844 478.0
1775	2 369 366.0	2800 BC	698 381.0
1750	2 360 234.0	3200 BC	552 284.0
1725	2 351 103.0	3600 BC	406 187.0
1700	2 341 972.0	4000 BC	260 090.0
1675	2 332 841.0	4400 BC	113 993.0
1650	2 323 710.0	4714 BC (24 Nov)	0.0
1625	2 314 579.0		

For days subsequent to 1 January, the following number of days should be added to the Julian day number.

Date (noon)	Number of days (non-leap years)	Number of days (leap years)
1 Feb	31	31
1 Mar	59	60
1 Apr	90	91
1 May	120	121
1 June	151	152
1 July	181	182
1 Aug	212	213
1 Sept	243	244
1 Oct	273	274
1 Nov	304	305
1 Dec	334	335
1 Jan	365	366

Appendix 3

Answers to the exercises

1.1.1 69.21″; and 21.4″ (don't forget the Purkinje effect)

1.1.2 1.834 m and −3.387 m

1.1.3 0.31 mm

1.1.4 94 mm, 850× (for the exit pupil to be smaller than the eye's pupil)

1.1.5 0.014″

1.2.1 84 m along the length of its arms

1.4.1 $0.002\,\text{m}^3$

1.4.3 250 per second

1.5.2 1.6×10^{-23}

1.5.3 About 44 $^{37}_{18}\text{Ar}$ atoms

1.6.2 The calculated data are tabulated below:

Planet	Mass $(M_\text{p}/M_\text{Sun})$	Period (days)	L_G (W)
Mercury	1.6×10^{-7}	88	62
Venus	2.4×10^{-6}	225	600
Earth	3.0×10^{-6}	365	180
Mars	3.2×10^{-7}	687	0.25
Jupiter	9.6×10^{-4}	4 330	5 300
Saturn	2.9×10^{-4}	10 760	20
Uranus	4.3×10^{-5}	30 700	0.015
Neptune	5.3×10^{-5}	60 200	0.002
Pluto	2.5×10^{-6}	90 700	10^{-6}

1.6.3   400 pc

2.4.1   (a) 35 mm (b) 52 mm

2.4.2 3.1×10^{16} m (or 1 pc)

2.4.3 5×10^{17} m (or 16 pc)

2.8.1   28 mW, 72 km

3.1.1   +25.9, −20.6

3.1.2   3050 pc

3.1.3 $(U - B) = -0.84$, $(B - V) = -0.16$
$Q = -0.72$
$(B - V)_0 = -0.24$, $E_{B-V} = 0.08$
$E_{U-B} = 0.06$, $(U - B)_0 = -0.90$
Spectral type B3 V
Temperature 20 500 K
Distance (average) 230 pc
$U_0 = 2.82$, $B_0 = 3.72$
$V_0 = 3.96$, $M_V = -2.8$

4.1.1   −80.75 km/s

4.1.2   One prism, 0.79 nm mm^{-1}

4.2.1 $f' = 25$, $W_\lambda = 10^{-12}$ m, $\lambda = 625$ nm
$R = 6.3 \times 10^5$, $L = 0.42$ m, $D = 0.38$ m
$d\theta/d\lambda = 1.66 \times 10^6$ rad m^{-1}
$f_1 = 13.3$ m, $f_2 = 12.1$ m
$D_1 = 0.533$ m, $D_2 = 0.533$ m
$S = 22$ μm
Problems to be solved if possible in the next iteration through the exercise. Grating is too large: a normal upper limit on L is 0.1 m. Slit width is too small. The overall size of the instrument will lead to major thermal control problems.

4.2.2   +5.3 (taking $q = 0.002$ and $\alpha = 5 \times 10^{-6}$)

4.2.3   64.96°

5.1.2   0.26 mm

5.3.1   50.6 mm, 0.79 mm

Appendix 4

Acronyms

Previous editions of this book have tried to avoid the use of acronyms, believing that except to those people working in the particular field, they obscure understanding without actually saving much time or effort. Unfortunately few other astronomers seem to follow this belief, instead appearing to delight in inventing ever more tortuous acronyms to name their instruments, techniques etc. Thus TAURUS, obtained from 'Taylor Atherton Uariable resolution Radial Uelocity System', is an extreme but not unusual example of the desperate attempt to get a 'nice' acronym. This particular example is so obscure that it is almost always referred to in the literature as the 'TAURUS tunable filter' thus eliminating any advantage in terms of brevity that the use of the acronym might have conferred. Second and third level (i.e. an acronym that contains an acronym that contains yet another acronym) acronyms are now being encountered. The use of acronyms in astronomy is now so prevalent, and the distortions so often introduced to the actual names, that they have now reluctantly been included in this fourth edition. The acronyms are defined when first encountered or within their main section, but are also defined below, so that their meaning can be found after the original definition can no longer be found within the text.

A list of thousands of mostly astronomical acronyms can be found at http://www.ipa.rssi.ru/PAGE/DEPFUND/GEOzm/zm_acre.htm, while there are some 200 000 general acronyms listed at http://www.acronymfinder.com/. There are also numerous lists of acronyms used within special projects such as the HST, AAO, ESO etc. that may easily be found by a web search.

2dF	2-degree Field (AAT)
2MASS	2 Micron All Sky Survey
6dF	6-degree Field (UKST)
AAT	Anglo-Australian Telescope
AC	Astrographic Catalogue
AGASA	Akeno Giant Air Shower Array
AGK	Astronomische Gesellschaft Katalog

AIPS	Astronomical Image Processing System
AMANDA	Antarctic Muon And Neutrino Detector Array
ANTARES	Astronomy with a Neutrino Telescope and Abyssal Environmental RESearch
AOS	Acousto-Optical radio Spectrometer
APD	Avalanche Photo-Diode
APM	Automatic Plate Measuring Machine
ATST	Advanced Technology Solar Telescope
AVO	Astrophysical Virtual Observatory
BAT	Burst Alert Telescope
BIB	Blocked Impurity Band device
BISON	Birmingham-Izaña Solar Oscillations Network
BWFN	Beam Width at First Nulls
BWHP	Beam Width at Half-Power points
CANGAROO	Collaboration of Australia and Nippon for a GAmma Ray Observatory in the Outback
CCD	Charge-Coupled Device
CDS	Centre de Données astronomiques de Strasbourg
CELT	California Extremely Large Telescope
CFHT	Canada–France–Hawaii Telescope
CHARA	Centre for High Angular Resolution Astronomy
CIAO	Coronagraphic Imager with Adaptive Optics
CID	Charge Injection Device
CMT	Carlsberg Meridian Telescope
COAST	Cambridge Optical Aperture Synthesis Telescope
COBE	Cosmic Background Explorer Satellite
COSMOS	CoOrdinates, Sizes, Magnitudes, Orientations, and Shapes
CW	Continuous Wave (radar)
CZT	Cadmium–Zinc–Tellurium detectors
DENIS	DEep Near Infrared Survey
DPOSS	Digital Palomar Observatory Sky Survey
EBCCD	Electron Bombarded Charge-Coupled Device
EBS	Electron Bounce Silicon (TV camera)
EGRET	Energetic Gamma Ray Experiment Telescope
EISCAT	European Incoherent SCATter
ESA	European Space Agency
ESO	European Southern Observatory
EUV	Extreme Ultra-Violet (also XUV)
FAME	Full-sky Astrometric Mapping Explorer
FCRAO	Five College Radio Astronomy Observatory
FGS	Fine Guidance Sensors
FIR	Far InfraRed
FIRST	Far InfraRed and Sub-millimetre Telescope (now called Herschel)

FITS	Flexible Image Transport System
FK	Fundamental Katalog
FLAIR	Fibre-Linked Array-Image Reformatter
FOC	Faint Object Camera
GAIA	Global Astrometric Interferometer for Astrophysics
GI2T	Grand Interféromètre à 2 Télescopes
GIF	Graphic Interchange Format
GLAST	Gamma-ray Large Area Space Telescope
GMOS	Gemini Multi Object Spectroscopes
GNO	Gallium Neutrino Observatory
GODs	Giant Optical Devices
GONG	Global Oscillation Network Group
GSC	(Hubble) Guide Star Catalogue
GSMT	Giant Segmented Mirror Telescope
GTC	Gran Telescopio Canarias
GZK	Greisen Zatsepin Kuzmin cut-off
HARPS	High Accuracy Radial velocity Planet Searcher
HEB	Hot Electron Bolometer
HEMT	High Electron Mobility Transistor
HFET	Heterostructure Field Effect Transistor
Hipparcos	HIgh-Precision PARallax COllecting Satellite
HST	Hubble Space Telescope
IBC	Impurity Band Conduction device
ICCD	Intensified Charge-Coupled Device
ICRS	International Celestial Reference System
IF	Intermediate Frequency
IMB	Irvine–Michigan–Brookhaven
INTEGRAL	INTErnational Gamma Ray Astrophysics Laboratory
IPCS	Image Photon Counting System
IRAF	Image Reduction and Analysis Facility
IRAM	Institut de Radio Astronomie Millimétrique
IRAS	InfraRed Astronomy Satellite
ISAAC	Infrared Spectrometer And Array Camera
ISI	Infrared Spatial Interferometer
ISO	Infrared Space Observatory
IUE	International Ultra-violet Explorer
JCG	Johnson–Cousins–Glass photometric system
JCMT	James Clerk Maxwell Telescope
JPEG	Joint Photographic Experts Group
JWST	James Webb Space Telescope
Kamiokande	Kamioka Neutrino DEtector
KAO	Kuiper Airborne Observatory
KAP	potassium (K) Acid Phthalate
LAMA	Large Aperture Mirror Array

LAMOST	Large sky Area Multi-Object fibre Spectroscopic Telescope
LAT	Large Area Telescope
LBT	Large Binocular Telescope
LENS	Low Energy Neutrino Spectroscope
LIGO	Laser Interferometer Gravitational-wave Observatory
LISA	Laser Interferometer Space Antenna
LZT	Large Zenith Telescope
MAMA	(1) Multi-Anode Micro-channel Array (2) Machine Automatique à Mesurer pour l'Astronomie
MAP	(1) Microwave Anisotropy Probe (2) Multichannel Astrometric Photometer
mas	milli-arc second
MAST	Multi mission Archive for the Space Telescope
MCAO	Multi-Conjugate Adaptive Optics
MEM	Maximum Entropy Method
MERLIN	Multi-Element Radio Linked INterferometer
MINOS	Main Injector Neutrino Oscillation Search
MIR	Mid InfraRed
MMIC	Monolithic Microwave Integrated Circuits
MOF	Magneto-Optical Filter
MOS	Metal Oxide–Silicon transistor
MTF	Modulation Transfer Function
NA	Numerical Aperture
NASA	National Aeronautics and Space Administration
NESTOR	Neutrino Extended Submarine Telescope with Oceanographic Research
NIR	Near InfraRed
NNLS	Non-Negative Least Squares
NOAO	National Optical Astronomy Observatories
NODO	NASA Orbital Debris Observatory
NPOI	Navy Prototype Optical Interferometer
NRAO	National Radio Astronomy Observatory
NSSDC	National Space Science Data Centre
NTD	Neutron Transmutation Doping
NVO	National Virtual Observatory
OASIS	Optically Adaptive System for Imaging Spectroscopy
OTCCD	Orthogonal Transfer CCD
OWL	OverWhelmingly Large telescope
PMM	Precision Measuring Machine
PSF	Point Spread Function
PZT	Photographic Zenith Tube
QWIPS	Quantum Well Infrared Photo-DetectorS
RAND	Radio Array Neutrino Detector

RL	Richardson Lucy algorithm
SAGE	Soviet–American Gallium Experiment
SALT	South African Large Telescope
SAR	Synthetic Aperture Radar
SCUBA	Submillimetre Common User Bolometer Array
SDSS	Sloan Digital Sky Survey
SEC	Secondary Electron Conduction (TV camera)
SETI	Search for Extra-Terrestrial Intelligence
SIM	Space Interferometry Mission
SIMBAD	Set of Identifications, Measurements and Bibliography for Astronomical Data
SIRTF	Space InfraRed Telescope Facility
SIS	Superconductor–Insulator–Superconductor device
SKA	Square Kilometre Array
SLR	Single Lens Reflex (camera)
SMM	Solar Maximum Mission
SNO	Sudbury Neutrino Observatory
SNU	Solar Neutrino Unit
SOFIA	Stratospheric Observatory For Infrared Astronomy
SOHO	SOlar and Heliospheric Observatory
SPIRAL	Segmented Pupil Image Reformatting Array Lens
SPIRE	Spectral and Photometric Imaging REceiver
ST-ECF	Space Telescope European Coordinating Facility
STIS	Space Telescope Imaging Spectrograph.
STJ	Superconducting Tunnel Junction
SUSI	Sydney University Stellar Interferometer
TAC	Twin Astrographic Catalog
TASS	The Amateur all-Sky Survey
TAURUS	Taylor Atherton Uariable resolution Radial Uelocity System
TDI	Time Delayed Integration
TES	Transition Edge Sensor
THEMIS	Télescope Héliographique pour l'Etude du Magnétisme et des Instabilités Solaires
TIFF	Tagged Image File Format
TPF	Terrestrial Planet Finder
TRACE	Transition Region And Corona Explorer
UCAC	USNO CCD Astrograph Catalog
UIST	UKIRT Imager SpecTrometer
UKIRT	United Kingdom Infrared Telescope
UKST	United Kingdom Schmidt Telescope
ULE	Ultra-Low Expansion fused silica
USNO	US Naval Observatory
UVES	Ultra-violet/Visual Echelle Spectroscope

VISTA	Visible and Infrared Survey Telescope for Astronomy
VLA	Very Large Array
VLBA	Very Long Baseline Array
VLOT	Very Large Optical Telescope
VLT	Very Large Telescope
VLTI	Very Large Telescope Interferometer
VPHG	Volume Phase Holographic Grating
VTT	Vacuum Tower Telescope
WARC	World Administrative Radio Conference
WET	Whole Earth Telescope
WFPC2	Wide Field Planetary Camera 2
WHT	William Herschel Telescope
WRC	World Radio Conference
WSRT	Westerbork Synthesis Radio Telescope
WUPPE	Wisconsin Ultra Violet Photo-Polarimeter Experiment
XUV	eXtreme Ultra-Violet (also EUV)

Appendix 5

Bibliography

Some selected journals, books and articles that provide further reading for various aspects of this book, or from which further information may be sought, are listed below.

Research journals

Astronomical Journal
Astronomy and Astrophysics
Astrophysical Journal
Icarus
Monthly Notices of the Royal Astronomical Society
Nature
Publications of the Astronomical Society of the Pacific
Science
Solar Physics

Popular journals

Astronomy
Astronomy Now
Ciel et Espace
New Scientist
Scientific American
Sky and Telescope

Ephemerises

Astronomical Almanac (published for each year), HMSO/US Government
    Printing Office
Handbook of the British Astronomical Association (published for each
    year), British Astronomical Association
Yearbook of Astronomy (published each year) Macmillan

Star and other catalogues, atlases, sky guides

Arnold H, Doherty P and Moore P 1999 *Photographic Atlas of the Stars* (IoP Publishing)

Bakich M E 1995 *The Cambridge Guide to the Constellations* (Cambridge University Press)

Coe S R 2000 *Deep Sky Observing* (Springer)

Hirshfield A and Sinnott R W 1992 *Sky Catalogue 2000 Volumes 1 and 2* (Cambridge University Press)

Inglis M 2001 *Field Guide to the Deep Sky Objects* (Springer)

Jones K G 1991 *Messier's Nebulae and Star Clusters* (Cambridge University Press)

Karkoschka E 1999 *Observer's Sky Atlas* (Springer)

Kitchin C R 1997 *Photo-Guide to the Constellations* (Springer)

Kitchin C R and Forrest R 1997 *Seeing Stars* (Springer)

Luginbuhl B and Skiff B 1990 *Observing Handbook and Catalogue of Deep Sky Objects* (Cambridge University Press)

Moore P 1998 *Observer's Year* (Springer)

Moore P 2002 *Astronomy Encyclopedia* (Philips)

Privett G and Parsons P 2001 *Deep Sky Observer's Year 2001* (Springer)

Ratledge D 2000 *Observing the Caldwell Objects* (Springer)

Ridpath I (ed) 1998 *Norton's 2000* 19th Edition (Longman)

Tirion W 2000 *Sky Atlas 2000* (Sky Publishing Corporation)

Reference books

Allen C W 2001 *Allen's Astrophysical Quantities* (Springer)

Kitchin C R 2002 *Illustrated Dictionary of Practical Astronomy* (Springer)

Lang K R 1998 *Astrophysical Formulae: Space, Time, Matter in Cosmology* (Springer)

Lang K R 1998 *Astrophysical Formulae: Radiation, Gas Processes and High-energy Astrophysics* (Springer)

Lang K R 2001*Cambridge Encyclopedia of the Sun 2001* (Cambridge University Press)

Murdin P (ed) 2001 *Encyclopedia of Astronomy and Astrophysics* (Nature and IoP Publishing)

Ridpath I 1997 *Oxford Dictionary of Astronomy* (Oxford University Press)

Shirley J H and Fairbridge R W 2000 *Encyclopedia of Planetary Sciences* (Kluwer Academic Publishers)

Introductory astronomy books

Engelbrektson S 1994 *Astronomy through Space and Time* (WCB)

Freedman R A and Kaufmann W J III 2001 *Universe* (Freeman)

Halliday K 1999 *Introductory Astronomy* (Wiley)

Moche D L 1993 *Astronomy; A Self-Teaching Guide* (Wiley)
Montenbruck O and Pfleger T 2000 *Astronomy on the Personal Computer* (Springer)
Moore P 1997 *Eyes on the Universe* (Springer)
Nicolson I 1999 *Unfolding our Universe* (Cambridge University Press)
Tonkin S F 2000 *AstroFAQs* (Springer)
Zeilik M 1994 *Astronomy; The Evolving Universe* (Wiley)
Zeilik M, Gregory S A and Smith E v P 1992 *Introductory Astronomy and Astrophysics* (Saunders)

Introductory practical astronomy books

Bone N 1999 *Observing Meteors, Comets and Supernovae* (Springer)
Covington M A 2002 *How to use a Computerized Telescope* (Cambridge University Press)
Dragesco J and McKim R (trans) 1995 *High Resolution Astrophotography* (Cambridge University Press)
Good G A 2002 *Observing Variable Stars* (Springer)
Kitchin C R 2001 *Solar Observing Techniques* (Springer)
Kitchin C R 2003 *Telescope & Techniques* 2nd edition (Springer)
James N and North G 2002 *Observing Comets* (Springer)
Jones B 1991 *An Introduction to Practical Astronomy* (Apple Press)
Lena P 1997 *A Practical Guide to CCD Astronomy* (Cambridge University Press)
Manly P L 1994 *The 20-cm Schmidt–Cassegrain Telescope* (Cambridge University Press)
Mobberley M 1999 *Astronomical Equipment for Amateurs* (Springer)
Mollise R 2001 *Choosing and Using a Schmidt–Cassegrain Telescope* (Springer)
Ratledge D 1997 *Art and Science of CCD Astronomy* (Springer)
Ratledge D 1999 *Software and Data for Practical Astronomers* (Springer)
Roth G D 1993 *Compendium of Practical Astronomy* (Springer)
Tonkin S F 1999 *Amateur Telescope Making* (Springer)
Tonkin S F 2001 *Astronomy with Small Telescopes* (Springer)
Tonkin S F 2002 *Practical Amateur Spectroscopy* (Springer)
Wlasuk P 2000 *Observing the Moon* (Springer)

Advanced practical astronomy books (or books with a significant section on practical astronomy)

Birney D S 1991 *Observational Astronomy* (Cambridge University Press)
Bode M F 1995 *Robotic Observatories* (Wiley)
Bray D F 1992 *The Observation and Analysis of Stellar Photospheres* (Cambridge University Press)
Budding E 1993 *Astronomical Photometry* (Cambridge University Press)

Butler C J and Elliott I 1993 *Stellar Photometry—Current Techniques and Future Developments* (Cambridge University Press)

Chapman A 1990 *Dividing the Circle: Development and Critical Measurement of Celestial Angles (1500–1850)* (Ellis Horwood)

Comte G and Marcelin M (ed) 1995 *Tridimensional Optical Spectroscopic Methods in Astrophysics* (Astronomical Society of the Pacific Conference Series Vol 71)

Cornwell T J and Perley R A (ed) 1991 *Radio Interferometry: Theory, Techniques and Applications* (Astronomical Society of the Pacific Conference Series Vol 19)

Davies J K 1997 *Astronomy from Space: The Design and Operation of Orbiting Observatories* (Praxis)

Davis Phillip A G *et al* (ed) 1990 *CCDs in Astronomy II* (L Davis Press)

Emerson D T and Payne J M (ed) 1995 *Multi-Feed Systems for Radio Telescopes* (Astronomical Society of the Pacific Conference Series Vol 75)

Filippenko A V (ed) 1992 *Robotic Telescopes in the 1990s* (Astronomical Society of the Pacific Conference Series Vol 34)

Glass I 1999 *Handbook of Infrared Astronomy* (Cambridge University Press)

Goddard D E and Milne D K (ed) 1994 *Parkes: Thirty Years of Radio Astronomy* (CSIRO)

Grey P M (ed) 1993 *Fiber Optics in Astronomy II* (Astronomical Society of the Pacific Conference Series Vol 37)

Guyenne T (ed) 1995 *Future Possibilities for Astrometry in Space* (European Space Agency)

Hearnshaw J B 1996 *Measurement of Starlight* (Cambridge University Press)

Henry G W and Eaton J A (ed) 1995 *Robotic Telescopes: Current Capabilities, Present Developments and Future Prospect for Automated Astronomy* (Astronomical Society of the Pacific Conference Series Vol 79)

Howell S B 2000 *Handbook of CCD Astronomy* (Cambridge University Press)

Kitchin C R 1995 *Optical Astronomical Spectroscopy* (Institute of Physics Publishing)

Korsch D 1991 *Reflective Optics* (Academic Press)

Laikin M 1995 *Lens Design* (Dekker)

Manly P L 1991 *Unusual Telescopes* (Cambridge University Press)

Mattok C (ed) 1993 *Targets for Space-Based Interferometry* (European Space Agency)

McClean I S 1994 *Electronic and Computer Aided Astronomy* (Ellis Horwood)

Mizon R 2001 *Light Pollution: Strategies and Solutions* (Springer)

Rieke G H 2002 *Detection of Light: From the Ultraviolet to the Submillimeter* (Cambridge University Press)

Robertson J G (ed) 1993 *Very High Angular Resolution Imaging* (Kluwer)
Sterken C and Manfroid 1992 *Astronomical Photometry* (Kluwer)
Stix M 2002 *The Sun* 2nd edition (Springer)
Wall J V 1993 *Optics in Astronomy* (Cambridge University Press)
Wall J V and Boksenberg A 1990 *Modern Technology and its Influence on Astronomy* (Cambridge University Press)
Zensus J A *et al* (ed) 1995 *Very Long Baseline Interferometry and the VLBA* (Astronomical Society of the Pacific Conference Series Vol 82)

Data reduction and analysis

Bruck M T 1990 *Exercises in Practical Astronomy using Photographs* (Institute of Physics)
di Gesù V *et al* (ed) 1992 *Data Analysis in Astronomy IV* (Plenum)
Emerson D 1996 *Interpreting Astronomical Spectra* (Wiley)
Grandy W T Jr and Schick L H 1991 *Maximum Entropy and Bayesian Methods* (Kluwer)
Jaschek C and Murtagh F (ed) 1990 *Errors, Bias and Uncertainties in Astronomy* (Cambridge University Press)
Taff L G 1991 *Computational Spherical Astronomy* (Krieger)
Wall J and Jenkins C 2003 *Practical Statistics for Astronomers* (Cambridge University Press)

Index

Note: page numbers shown in italic mark the start of a chapter or major section on the topic.